Charles Hood

A Practical Treatise on Warming Buildings

By Hot Water, Steam and Hot Air

Charles Hood

A Practical Treatise on Warming Buildings
By Hot Water, Steam and Hot Air

ISBN/EAN: 9783743465084

Manufactured in Europe, USA, Canada, Australia, Japa

Cover: Foto ©Lupo / pixelio.de

Manufactured and distributed by brebook publishing software (www.brebook.com)

Charles Hood

A Practical Treatise on Warming Buildings

A PRACTICAL TREATISE

ON

WARMING BUILDINGS

BY

HOT WATER, STEAM, AND HOT AIR;

ON

VENTILATION,

AND THE

VARIOUS METHODS OF DISTRIBUTING ARTIFICIAL HEAT, AND THEIR
EFFECTS ON ANIMAL AND VEGETABLE PHYSIOLOGY.

TO WHICH ARE ADDED AN

INQUIRY INTO THE LAWS OF RADIANT AND CONDUCTED HEAT,
THE CHEMICAL CONSTITUTION OF COAL, AND
THE COMBUSTION OF SMOKE.

BY

CHARLES HOOD, F.R.S., F.R.A.S., &c.

SIXTH EDITION.

E. & F. N. SPON, 125, STRAND, LONDON.

NEW YORK: 35, MURRAY STREET.

1885.

PREFACE

TO THE FIFTH EDITION.

———•———

A NEW Edition of the TREATISE ON WARMING AND VENTILATING BUILDINGS being required, it has been thought desirable not only to revise the former edition, but also to enlarge it with a considerable quantity of new matter.

The various inquiries addressed to the Author from time to time, have led to the conclusion that fuller information on certain subjects treated in this work would be desirable, and it has been his endeavour to supply this information in the present edition.

The ever-varying conditions under which the Warming and Ventilating of Buildings are sought to be applied, necessarily produce new combinations. · In the present edition considerable additions have been made with the view of meeting, as far as possible, these requirements. A chapter on heating large bodies of water for Baths and other purposes has been added; and considerable additions have been made to the chapters on forced and spontaneous ventilation, and on the action of the various forms of ventilators and chimney cowls. The causes of failure in several forms of apparatus have also been pointed out, and remedies suggested; and generally, throughout the work, many additional points of practical experience have been investigated, and it is hoped the results may be found useful.

In the former editions of this work all the paragraphs were numbered consecutively. In the present edition many new paragraphs have been added without being numbered, and in other cases duplicate numbers, with an asterisk prefixed, have been inserted. This plan has been adopted advisedly. Not only was it deemed undesirable to disturb the original numbering of the paragraphs as given in the former editions; but further, the references throughout the work, from one part to another, are so numerous, that it was found to be scarcely practicable to alter the original numbering of the paragraphs, without great risk of error, and an amount of additional labour which would be hardly justified.

All that was given in the former editions will be found in the present work. But the additions which have been made will, it is hoped, throw some further light on various points of practice which have occasionally caused difficulties.

Very copious extracts from the former editions of this work have for many years past been made both at home and abroad, by other authors. Not only paragraphs and tables, but whole pages, and even entire chapters, have been thus appropriated without acknowledgment. It may therefore be well to state that the first edition of this treatise was published in 1837. The extensive appropriation of parts of this work by other authors, without proper acknowledgment, has sometimes led to the borrowers being credited with the original production, and the real author esteemed to be the plagiarist. This is the excuse for noticing what might otherwise lead some persons to suppose an unfair use had been made by the author of what properly belonged to others; but he believes that he has in every instance quoted the authors from whom he has derived any of the information contained in the following pages.

In this edition, as in those which preceded it, the plan of throwing into Notes a good deal of the information which is explanatory of the subject-matter of the work, has been followed. The object of this is to render the text more simple for those who only wish for the practical information, apart from any scientific discussion.

C. H.

No. 1, Upper Thames Street,
         London; and
Leinster Gardens, Hyde Park.

This sixth edition is a reprint of the fifth.

CONTENTS.

PART FIRST.

ON WARMING BUILDINGS BY HOT WATER.

CHAPTER III.

ON THE RELATIVE SIZES OF PIPES, AND THE USE OF STOP-COCKS AND VALVES.

CHAPTER IV.

ON TEMPERATURE, PIPES AND BOILERS, DURABILITY OF MATERIAL, AND FUEL.

CHAPTER V.

ON FURNACES, THEIR CONSTRUCTION, AND MODES OF FIRING.

CHAPTER VI.

ESTIMATE OF THE HEATING SURFACE REQUIRED TO WARM ANY
DESCRIPTION OF BUILDINGS.

CHAPTER VII.

ON VARIOUS HOT-WATER APPARATUS.

CHAPTER VIII.

GENERAL REMARKS ON HOT-WATER APPARATUS.

CHAPTER IX.

ON APPARATUS FOR BATHS AND DOMESTIC SERVICE.

CHAPTER X.

ON HEATING BY STEAM.

CHAPTER XI.

ON HEATING BY HOT AIR.

CHAPTER XII.

ON THE LAWS AND PHENOMENA OF HEAT.

CHAPTER XIII.

EXPERIMENTS ON COOLING.

PART SECOND.

ON THE WARMING AND VENTILATION OF BUILDINGS BY THE COMBUSTION OF FUEL. ·ᵉ

CHAPTER I.

ON THE VARIOUS METHODS OF WARMING AND VENTILATING BUILDINGS, THE COMBUSTION OF FUEL, ETC.

CHAPTER II.

ON THE VARIOUS FORMS OF FIREPLACES AND STOVES.

CHAPTER III.

ON THE CHANGES PRODUCED IN ATMOSPHERIC AIR BY HEAT, COMBUSTION, AND RESPIRATION.

CHAPTER IV.

ON THE VARIOUS METHODS OF PRODUCING VENTILATION.

CHAPTER V.

ON THE THEORY OF GASEOUS EFFLUX.

CHAPTER VI.

ON THE CHEMICAL CONSTITUTION OF COAL, AND THE COM-
BUSTION OF SMOKE.

INTRODUCTION.

THE practice of employing hot water, circulating through iron pipes, for diffusing artificial heat, is an invention of acknowledged utility; and the present extensive and extending use of this invention renders it extremely desirable that its principles should be clearly defined, and the rules for its practical application laid down with precision. Without this knowledge its success will be uncertain, its application limited, and its results unsatisfactory.

It can scarcely excite surprise, that prejudices should formerly have existed against this invention, while its merits and its principles were alike imperfectly known. Even at the present time they are but partially understood; and, therefore, to investigate these two subjects is the proposed object of the present treatise, with the view of facilitating its application, and extending the sphere of its utility.

There is scarcely any branch of science, or of art, in which an acquaintance with the laws of Nature does not enable us to derive greater advantages in its application than we could otherwise possess. There is no art, however humble, in which a knowledge of the laws that regulate matter does not open a wide and extensive field of useful improvement; and no man can hope to

B

advance the sphere of useful invention, in any considerable degree, without some knowledge of the principles on which it depends. But, having this knowledge, he may, to use the words of a well-known writer, "if he have only a pot to boil, be sure to learn from science lessons which will enable him to cook his morsel better, save his fuel, and both vary his dish and improve it." Although it is true that we are still ignorant of the more subtile agents which exist in the vast chain of causation, the laws which regulate the various phenomena of Nature are sufficiently known to afford the most beneficial assistance to every branch of the arts and sciences: and the most recondite of scientific discoveries, as well as the most valuable inventions and improvements in the arts, are not more demonstrative of the truth of this assertion, than those which are the most simple and inartificial.

For an illustration of the utility of this knowledge, we may refer to the law of gravity, not only because it is, of all natural phenomena, the most constant in its operation, and the most universal in extent, but because its influence is closely connected with the present subject of inquiry.

That all falling bodies gravitate with the same velocity, and, therefore, descend through a certain definite space in a given time, is, we know, an *effect*, of which gravity is the *cause*. It is on the operation of this invariable law that many of our most valuable inventions depend. Its influence is equally exerted on all objects alike;—the most mighty as well as the most minute. It is this which gives stability to the grandest productions of Nature, as well as to the most minute or artificial of our own works. It is from this cause that we obtain the unerring action of our pendulums and clocks; and it is by this we obtain the circulation of hot water, with which we warm our dwellings. By a knowledge of the *cause* of these effects, of the extent of its operations, and of the laws by which it acts, we can, by varying the circumstances of a gravitating body, alter also the velocity of its descent. We accomplish this by bringing other causes into operation, which modify the result, notwithstanding the immutability of the laws of gravity: and thus, by a knowledge of the physical laws, we can modify and subject to our will one of the most constant and universal agents in Nature.

The study of the laws which govern natural phenomena—which in all cases are so simple, so beautiful, so perfect—is, therefore, one of the most fruitful sources of inquiry which the mechanician can pursue. It opens to him new fields of useful inquiry; new applications of known inventions; new and simpler means of accomplishing known effects. And while it points to improvement in every direction, it restrains the judgment from false principles. Where it exists not, it is almost certain that the plans of the mechanician will either be only modified copies of existing inventions, or they will degenerate into wild speculations, unsupported on any reasonable foundation.

This is particularly observable in the case before us. The numerous failures which have occurred in the practical application of the invention of heating buildings by the circulation of hot water, are all distinctly referrible to the want of this kind of knowledge, and not to the object aimed at being itself unattainable. For whenever the physical laws are intended to be employed as the principal agents in producing any mechanical effect, it is an indispensable condition that simplicity of action be kept in view. While it may further be observed,

that the endeavours to trace and elucidate the
operating causes of the various phenomena, which
occur in the course of practical experiments, are
the surest means of facilitating original discoveries,
as well as of promoting new adaptions of recog-
nized principles.

The origin of the invention of employing hot
water for diffusing artificial heat appears to be hid
in considerable obscurity. It is not improbable
that, like many other discoveries, it has been re-
produced at various periods. It seems, however,
to have been first used in France, by M. Bonnemain,
in the year 1777, and was employed by him during
several years for hatching chickens by artificial
heat. The French Revolution, which followed
shortly afterwards, put a stop to this, as well as
many other useful and scientific inventions, in that
country; and for several years the invention appears to have been entirely dormant; nor indeed,
does it appear ever to have been used by M. Bonnemain except for the purpose above mentioned.
About the year 1817, the Marquis de Chabannes
introduced a similar apparatus into this country
for heating a conservatory, and also heating some
rooms in a private house, by pipes leading from the
kitchen boiler. In the following year he published,
in London, a pamphlet describing this apparatus,
and some ingenious modifications of hot-air stoves.
The invention appears to have made very little progress for some years. In 1822, Mr. Bacon, a
gentleman of fortune, introduced the use of hot
water into his forcing-houses, using for the purpose
a single pipe of large diameter, communicating with
the boiler; and, by giving a slight elevation to the
pipe from the horizontal line, he was thus enabled
to produce a circulation of the water, the hot water
slowly passing along the upper part of the nearly
horizontal pipe, and the colder water returning to the

boiler along the lower part of the same pipe. The circulation in this apparatus was very imperfect; and Mr. Atkinson, an architect, almost immediately afterwards suggested the addition of a second pipe to bring back the colder water to the boiler; and thus at once the apparatus assumed the form which it has ever since retained. By this alteration the apparatus was brought very nearly to the same form as that contrived by M. Bonnemain more than forty years before; the principal difference being that M. Bonnemain used only very small pipes of gun-barrel size, while Mr. Atkinson used pipes of four or five inches diameter.

The honour of this invention has been claimed for Mr. Watt, prior to the time of M. Bonnemain using it in France; but there appear no grounds for supposing he ever employed it without the intervention of steam, as a distributor of heat by circulation, in the manner in which it is now used. The mere motion of hot water in pipes is an invention of far greater antiquity than the time either of Watt or Bonnemain. Seneca has accurately described the mode of heating the water in the Thermæ of Rome, of which Castell has given drawings;[*] and which show that the method of heating baths by passing the water through a coil of pipes which passed through the fire was known and practised previous to the Christian era. And, except that these tubes were of brass, instead of iron, they were precisely similar, both in form and arrangement, to those occasionally used at the present day for the like purpose; the lapse of nineteen centuries having apparently added nothing to our knowledge on this subject.

Since the first introduction of the hot-water ap-

---

[*] Castell's "Illustrations of the Villas of the Ancients," p. 10.

paratus for warming buildings, the variations made in its more complicated arrangements appear to have been very gradually adopted. Each time that an apparatus has been erected, experimentalists have deviated in some small degree from the model of that which preceded; apparently afraid of venturing on too great a variation, yet requiring, from contingent circumstances, some alteration of its form and application. This mode of proceeding, though natural while the principles were not thoroughly understood, has frequently led to both inconvenience and loss, in consequence of the numerous failures to which it has given rise, by unintentional deviations from the true principles. In the present attempt to elucidate the subject, it will, however, be shown that success needs not be uncertain, provided only that the laws of physics be justly applied and strictly adhered to.

So numerous have been the failures which have occurred in this method of heating buildings, that nothing but the intrinsic merits of the invention could possibly have made it retain its hold on the public favour. Every imaginable kind of mistake has been made in apparatus erected on this plan. And these mistakes still continue to be made almost as frequently as ever, notwithstanding the vast number of buildings in which it has been successfully applied, and which might be consulted for correct information.

Neither the capabilities of this method of warming, nor the various useful purposes to which it is applicable, are even yet fully appreciated. There are no buildings, however large, to which it cannot be advantageously adapted, nor any that present insurmountable difficulties in its practical application. In many useful purposes connected with arts and manufactures, it can be most advantageously employed, though its application to these purposes has hitherto been greatly overlooked. Its merits, however, will best appear by the plain statement of facts in the following pages.

Since the publication of the first edition of this work, in 1837, ample opportunities have occurred for testing in every variety of form the accuracy of the rules and calculations which were then given, for constructing and apportioning the hot-water apparatus to the varying circumstances under which it could be applied. The most important of the calculations are those on the heated surfaces required to warm a given building, and the proper size of the boiler and the furnace. The data on which were founded the rules, and tables for calculating these proportions, were carefully compared, both experimentally and practically; and the extensive use which has been made of these rules, with perfect success, leaves no doubt whatever as to their complete accuracy.

The object which has been aimed at in this treatise, has been to render the work as clear as possible to practical men; while at the same time, it should not be below the notice of those who might desire either to acquaint themselves with the scientific principles of the invention, or with the general bearing of the subject, on the health, the comfort, the physical development, and even the duration of life, of a large portion of the human race. The effects of unwholesome air on the animal economy from imperfect ventilation and deleterious methods of producing artificial warmth, is a subject which has hitherto excited far too little attention. The laws which connect man with the physical universe are of far higher generalisation than many are inclined to believe. They bind together not alone the phenomena of our own world, but of the entire material universe. And yet mankind vainly suppose they can act as they

please with relation to these immutable laws,—
that they can fashion them to their own will, and
to their own imagined wants; and instead of
endeavouring to make themselves more thoroughly
acquainted with these laws, devised by unerring
Wisdom for the universal good, they choose rather
to act either in direct opposition, or, at least, in
total neglect of those great truths which philosophy
has been permitted to discover and unfold. But it
is certain that they cannot with impunity neglect
the great truths which physiology discloses; nor
can they place themselves in opposition to those
fundamental laws which bind together the whole
physical universe, without entailing upon themselves the penalties which the Great Author of
those laws has made the inseparable condition of
their infringement.

It is hoped the following pages will place this
matter in a clear point of view, in a sufficiently
popular manner.

A PRACTICAL TREATISE,

ETC.

PART FIRST.

ON WARMING BUILDINGS BY HOT WATER, AND ON

THE LAWS OF HEAT, ETC.

CHAPTER I.

Cause of Circulation of the Water—Inclination of the Pipes—
Necessity for Air-vents—Open and Close Boilers—Pressure
of Water—Expansive Power of Steam—Effect produced
on the Circulation by increased Height of the Pipe—
Compression of Water—Branch Pipes—Variations in
Level of Pipes.

(ART. 1.) In endeavouring to explain the principles of the various forms of apparatus in which
hot water, circulating through iron pipes, is employed as a means for distributing artificial heat,
the first object should be to point out, as clearly as
possible, the power which produces the circulation
of the water ; for without a clear perception of this
part of the subject, there will always be an uncertainty as to the results which will obtain, when any
departure is made from the most simple form and
arrangement of the different parts of the apparatus. It is this circulation which causes all
the water in the apparatus to pass successively
through the boiler, and then communicates the
heat that is thus received from the fuel to the

various buildings or apartments which it is designed to warm. Without this circulation, those
parts of the apparatus which are remote from the
fire would not receive any heat; because water is
so bad a conductor that it is only when there
exists perfect freedom of motion among its particles, that it acts at all as a conductor of heat, so
far, at least, as regards any practical and useful
effect. It is in a complete and perfect circulation,
therefore, that the efficiency of a hot-water apparatus depends, and that the greatest amount of heat
is obtained by it from a given quantity of fuel.

(2.) The only treatise hitherto published, in
which any attempt has been made to explain the
cause of the circulation of the water in this description of apparatus, is Mr. Tredgold's work on
heating by steam; and the effect is there referred
entirely to an erroneous cause. In the Appendix
to that work, the cause of motion is thus explained.

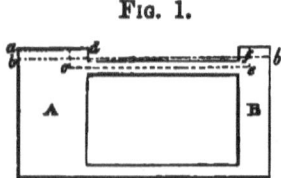

FIG. 1.

"If the vessels A B, and
pipes, be filled with water,
and heat be applied to the
vessel A, the effect of heat
will expand the water in
the vessel A, and the surface will, in consequence,
rise to a higher level, a d, the former general level
surface being b b. The density of the fluid in the
vessel A will also decrease in consequence of its
expansion; but as soon as the column c d (above
the centre of the upper pipe) is of a greater weight
than the column f e, motion will commence along
the upper pipe from A to B, and the change this
motion produces in the equilibrium of the fluid
will cause a corresponding motion in the lower
pipe from B to A."

(3.) Now it is certain that this theory will not
account for the circulation of the water under all

circumstances, and every variety of form of the apparatus; and as the cause of motion must be the same in all cases, any explanation which will not apply universally must necessarily be erroneous. Were this the true cause of motion, there would be no difficulty in obtaining a circulation in all cases; for, according to this reasoning, whenever the level of the water is higher in the boiler than in the pipes—or even if an upright pipe were placed on the top of a close boiler, by which the pressure on the surface would be increased—the water must of necessity circulate through the pipes: while, on the other hand, if this hypothesis were correct, the water in an apparatus constructed as in the following figure would not circulate at all.

(4.) Suppose the apparatus, fig. 2, to be filled with cold water, and the two stop-cocks, $f g$, to be closed: on applying heat to the vessel A, the water it contains will expand in bulk, and a part of it will flow through the small waste-pipe x, which

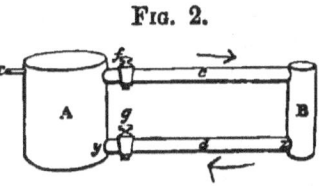

Fig. 2.

is so placed as to prevent the water rising higher in the vessel A than the top of the vessel B. The water which remains in the vessel A, after it has been heated, and a portion of it has passed through the waste-pipe x, will evidently be lighter than it was before, while its height will remain unaltered. Suppose, now, the two cocks, $f g$, to be simultaneously opened; the hot water in the boiler A will immediately flow towards B through the upper pipe, and the cold water in B will flow into A through the lower pipe; although, by the hypothesis previously alluded to, unless the water in the vessel A, above the pipe c, were heavier, or rose to a higher level than the water in the vessel B, no circulation could take place. In this case, therefore, we must find another explanation of the cause of motion.

(5.) The power which produces circulation of the water will be found to arise from a different cause than that which is here stated; for we see that this reason is insufficient to account for the effect, even in one of the simplest forms of the apparatus.

In order to explain this, let us suppose heat to be applied to the boiler A, fig. 2. A dilation of the volume of the water takes place, and it becomes lighter; the heated particles rising upwards through the colder ones, which latter sink to the bottom by their greater specific gravity, and they in their turn become heated and expanded like the others. This intestine motion continues until all the particles become equally heated, and have received as much heat as the fuel can impart to them. But as soon as the water in the boiler begins thus to acquire heat, and to become lighter than that which is in the opposite vessel B, the water in the lower horizontal pipe d is pressed by a greater weight at z than at y, and it therefore moves towards A with a velocity and force equal to the difference in pressure at the two points y and z.* The water in the upper part of the vessel B would now assume a lower level, were it not that the pipe c furnishes a fresh supply of water from the boiler to replenish the deficiency.

* To any person unacquainted with the science of Hydrostatics, this may probably appear erroneous, because the quantity of water contained in A is much greater than that in B. It is, however, one of the first laws of Hydrostatics that the pressure of fluids depends for its amount on the height of the fluid only, and is wholly irrespective of the bulk, or actual quantity of fluid: therefore, a pipe which is not larger than a quill will transmit the same amount of pressure as though it were a foot, or a yard, in diameter, provided the height be alike in both cases. (See Art. 10.)

By means of this unequal pressure on the *lower* pipe, the water is forced to circulate through the apparatus, and it continues to do so as long as the water in B is colder, and therefore heavier, than that which is in the boiler. And as the water in the pipes is constantly parting with its heat, both by radiation and conduction, while that in the boiler is as continually receiving additional heat from the fire, an equality of temperature never can occur; if it did, the circulation would cease.

(6.) We see, then, that the cause of the circulation is the unequal pressure on the lower pipe of the apparatus; and that it is not the result of any alteration which takes place in the level of the water, as has been erroneously supposed. Indeed, the truth of this appears so plain, that it would scarcely require explanation at such a length, were it not that false opinions in this matter appear to have led to many errors in practice.

As the circulation is caused by the water in the descending pipe being colder, and therefore heavier than that which is in the boiler, it follows, as a necessary consequence, that the colder the water in the descending pipe shall be, relatively to that which is in the boiler, so much the more rapid will be its motion through the pipes. In such an arrangement of pipes as fig. 3, the water in the descending pipe *e f*, having to travel farther before it descends to the

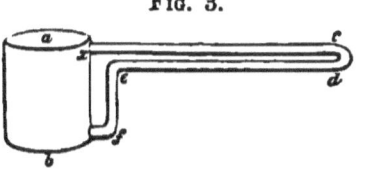

FIG. 3.

lower part of the boiler than when the pipes are arranged as in fig. 2, it will of course be colder at the time of its descent, in the case fig. 3 than in fig. 2, and therefore the circulation will be more rapid. The height of the descending pipe is sup-

posed to be alike in both cases, because *c d* and
e f are together equal to *a b*.

(7.) Some persons have imagined that if the
pipes be inclined, so as to allow a gradual fall of
the water in its return to the boiler, additional
power will be gained; as, for instance, by inclining the lower pipe of fig. 3, so as to make the
part *e* lower than *d*, and then reducing the vertical
height of the return pipe *e f*. This, at first, appears very plausible, particularly with regard to
some peculiar forms of the apparatus; but the
principle is, in fact, entirely erroneous. The
author of the Appendix to Tredgold's work,
already quoted, in consequence of adopting the
erroneous hypothesis, that the motion of the water
commences in the upper pipe instead of the lower
one, as already described, appears to recommend
an inclination being given to the pipes in this
manner; and he has described an apparatus that
he erected, to which a fall of four feet was given
to the water by this method.

This error appears to arise from treating the
subject as a simple question of hydraulics, instead
of a compound result of hydrodynamics. But, in
order to ascertain what is the effect of thus inclining the pipes, let us suppose an extreme case.

It is evident that the farther the water flows,
the colder it becomes. It must, therefore, be
hotter at A (fig. 4) than it is at B, and hotter at B
than C, and so on. Let us, now, suppose any
arbitrary number to represent the specific gravity
of the water at A; say, for instance, ·94. The
water at B, in consequence of having flowed farther,
and therefore become colder and heavier, will be,
we will suppose, of the specific gravity of ·95; at
C, for the same reasons, it will be ·96, and so on
to F, where, from having run the greatest distance from the boiler, it will be the heaviest of

all; * and the sum of all these numbers represents
the pressure at F. But had the pipe, instead of
inclining gradually from the boiler, continued on a
level to *a*, as represented by the dotted lines, the
water would have been as cold, and therefore as
heavy, at *a* as by the former arrangement it is at
F, and therefore its specific gravity would be the
same, namely, ·99. Now, as the pressure of water

Fig. 4.

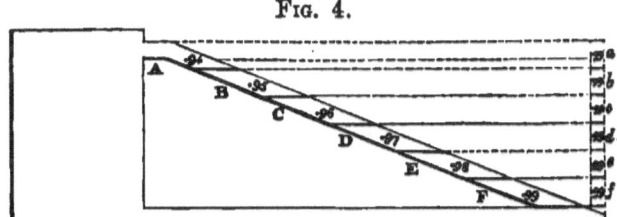

is as its vertical height, by dividing the vertical
pipe, *a f*, in the same manner as we have done
with the inclined pipe, we shall have *a*, *b*, *c*, *d*, *e*,
f, each equal in altitude to the corresponding divisions of the inclined pipe; and as the specific
gravity of each division is equal to ·99, the total
number representing the sum of all these will show
the pressure at the point *f*. We shall hence find
the pressure of the vertical pipe, compared with
that of the inclined pipe, will be as 5·94 is to
5·79.†

(8.) It is evident from this that there must be
a considerable loss on the effective pressure by
making the return pipe incline below the horizontal
level. Nor can this loss be compensated in any

* The real specific gravities could not conveniently be used
in this illustration, as they would require several decimal
places of figures. (See Table IV., Appendix.)

† If the strict analogy were carried out, the difference ought
to be greater than is here represented, because it is evident
that instead of *a*, *b*, *c*, *d*, &c., being each of equal density, *b*
will be heavier than *a*, and *c* heavier than *b*, and so on; but
the illustration, as now given, will be sufficient to show the
principle.

manner; for the total height being the same, whether the water descends through a vertical or through an inclined pipe, the force or pressure will only be equal to the specific gravity of the matter. And as there is actually more matter in a pipe filled with cold water than in a similar pipe filled with hot water, the gravitating force will be inversely proportional to the temperature; that is, it will be less in proportion as *the temperature* of the water is greater. There must, therefore, under all circumstances, be a positive loss of effect by inclining the pipe in the manner stated.*

(9.) In such a form of apparatus as fig. 3, there would be no circulation of the water, unless some plan were adopted by which the air would be dislodged from the pipes, and a ready escape provided for it. Nothing is more necessary to be attended to than this. In the more complicated forms of the apparatus, the want of an efficient means of discharging the air has been the cause of innumerable failures. Suppose we require the apparatus fig. 3 to be filled with water: by pouring it in at the boiler, the pipe *e f* will of course be filled simultaneously with it, and then the lower pipe *d*; and the water will then gradually rise higher in the boiler until it partially fills the upper pipe. At last the orifice of the pipe *x* will become full, and the air which is in the pipe *c x*, being thus prevented from escaping, will be forced towards *c* by the weight of water behind it; and if the quantity of air be sufficiently large, it will entirely prevent the junction of the water at *c*, and cut off the communication between the two pipes at

---

* It must not be supposed that this reasoning at all applies to any case of pure hydraulics. If the question were only as regards a fluid of uniform temperature, then the greatest effect would be obtained by using an inclined pipe; but the fluid which we are now regarding is one of a varying density and temperature, which materially alters the conditional results.

c d. If an opening be now made in the pipe at *c*, the air will immediately escape, being forced out by the greater density of the water; and therefore, either a valve or a cock must be placed there, to allow of its discharge, for otherwise no circulation of the water can ensue. As water, while boiling, always evolves air, it is not sufficient merely to discharge the air from the pipes on first filling them with water, because it is continually accumulating: * and in many instances, particularly with a close-topped boiler, it is desirable to have the air-vent self-acting, either by using a valve or a small open pipe; in others, a cock will often be found most convenient.

The size of the vent is not material, as a very small opening will be sufficient to allow the air to escape. For the rapidity of motion in fluids when pressed by equal weights being inversely proportional to their specific gravities, as water is 827 times more dense than air, an aperture which is sufficiently large to empty a pipe in fourteen minutes if it contained water, would, if it contained air, empty it in about one second.† Air being so very much lighter than water, it is of course necessary that the vents provided for its escape be placed in the highest part of the apparatus, for it is there it will always lodge; and sometimes it will be found necessary to have several vents in different parts of the apparatus.

Though it is perfectly easy, as far as the mere mechanical operation is concerned, to provide for the discharge of the air from the pipes, it requires

---

* If the water were always kept boiling, the air, after being once expelled, would not again accumulate. But when the water cools, it again imbibes air; and thus a continual discharge of air occurs in a hot-water apparatus.

† *Manchester Memoirs*, vol. v., p. 398; and *Nicholson's Journal*, vol. ii., p. 269, also *Robinson's Philosophy*, vol. iii., pp. 682–696. Regnault (*Ann. de Chimie*) states this to be 813·67 to 1.

much consideration and careful study to direct the application of those mechanical means to the exact spot where they will be useful. The subject will, therefore, be again adverted to in a subsequent chapter, when we have investigated the principles of the apparatus in some of its more complex forms of arrangement.

(10.) The plan of the boiler and pipes which has been given in the preceding figures is applicable to comparatively but few purposes; for, in consequence of the boiler being open at the top, the pipes must be laid level with it, otherwise the water would overflow. When the pipes are required to rise higher than the boiler, the latter must be closed at the top, and the pipes can then be carried upwards to any required height. This arrangement possesses considerable advantages: for the higher we make the ascending and descending pipes, the more rapid is the circulation of the water.* This effect will necessarily result from the principles already explained: because, as motion is obtained in consequence of the difference in weight of the ascending and descending columns of water, the greater the height of these columns, the greater must be the difference in their weight, and therefore the greater must be the force and velocity of motion.

The advantages which may be derived from an increased height in the ascending pipe cannot, however, be applied in an unlimited manner, because it might lead to inconvenience, and even be attended with some degree of danger to the apparatus, if the increased height were not regulated by

---

* In this and the preceding figures, the pipes are drawn so as to show the flow and return pipes lying one above the other. A moment's consideration will satisfy anyone that the effect will be the same if they were placed side by side on the same level; and frequently this arrangement is far more convenient.

certain rules, and these, when ascertained, applied with judgment.

The pressure produced by water is calculated by its columnar height, reckoned from the bottom of the vessel in which it is contained. Whether the vessel be open at the top and very deep, or closed at the top and very shallow, but with a pipe attached to the top, like the boiler and pipe A B, fig. 5, the pressure will be exactly alike in either case, if the deep open boiler be equal in height to that of the shallow boiler and upright pipe conjointly; notwithstanding *the quantity* of water may be ten times, or 100 times, larger in the one case than the other. Neither is the pressure increased, however large may be the diameter of the pipe which is used ; nor is it lessened if the pipe be inserted at the side of the boiler, as in the dotted lines *y z*, fig. 5, instead of being placed on the top.

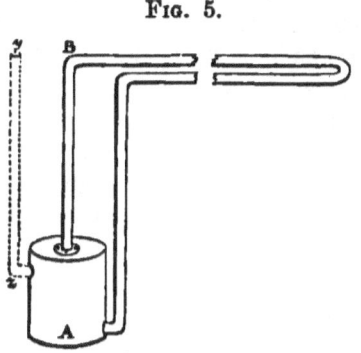

Fig. 5.

As the pressure of water on each square inch of surface increases at the rate of about half a pound * for every foot of perpendicular height, if the height from the bottom of the boiler to the top of the pipe be six feet, the pressure on the bottom will be three pounds on every square inch of surface; but if the boiler be two feet high, the pressure on the top—

* The exact weight of a perpendicular foot of water with a base in one square inch, is 3030·24 *grains*, at the temperature of 60° ; which is therefore only ·4328 of a pound avoirdupois. A column of water 30 feet high only gives a pressure of 12·68 lbs., instead of 15 lbs., as usually reckoned ; and therefore the real height of a column of water, which will give a pressure equal to one atmosphere, must be 34½ feet.

which will be a pressure upwards—will be only two pounds on every square inch of surface, because it will only have four perpendicular feet of water above it. If the height of the pipe be increased to 28 feet, and the depth of the boiler be two feet, as before, making 30 feet together, the pressure will be 15 lbs. on each square inch of the bottom, 14 lbs. on each square inch of the top, and an average pressure of 14½ lbs. on each square inch of the sides of the boiler. Suppose, now, a boiler to be three feet long, two feet wide, and two feet deep, with a pipe 28 feet high from the top of the boiler; when the apparatus is filled with water there will be a pressure on the boiler of 66,816 lbs., or very nearly 30 tons.*

(11.) When a great pressure is used in a hot-water apparatus, in the manner here described, it is necessary that the materials of the boiler should be stronger than they otherwise need be; and more care is also required in making the joints very sound, for attaching the pipes to the boiler so as to prevent any leakage. But when these mechanical difficulties are overcome, the amount of danger arising from a great pressure of water must not be overrated, for it might otherwise deter some persons from adopting this form of the apparatus, notwithstanding its numerous advantages.

(12.) The great danger that arises from the bursting of a steam apparatus is in consequence of the elastic force of steam, which, at very high temperatures, is immense. But water possesses

---

* This enormous pressure on vessels which contain water does not occur in the case of pipes merely used for *the conveyance* of water; for in this case, when the water runs out of the lower end of a long vertical pipe as fast as it runs in at the top, although it be always kept perfectly full, still there may probably be no pressure whatever on any part of the pipe, however great its length may be.—*Robinson's Mechanical Philosophy*, vol. ii., p. 580, *et seq.*

very little elasticity compared with steam, its expansive force being almost inappreciable under ordinary circumstances. At the pressure of 15 lbs. per square inch, the water in the boiler last described, which holds about 75 gallons, would be compressed rather less than *one cubic inch*, or about 1–35th part of a pint.* The expansive force of the water in this apparatus, therefore, even supposing it were to burst, would be perfectly harmless; for it could only expand as much as it had been compressed; namely, *one cubic inch*. The effect on a boiler, by the pressure of the water, will be precisely similar to a weight pressing upon it equal to the estimated pressure of the water, which is quite different from the sudden and violent force produced by the expansive power of steam. As an apparatus of this kind could never be forced asunder, as in the explosion of a steam-boiler, the only result, under the worst circumstances that could occur, would be a leakage of the water, in consequence of the cracking of some part of the boiler.

Neither the principle nor the practical working of the apparatus is in the least affected by having any additional number of pipes leading out of or into the boiler. The effect is the same whether there are more flow-pipes than return-pipes, or, conversely, more return-pipes than flow-pipes. If there be two or more flow-pipes, whether they lead from the boiler separately, or branch from one main pipe, or whether they lead from op-

* According to the experiments of Professor Oerstead, the compression of water is ·0000461 by a pressure of 15 lbs. per square inch; and he has found that it proceeds *pari passu* as far as 65 atmospheres, which was the limit to which his experiments extended. This compression is about equal to reducing a given bulk of water $\frac{1}{16}$ of its volume by a pressure of 20,000 lbs. per square inch.—*Report British Scientific Association*, vol. ii., p. 353.

posite sides of the boiler, or all from one side, each range of pipes will act separately and have a velocity of circulation peculiar to itself. One range of pipes may act efficiently, while another, though attached to the same boiler, may have no circulation whatever through it; and this effect will not be altered whether the pipes return into the boiler separately, or all unite into one main pipe. The pressure, supposing the pipes to rise vertically from the boiler, will likewise be precisely the same, however numerous the pipes may be. This circumstance is one of the peculiarities which distinguish fluids from solids. For if the fluid in any close vessel be pressed by the fluid contained in an upright pipe, so as to produce a pressure of 10 lbs. on a square inch; if a second pipe, capable of exerting a similar pressure with the first, be placed upon the same vessel, the united pressure will still be only 10 lbs. per square inch; and it would be no more, though ever so large a number of pipes were added, provided the vertical height were not increased.

(13.) One advantage may be obtained by causing the water to rise from the boiler by an ascending pipe, as in fig. 5, which cannot be accomplished by any other means; and it is of considerable importance to ascertain its true effect, as it has produced consequences which are not generally attributed to the right cause.

The force and velocity of motion of the water, being proportional to the vertical height of the ascending and descending pipes; by increasing this height, a facility is afforded for taking the pipes below the horizontal level, as, for instance, when it is required to pass them under a doorway, or other similar obstruction, before they finally return to the bottom of the boiler. Innumerable failures have occurred in attempting to make the water descend and again to ascend in this manner, the success of the experiment depending entirely upon the vertical height of the ascending pipe above the boiler. These alterations of the horizontal level, which are frequently very desirable, have, of course, their limits, beyond which they cannot be carried. It is from not having ascertained what are these limits, and what the cause of the limitation, that such uncertainty has hitherto prevailed with regard to this experiment; for it frequently succeeds, but more frequently fails, in practice. It will be most convenient, however, to consider this subject after we have ascertained what is the amount of *motive power* of the water in this kind of apparatus.

CHAPTER II.

On the Motive Power of the Water—On increasing the Motive Power—Velocity of Circulation—Circulation of Water below the Boiler—Direct and Reversed Circulation—Air-Vents—Supply Cisterns—Expansion of Pipes, &c.

(14.) IT has already been mentioned that the power which produces circulation of the water is the unequal pressure on the return-pipe, in consequence of the greater specific gravity of the water in the descending pipe, above that which is in the boiler. Whether this force acts on a long length of return-pipe, as $y\,z$, fig. 2, or only on a very short length, as f, fig. 3, the result will be precisely similar.

Now, it is evident that, if this unequal pressure is the *vis viva*, or motive power, which sets in motion the whole quantity of water in the apparatus, it is only necessary, in order to ascertain the exact amount of this force, that we know the specific gravities of the two columns of water; and the difference will, of course, be the effective pressure, or motive power. This can be accurately determined when the respective temperatures of the water in the boiler and in the descending pipe are known.*

* A thermometer suitable for this purpose was long since proposed by M. Fourier, and called by him the thermometer of contact. It consists of a very small iron cup, just large enough to hold the bulb of the thermometer, but without a bottom to it. Over the bottom, a piece of goldbeaters' skin is to be tied, and the cup is then to have a little mercury put in it, into which the bulb of the thermometer is to dip. The

As this difference of temperature rarely exceeds a very few degrees in ordinary cases, the difference in the weight of the two columns must necessarily be very small. But, probably, the very trifling difference which exists between them, or, in other words, the extreme smallness of the motive power, is very imperfectly comprehended, and will, perhaps, be regarded with some surprise when its amount is shown by exact computation.

(15.) In order to ascertain, without a long and troublesome calculation, what is the amount of motive power for any particular apparatus, the following Table has been constructed. An apparatus is assumed to be at work, having the temperature in the descending pipe 170°; and the difference of pressure upon the return-pipe is calculated, supposing the water in the boiler to exceed this temperature by from 2° to 20°. This latter amount exceeds the difference that usually occurs in practice.

By refering to the annexed Table, it will be found that when the difference between the temperature of the ascending and the descending columns amounts to 8°, the difference in weight is 8·16 *grains* on each square inch of the section of the return-pipe, supposing the height of the boiler A, fig. 2, to be 12 inches. This height, however, is only taken as a convenient standard from which to calculate; for, probably, the actual height will seldom be less than about 18 inches, and, in many cases, it will be considerably more.

cup, when placed on any hot surface, will accurately show the temperature, the contact between the skin and the surface being extremely perfect. As a permanent adjunct, however, to the hot-water apparatus, it is much better to use a thermometer fitted with a hollow screw-nut, by which it can be accurately fixed to the apparatus, and the bulb actually dips into the water of the apparatus.

TABLE I.

Difference in Weight of Two Columns of Water, each One
Foot high, at various Temperatures.

Difference in Temperature of the Two Columns of Water: in Degrees of Fahrenheit's Scale.	Difference in Weight of Two Columns of Water contained in different-sized Pipes, each One Foot in height.				
	1 in. diam.	2 in. diam.	3 in. diam.	4 in. diam.	per sq. in.
	grs. weight.	grs. weight.	grs. weight.	grs. weight.	grs. weight.
2°	1·5	6·3	14·3	25·4	2·028
4°	3·1	12·7	28·8	51·1	4·068
6°	4·7	19·1	43·3	76·7	6·108
8°	6·4	25·6	57·9	102·5	8·160
10°	8·0	32·0	72·3	128·1	10·200
12°	9·6	38·5	87·0	154·1	12·264
14°	11·2	45·0	101·7	180·0	14·328
16°	12·8	51·4	116·3	205·9	16·392
18°	14·4	57·9	131·0	231·9	18·456
20°	16·1	64·5	145·7	258·0	20·532

₄ The above Table has been calculated by the formula
given with Table IV., in the Appendix, for ascertaining the
specific gravity of Water at different temperatures. The
assumed temperature is from 170° to 190°.

Now, suppose, in such a form of apparatus as
fig. 2, the boiler to be two feet high; the distance
from the top of the upper pipe to the centre of the
lower pipe to be 18 inches; and the pipe four
inches diameter;—if the difference of temperature
between the water in the boiler and in the descending pipe be 8°, the difference of pressure on
the return-pipe will be 153 *grains*, or about one-
third part of an ounce weight: and this will be the
whole amount of *motive power* of the apparatus,
whatever be the length of pipe attached to it. If
such an apparatus have 100 yards of pipe, four
inches diameter, and the boiler contains, suppose
30 gallons, there will be 190 gallons, or 1,900 lbs.

weight of water kept in continual motion by a force only equal to one-third of an ounce.* This calculation of the amount of the motive power, in comparison with the weight moved, will vary under different circumstances; and in all cases the velocity of the circulation will vary simultaneously with it.

(16.) It will be observed in the foregoing Table that the amount of motive power increases with the size of the pipe: for instance, the power is four times as great in a pipe of four inches dia- meter as in one of two inches. The power, how- ever, bears exactly the same relative proportion to the resistance or weight of water to be put in motion in all the sizes alike. For although the motive power is four times as great in pipes of four inches diameter as in those of two inches, the former contains four times as much water as the latter; the power and the resistance, therefore, are relatively the same.

(17.) As the motive power is so small, it is not at all surprising that, by an injudicious arrange- ment of the different parts of an apparatus, the

* M. Dutrochet made some experiments on the influencing causes of the motion of currents of liquids. He found that a difference of temperature of 1-800th of a degree was sufficient to produce currents when aided by light, but the motions ceased on light being excluded. In the absence of light (except what was necessary to distinguish the object) the sound of a violoncello or of a bell produced circulation in the liquid. He therefore concluded that the most minute differences of tem- perature will produce motion among the particles of a fluid when aided by light or any other cause which produces feeble vibrations to the particles of the fluid. *Quart. Journal of Science*, vol. xxix., p. 194. It is not stated how this small excess of heat was ascertained. The effects of sound may possibly be in some way connected with a fact which Biot at- tempted to demonstrate by experiment—that every vibration of a sonorous body in elastic media is accompanied with a change of temperature. *Mémoires de la Société d'Arcueil*, 1809; and *Retrospect of Science*, vol. v., p. 429.

resulting motion may frequently be impeded, and
sometimes even totally destroyed; for the slower
the circulation of the water, the more likely is it
to be interrupted in its course.   There are two
ways by which the amount of the motive power
may be increased; one, by allowing the water to
cool a greater number of degrees between the time
of its leaving the boiler and the period of its return
through the descending pipe; the other, by increasing the vertical height of the ascending and
descending columns of water.   The effects produced by these two methods are precisely similar;
for, by doubling the difference of temperature
between the flow-pipe and the return-pipe, the
same increase in power is obtained as by doubling
the vertical height; and tripling the difference in
temperature is the same as tripling the vertical
height.*   This can be ascertained by referring to
the preceding Table.   Thus, suppose, when the
difference of temperature is 8°, and the vertical
height four feet, that the motive power is 32·6
grains per square inch : if the difference of temperature be increased to 16°, while the height
remains the same, or if the height be increased to
eight feet while the temperature remains as at
first, the pressure, in either case, will be 65·2
grains per square inch, or twice the former
amount.   The same rule applies to other differences, both of height and temperature.

(18.) Almost the only two methods of increasing
the difference of temperature between the ascending and the descending columns, are, either by increasing the quantity of pipe, so as to allow the
water to flow a greater distance before it returns
to the boiler; or, by diminishing the diameter of
the pipe, so as to expose more surface in proportion

* This is without reference to friction : the effect will therefore be a little modified by this cause. (See Art. 22 and 13.)

to the quantity of water contained in it, and by this means to make it part with more heat in a given time. (See Art. 61.) The first of these two methods, however, is necessarily limited by the extent of the building that is to be heated, to which the quantity of pipe must be adjusted in order to obtain the required temperature: and, as to the second, there are many objections against reducing the size of the pipes, which will be considered presently. The increase of motive power to be obtained by increasing the height of the ascending column of water is, therefore, what must principally be depended on, when additional power is required to overcome any unusual obstructions.

(19.) In all cases the rapidity of circulation is proportional to the motive power; and, in fact, the former is the index of the latter, and the measure of its amount. For if, *while the resistance remains uniform*, the motive power be increased in any manner or in any degree, the rate of circulation will increase in a relative proportion. Now the motive power, irrespective of retardation by friction, may be augmented, as we have already seen, either by increasing the vertical height of the pipe, by reducing its diameter, or by increasing its length. If by any of these means the circulation be doubled in velocity, then, as the water will pass through the same length of pipe in half the time it did before, it will only lose half as much heat as in the former case, because the rate of cooling is not proportional to the distance through which the water circulates, but to the time of transit. If, then, by sufficiently increasing the vertical height, and doubling the velocity of the circulation, the difference between the temperature of the flow-pipe and the return-pipe be diminished one-half, it might be supposed that the

motive power of the apparatus would remain the
same, and no advantage would appear to be gained
by this means. But this is not exactly the case.
For although, whether we double the vertical
height, or double the horizontal length, we shall,
in either case, increase the velocity of motion;
yet it will require a quadruple increase of vertical height, or a quadruple increase of horizontal
length, to obtain double the original rate of circulation. (See Art. 21.) The increased velocity
is, therefore, indicative of increased power; and,
in a hot-water apparatus, it is the increased velocity of circulation which overcomes any obstructions of a greater amount than ordinary.

(20.) The velocity with which the water circulates in this kind of apparatus, although continually subject to variation, can nevertheless be
calculated theoretically, when certain data are
agreed upon, or are ascertained to exist.

When the two legs of an inverted siphon A,
fig. 6, are filled with liquids of unequal density,
if the stop-cock, z, be turned, so as
to open the communication between
them, the lighter liquid will move
upwards with a force proportional
to the difference of weight of the
two columns, provided the bulk of
the two liquids be equal. If one
leg contains oil and the other contains water, the relative weights
will be about as nine to ten; therefore it will require 10 inches of vertical height of oil to balance
9 inches of water, and no motion will in that case
take place. But when equal bulks of the two
fluids are used, the velocity of motion with which
the lighter fluid is forced upwards is equal to the
velocity which a solid body would acquire in falling, by its own gravity, through a space equal to

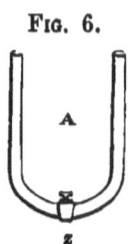

Fig. 6.

the additional *height* which the lighter body would occupy in the siphon, supposing a similar *weight* of each fluid be used. This velocity is easily calculated. A gravitating body falls 16 feet in the first second of time of its descent, 64 feet in two seconds, and so on, the velocity increasing as the square of the time; therefore the relative velocities are as the *square roots of the heights.* Now, in the case of the siphon, which we have supposed to contain a column of water and a column of oil, as the oil ought to be 10 inches high to balance the 9 inches of water, the oil in the one leg will be forced upwards with a velocity equal to that which the water (or any other body) would acquire by falling through one inch of space; and this velocity, we shall find, is equal to 138 feet *per minute.**

(21.) To estimate the velocity of motion of the water in a hot-water apparatus, the same rule will apply. If the average temperature be 170°, the difference between the temperature of the ascending and the descending columns 8°, and the height

* The velocity will be as the square root of 16 feet *per second* to the square root of the additional height which an equal weight of the lighter liquid would occupy, reduced to the decimal of a foot; and as the acquired velocity of a body at the end of a given time is twice as much as the distance it passes through in arriving at any given velocity by accelerated motion, or, in other words, as a body which falls through 16 feet of space, in one second, will proceed at the rate of 32 feet per second afterwards, without receiving any additional impulse; so the velocity found by this rule will be only half the real velocity; and the number thus obtained must be multiplied by 2. The velocity will, therefore, be found by multiplying the *square root* of the difference between the height of the two columns in decimals of a foot by the *square root* of 16, and then, multiplying that product by 2, will give the real velocity *per second.*

The discharge through a siphon, employed to empty casks and other vessels, can also be calculated by this rule: the velocity of motion will be equal to the difference in length of the two legs.

10 feet; when similar weights of water are placed
in each column, the hottest will stand ·331 of an
inch higher than the other; * and this will give a
velocity equal to 79·2 feet *per minute*. If the height
be 5 feet, the difference of temperature remaining
as before, the velocity will be only 55·2 feet *per
minute;* but if the difference of temperature in this
last example had been double the amount stated—
that is, had the difference of temperature been 16°,
and the vertical height of the pipe 5 feet—then the
velocity of motion would have been 79·2 feet *per
minute,* the same as in the first example, where the
vertical height was 10 feet, and the difference of
temperature 8°. This, therefore, proves, in cor-
roboration of what has been already stated (Art.
19), that reducing the temperature of the water,
either by using smaller pipes, or by increasing the
length through which it flows, has the same effect
on the circulation as increasing the vertical height,
leaving out of consideration the question of friction.
The velocity for 3 feet of vertical height, by the
same rule, will be 43·2 feet *per minute*, for 2 feet of
vertical height it will be 36 feet *per minute*, and
for 18 inches of vertical height it will be 30·7 feet
per minute, if the difference of temperature between
the two columns be in each case 8°, the same as in
the former examples. It must here be observed,
however, that, although it appears by these calculations that increasing the vertical height of the
pipe fourfold will produce a double velocity of
circulation, as the water will then pass through
the pipe in half the time, the difference between
the temperature of the flow-pipe and the return-
pipe will be lessened, and the velocity will at last
become a mean rate; so that the mere quadruple
increase of vertical height, without the horizontal

---

* The expansion of water will be found in Table IV.,
Appendix.

length be at the same time increased, will only produce a rate of circulation about one and a half times the original velocity.

(22.) Such is the result of theory; but, although this be true in itself, we shall, in practice, find but few cases that in any way agree with these results, in consequence of other causes modifying the effects. Even in an apparatus in which the length of pipe is not very considerable, where the pipes are of large diameter, and the angles few, a large deduction from the theoretical amount must be made, to represent, with tolerable accuracy, the true velocity. And in more complex apparatus the velocity of circulation is so much reduced by friction, that it will sometimes require from 50 to 90 per cent. and upwards to be deducted from the calculated velocity, in order to obtain the true rate of circulation.* The calculation of the friction of water

---

* It has been found by experiment (*Robinson's Philosophy*, vol. ii., p. 336), that a smooth pipe 4½ inches in diameter, and 500 yards long, yields but one-fifth of the quantity of water which it ought to do, independent of friction. And Mr. George Rennie found (*Philosophical Transactions*, 1831) that the velocity of a half-inch pipe was reduced nearly three-fourths (that is, from 3·7 to 1) by increasing its length from 1 foot to 30 feet; that three semicircular bends reduced the velocity $\frac{1}{30}$ in a short pipe, and 14 such bends reduced it $\frac{1}{10}$ of its velocity, while 24 right-angled bends reduced the velocity nearly two-thirds. The results of M. Prony's experiments led him to adopt the formula $V = 26·79 \frac{\sqrt{DZ}}{L}$ for the discharge through straight pipes : D being the diameter of the pipe; Z the altitude of the head of water; L the length of the pipe in mètres; and V the mean velocity. M. Dubuat's formula for diminution by flexure is $R = \frac{V^2 S^2 n}{3000}$; where R is the resistance; V the mean velocity; S the sine of the angle of incidence; n the number of equal rebounds. Dr. Young (*Philosophical Transactions*, 1808) objects to this theory, and gives a different one, which he considers more nearly to represent the true result.

passing through pipes is alike complicated and unsatisfactory. Though the question has been investigated by some of the most able philosophers and mathematicians, a simple and correct formula on this subject is still a desideratum ; and in the present state of knowledge of the subject it would be almost impossible to determine what would be the resulting velocity of circulation in a hot-water apparatus of complicated construction.*

(23.) In addition to these ordinary causes which impede the circulation, and which are common to all hydraulic experiments, there is another that is still more important, and is peculiar to the hot-water apparatus. The vertical angles in the pipe, or those angles which carry the pipe below the horizontal level, increase the resistance in this case to a very considerable extent, for they oppose not merely a passive resistance by friction, but they engender a force of their own, tending in an opposite direction to that of the prime moving power.

The motion of the heated particles of water is very different in passing through an ascending pipe, compared with that which takes place in a descending pipe. The heated particles rise upwards through an ascending pipe with great rapidity, and when the space occupied by the displaced particles is supplied by water from

---

* In *Robinson's Mechanical Philosophy*, vol. ii., pp. 261–627, will be found much information on this subject, with the results of nearly all the experiments that have been made. Also see Dr. T. Young, *Philosophical Transactions*, 1831 ; *Nicholson's Journal*, vol. xxii., p. 104, and *Philosophical Magazine*, vol. xxxiii., p. 123 ; Mr. G. Rennie, *Philosophical Transactions*, 1831, and *Reports Brit. Sci. Assoc.*, vol. ii., p. 153, and vol. iii., p. 415. In these several works are given the experiments of Bossut, Prony, Dubuat, Eytelwein, Venturi, Borda, and others, which comprise nearly all that is known on this difficult subject.

below, the motion becomes general in one direction, being most rapid in the centre, and gradually decreasing towards the circumference, where, on account of the friction, it becomes comparatively slow. But in a descending pipe the circumstances are very different, the motion being much more like that of a solid body. For as the heated particles are unable to force their way downwards through those which are colder and heavier than themselves, the only motion arises from the cold water flowing out at the bottom, its place being then supplied at the top by that which is warmer, the whole apparently moving together, instead of the molecular action which has been described as the proper motion in an ascending pipe.

(24.) In an apparatus constructed as fig. 7, the motion through the boiler and pipe A B, and through the descending pipe c D, takes place according to the two methods here described. But it is evident that, on motion commencing in the return pipe $y\ z$, in consequence of the greater

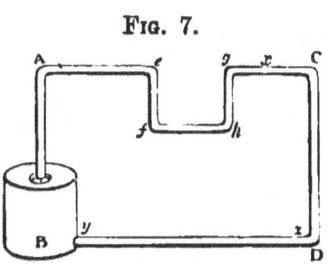

FIG. 7.

pressure of c D than of A B, the water from A will be forced towards e, at the same time that the water in e, f, g, h flows towards c. But when a very small quantity of hot water has passed from the pipe and boiler, A B, into the pipe $e\ f$, the column of water $g\ h$ will be heavier than the column $e\ f$, and therefore there will be a tendency for motion to take place along the upper pipe *towards* the boiler, instead of from it. This force whatever be its amount, must be in opposition to that which occurs in the lower or return pipe, in

consequence of the pressure of C D being greater
than A B; and unless, therefore, the force of motion
in the descending pipe C D be sufficient to overcome this tendency to a retrograde motion, and
leave a residual force sufficient to produce direct
motion, no circulation of the water can take
place.

(25.) An extremely feeble power, as we have
already seen (Art. 15), will produce circulation of
the water in an apparatus where there are no unusual obstructions; but it is a necessary result of
the motive power being so very small, that it is
easily neutralized. So trifling a circumstance as a
thin shaving planed off a piece of wood, and accidentally getting into a pipe, has been known effectually to prevent the circulation in an apparatus
otherwise perfect in all its parts.

It is not sufficient, then, when such an obstruction as the vertical dip from the horizontal level,
shown by the last figure, has to be surmounted,
merely to make the *direct* force of motion sufficient to overcome the antagonist force, and to
leave the smallest possible residual amount, for
the purpose of causing circulation; because an
amount which would be sufficient for this purpose,
as an undivided force, would not be found sufficient
as a residual force.

(26.) In estimating the additional height which
it is necessary to give to the ascending column,
in order to overcome such an obstruction as shown
in fig. 7, it will be necessary to take into account
what is the length and diameter of the pipe
through which the water will have to pass; for on
this depends the difference of temperature between the ascending and descending columns,
which we have seen materially affects the amount
of the motive power of the apparatus. If the
length of pipe be considerable, a somewhat smaller

increase of the vertical height of the ascending pipe will suffice ; but if the length of pipe be short, a greater height must be allowed.* The temperature to which the air surrounding the pipes is to be raised will also modify the result ; for on this will depend the quantity of heat given out by the pipes *per minute*, which likewise affects the temperature of the descending pipe. (Art. 222.)

(27.) Under such a great diversity of circumstances, it would be difficult to form a rule for estimating what ought to be the height of the ascending pipe in such cases ; because not only are these circumstances different in each apparatus, but they likewise differ, in some respects, in the same apparatus in the different stages of its working. The difficulty is also increased by not being able to fix on an absolute minimum measurement, which is sufficient, under all circumstances, to cause a circulation of the water in the common form of the apparatus. There have been instances where apparatus have succeeded, though con-

* This applies merely to the possibility of producing motion, and not to the resulting velocity of the circulation. For it must be borne in mind that, although in every case, by increasing the length of the pipe, or by reducing its diameter, we cause the water to assume a greater difference of temperature between the ascending and the descending columns, and thereby increase the circulation, still, in both these cases, we greatly increase the friction, which therefore considerably detracts from the advantages gained by this greater difference of temperature. And as the friction is a certain quantity, compounded of the square root of the length of the pipe directly, and the diameter of the pipe inversely, it follows that the friction may become so great, by increasing the length and reducing the diameter, as completely to neutralize all beneficial effect. Unless, therefore, the circulation is moderately active, the apparatus will be of such unequal temperature as to render it nearly useless. The utmost caution is necessary, in order that the friction may not become so great as to interfere with the due circulation of the water.

structed on the very worst principles, in consequence of various circumstances having favoured
the result. Thus in an apparatus constructed as
fig. 8, where the pipes
were not more than
three inches apart, the
water circulated with
perfect freedom; but
in this case, not only
was the pipe of considerable length, and without
angles or turns, but the size of the pipe was only
two inches diameter, so that the water cooled
twice as fast as it would have done had pipes of
four inches diameter been used (Art. 61). It is,
however, quite certain that such a distance between the pipes, at their insertion into the boiler,
as that which has just been described, is insufficient, under ordinary circumstances, to give a
steady and good circulation. But when the two
pipes are about 12 inches apart, at the place
of their insertion into the boiler x f, fig. 3 (or
16 inches from centre to centre when the diameter
of the pipe is four inches), it will be sufficient
to produce a good circulation for almost any ordinary length of pipe, when it is not required to
dip below the horizontal level. If this be considered as the minimum height, which, under
ordinary circumstances, will obtain a good circulation when the pipes are not required to dip
below the horizontal level, then an average height
can be estimated for enabling any vertical declination of the pipes to be made.

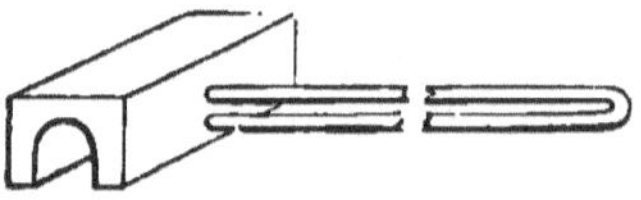

(28.) When the pipe dips below the horizontal
level, the height of the ascending pipe should
generally be just so much greater than the above
dimensions, as the depth which the circulating
pipe is required to dip below the horizontal level;
bearing in mind the circumstances mentioned

Fig. 8.

(Art. 26), which modify the general results.* Thus, suppose the depth of the dip, shown by the dotted line *a b*, fig. 9, to be 24 inches, then the distance *y z* ought to be 40 inches, if the pipes be four inches diameter; that is, 36 inches from centre to centre, or 40 inches from the top of the pipe *y* to the bottom of the pipe *z*; and with these dimensions, as good a circulation will be obtained (excepting the friction from the additional elbows) as when the distance between the top and bottom pipes is 16 inches from centre to centre, in the common form of the apparatus. It will be ob-

FIG. 9.

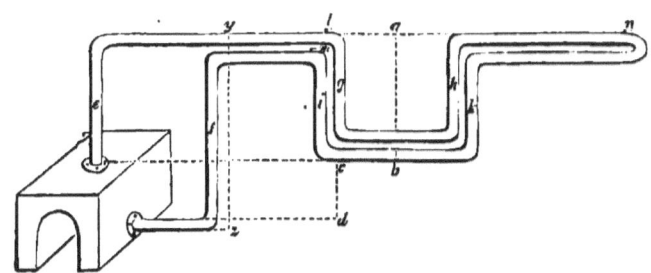

served that, by this arrangement, the distance *c d*, from the under side of the flow-pipe to the upper side of the return-pipe, is just 12 inches, which is the same height that was stated to be necessary to insure a good circulation, on the ordinary plan, without a vertical dip. The reason why this height is sufficient in the present case, notwith-

* So greatly, in fact, do these circumstances affect the general result that it is very possible, in particular circumstances, to make the water descend below the bottom of the boiler to a considerable depth without stopping the circulation. It is therefore evident that the dimensions which are here given for the height of the ascending pipe, relative to the dip, must not be taken as an absolute minimum, but simply as a general rule, which will succeed in all cases. See Art. 30 and 31.

standing the increased friction of the angles, is because there must always be a greater difference
between the temperature of *e* and *f* than between
either *g* and *h*, or between *i* and *k*, or even more
than between both these together; therefore the
tendency to *direct* motion is greater than towards
retrograde motion, in proportion to this difference,
and is sufficient to overcome the increased friction
caused by the vertical declination; while the additional height of 12 inches beyond the height of
the dip, possessed by the descending pipe *f*, is
sufficient to produce circulation of the water.   If
g and *h*, and also *i* and *k*, were very wide apart,
say 40 or 50 feet, instead of being, as usual, only
about three or four feet, the balance of effect,
though still in favour of *direct* motion, would not
be so great as in the last supposed case; because
there would be a greater difference in temperature
between *g* and *h* (that is, *h* would be heavier than
g in a greater degree), which would give a greater
tendency to retrograde motion.   In many cases,
therefore, it will be advisable to make the ascending pipe higher in proportion to the dip than is
here stated, particularly when there are several
such alterations required in the level of the pipes;
and, in all cases, as has been before observed, the
higher the ascending pipe is made, the more rapid
will be the circulation, and, therefore, the more
perfect the apparatus will become.

(29.) The remarks which have been made with
respect to the height of the ascending pipe, relatively to the vertical declination, or dip, below the
level of the horizontal pipe, applies to all the usual
forms which are given to the apparatus.   But
there are peculiar arrangements which may be
adopted that will allow the dip of the circulating
pipe to be much greater than the proportion which
has here been stated.   For, in some cases, the dip

pipes may even pass below the bottom of the boiler to a considerable depth, without destroying the circulation; and from the very extensive use that is now made of the hot-water apparatus for heating buildings of every description, it is very desirable to examine this part of the subject at some length, as its application will, in many cases, entirely depend upon the possibility of making the pipes descend below the boiler.

(30.) In an arrangement of pipes such as fig. 10, the circulation will depend entirely upon the quantity of heat given off by the coil c; for it is evident that when the boiler B and pipe a are heated, the *direct* motion will arise in consequence of the greater weight of the water in the coil c and pipe d, above that which is in the boiler and pipe, B a. But as the water in the pipe e, below the dotted line, will be lighter than that in the pipe f, the tendency in that part of the apparatus will be towards a retrograde motion. The result of these

Fig. 10.

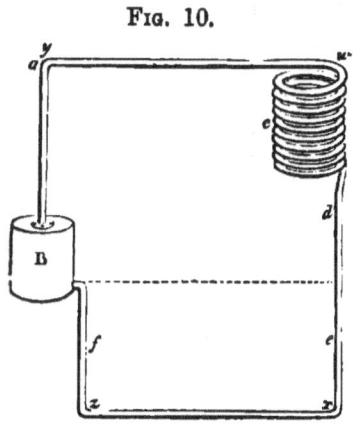

two forces will be that if the water in the whole length of pipe w x is heavier than that of the whole length, y z, in a sufficient degree to overcome the increased friction, circulation of the water will take place; and the velocity of motion will depend upon the amount of this difference in weight.

(31.) Another form, though somewhat more complicated, may be given to this arrangement

of the apparatus. In fig. 11, B represents the
boiler; and the effective or *direct* motion is, in
this case, caused by the water in the coil and pipe
c d being so much
heavier than that
in the boiler and
pipe, B *a*, that it
overcomes the re-
trograde motion
which is produced
by all the other
parts of the appa-
ratus. Thus the
water in *g h*, be-
ing *heavier* than
that in *i k*, and
that in *e f* (below
the dotted line) being *lighter* than that in *l m*,
has in both cases a tendency to retrogression;
and this will be more considerable in proportion
as the pipes *i k*, and *g h*, &c., are more distant
from each other. The motive power, therefore,
entirely depends upon the quantity of heat given
off by the coil; for the water must be cooled down
many degrees, in order to give it a sufficient preponderance over the water in B *a*, to cause a circulation; and the circulation must necessarily be
very slow, and, therefore, the temperature very
unequally diffused.

Fig. 11.

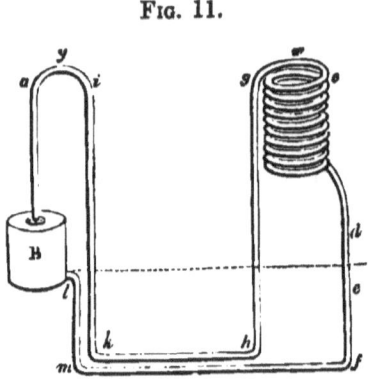

If the coil, in the last two figures, be placed in
any lower position than is here shown, the effect
will be proportionately less in producing circulation; and if placed below the dotted lines, it
would be scarcely possible to obtain any circulation
at all. Nor would there be any circulation if the
coil were omitted, because the mere descent of the
water through a straight pipe would not cool it
sufficiently to give the necessary preponderance

to the descending column, unless some other contrivance for the purpose of cooling the water to an
equal extent were adopted.

(32.) The principle which governs the circulation in these last-mentioned cases is capable of
many applications. And it must be remembered
that, as a coil of pipes produces an enormous
degree of friction in the fluid passing through it,
which must be overcome before circulation can be
produced, a smaller difference of temperature
between the ascending and descending columns

Fig. 12.

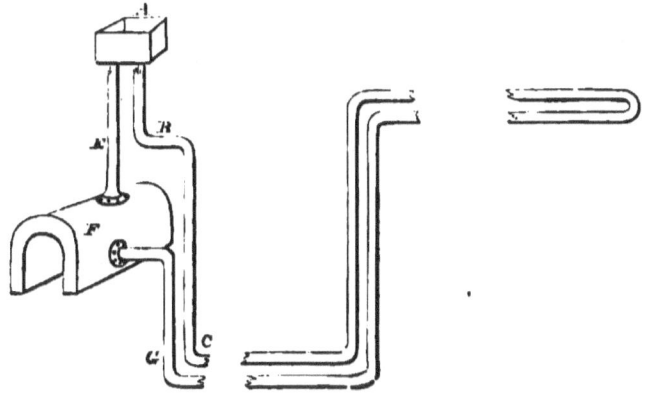

would produce circulation, if the apparatus were
contrived so as to cause less friction to the fluid
passing through it.

In an apparatus constructed as fig. 12 the water
rises directly from the boiler into an open cistern,
A; and it then descends through the pipe B, which
communicates with the bottom or the side of this
cistern. In cases of this kind it has been generally assumed that the water will descend as far
below the boiler as the rising pipe and the cistern
are *above* the boiler; and, practically, it is often
found that this is the case, though the explanation

of the fact must be sought for among a different class of phenomena than those which merely regard the height of the ascending pipe.

(33.) The advantage of conveying the water into an elevated cistern, as shown in the last figure, appears to be twofold.* It allows the freest escape of air and of steam, either of which would prevent the circulation, if it lodged in the part of the apparatus a, y, i, in fig. 11; and which part is in fig. 12 occupied by the open cistern. This cistern also facilitates the circulation, by increasing both the actual as well as the relative weight of the descending column of water; because no part of the descending pipe B can possibly contain steam, as the water will remain in the cistern A until it has become colder than that in the pipe E, and boiler F; and it is evident that, by such an arrangement as fig. 12, this difference of temperature must constantly increase, after heat is applied to the boiler, until it becomes sufficient to give a preponderance to the water in B. And even if the heat were sufficient to raise steam in the pipe E, this would only still further increase the effect, instead of diminishing or even wholly stopping the circulation, as would be the case with an apparatus like fig. 11, under similar circumstances.

(34.) Many other arrangements of the apparatus answering the same purpose as these last three figures, might be contrived; but while these forms are advantageous when difficulties of adaptation have to be surmounted, it must not be imagined that they are recommended above the more simple forms shown in figs. 3, 7, and 9. It requires great judgment in adopting some of these complicated

---

* The Marquis de Chabannes was the first to employ an elevated cistern in this way. His apparatus is described in his pamphlet, published in 1818, on "Warming and Ventilating Buildings."

arrangements ; for many causes may interfere to prevent complete success. It is sometimes very difficult to detect the various causes of interference, and the impediments which arise are often, apparently, so insignificant in their extent, that even when ascertained they are frequently neglected. Those, however, who bear in mind how small is the amount of motive power in any apparatus of this description, will not consider as unimportant, any impediment, however small, which they may detect; but, in the more complicated forms of the apparatus, so many causes become operative, that the reason of failure may sometimes elude the detection of even an experienced practitioner.

(35.) It has occasionally occurred that the circulation of the water in an apparatus has been reversed, the hot water passing along what should be the return-pipe, and the colder water following the course of the flow-pipe. This effect has sometimes been exceedingly puzzling; but it will be found to arise in those apparatus which have but small motive power, and in which the principle has not been followed out of making the water rise to the highest point of the apparatus as soon as possible, and allowing it, *in its return to the boiler*, to give out its heat to the various pipes, coils, or other distributing surfaces which it is intended to heat. If the opposite course to this be adopted, the friction in the flow-pipes often becomes so considerable that the direct and natural action of the ascending column is altered, and the hot water, meeting less resistance by passing through what ought to be the return-pipe, converts it at once into a flow-pipe, and entirely reverses the ordinary action of the apparatus. This is particularly liable to occur in boilers which have but little depth ; and it sometimes happens that the appa-

ratus, when constructed in this way, will operate
most capriciously, the circulation sometimes being
direct and sometimes retrograde, and sometimes
stopping altogether. Whenever, therefore, the
pipes rise to any considerable height above the
boiler, it is very desirable that the most direct
route should be provided for the water to flow
first to the highest elevation, after which in its
return to the boiler, it may be made to pass
through the various pipes, coils, and other heat-
distributing surfaces, which will thus secure its
most efficient action and the most ·perfect circulation.

(36.) The distance through which the water
will circulate in a hot-water apparatus is very considerable ; its limit has not yet been ascertained,
and probably never will be, as it must depend upon
many circumstances totally differing in almost
every apparatus. The higher the water rises
above the boiler, the greater is the length through
which it may afterwards be made to circulate ;
and many apparatus are successfully working
where the water circulates through several hundred
feet of pipe, in a continuous course, before it again
returns back to the boiler to be reheated. In
general, however, it is very desirable to shorten
this circulation as much as possible, and an apparatus will always be more efficient if it can be so
arranged, by altering the position of the boiler, or
the disposition of the pipes, that the water shall
run through two or more distinct and short cir-
culations, rather than through one long one. Nor
is this at all at variance with what has been previously stated about the velocity of circulation
being increased by lengthening the circulating
pipe (Art. 19). For while impediments to the
circulation may be overcome by a considerable
difference of temperature in the flow and return-

pipes, the apparatus will always be more efficient when the temperature of the various parts of the apparatus does not very widely differ.

(37.) When an apparatus is so constructed that the boiler is considerably below the pipes or other surfaces giving out heat, the circulation is sure to be very rapid, and the greatest effect will always be obtained by making the circulation as short as possible, so as to have as little difference as may be in the temperature of the flow and the return-pipe. But when, on the other hand, the pipes and boiler are nearly on a level, it is frequently necessary that a greater difference shall exist between the temperature of the flow and return-pipes, in order to produce a good circulation, and to overcome any obstructions arising from dips in the pipes or any other disturbing cause.

(38.) The necessity for making provision for the escape of the air from the pipes has already been mentioned. It may be observed that, in such forms of the apparatus as are described in the last four figures, the difficulty of its expulsion is much increased, as there are several points where it will collect and stop the circulation, unless proper means be taken to prevent this result. In the apparatus, fig. 9, the air will collect at three points, l, m, and n; and the nature of the outlets provided for its escape will depend, in some measure, upon the plan adopted for supplying the apparatus with water. In some cases open-air pipes will be the best: in other cases air-cocks will be necessary to prevent an over-flow of water through an open-air pipe. It frequently requires the greatest care and the closest attention to discover where the air is likely to lodge, as the most trifling alteration in the position of the pipes will entirely alter the arrangements with respect to the air-vents. Want of attention to this has been the cause of innumerable failures; and the discovery of the places where the air will accumulate is, occasionally, a matter of some difficulty. For although it be true, in a general sense, that the air will rise to the highest part of the apparatus, it will frequently be prevented getting to those parts by alteration in the level of the pipes, and by other causes. This is the case at *m*, fig. 9, where, it will be seen, the air that accumulates in that part of the apparatus is prevented from escaping to a higher level, by the vertical angle at *f* on the one side, and *i* on the other. In the apparatus, fig. 11, the air will accumulate at *y* and at *w*, and must be carried off by proper outlets; and in every case provision must be made for the air to pass, either by a level pipe or by ascending gradients, for in no case can it be made to pass *downwards* (however small the extent) in its passage to the vent provided for its escape.

(39.) When a boiler is open at the top, or merely has a loose cover laid on it, no particular care is necessary respecting the supply of water. It can generally be poured in at the boiler, taking care not to fill it quite full, so as to allow for the expansion of the water when heated, as otherwise it will overflow. But when, as figures 7, 9, 10, 11, and 12, the boiler is close at the top, it is necessary to place a supply cistern on a level with, or above the highest part of the apparatus, so as to keep it always full of water. But as water expands about $\frac{1}{24}$ part of its bulk, when it is heated from 40° (the point of its greatest condensation) to 212°, it is indispensably necessary to provide for a part of the water returning back to the supply cistern when this expansion takes place. The cistern, however, need not contain so much water as $\frac{1}{24}$ part of the whole contents of the apparatus;

for it is found in practice, that a less quantity than this returns back into the cistern on the apparatus being heated. This arises from the fact of the water not reaching to so high a temperature as 212°, and also in consequence of its being generally at a higher temperature than 40° before it is heated, and by both these causes the expansion is considerably lessened. For if the water be raised from 50° to 180°, the expansion will only be about $\frac{1}{36}$ part of its bulk ; and the expansion of the iron itself, by giving an increased capacity to the apparatus, will also tend still further to diminish the quantity of water returned back into the cistern. A very good proportion for general purposes is to make the supply cistern contain about $\frac{1}{30}$ of the whole quantity of water in the pipes and boiler; though, for the reasons above stated, a smaller size will answer in many cases, where economy or convenience requires it to be reduced. A table is given at the end of this volume, showing the contents of pipes of all sizes, and which will enable any one easily to ascertain the correct size of these expansion and supply cisterns for any apparatus.

(40.) The usual plan for a supply cistern is shown at A, in fig. 13. The cistern is placed in some convenient situation, and then attached, by a small pipe, to any part of the appara-tus—usually, to the lower pipe, and it is then less likely to allow of the escape of vapour than if it were fastened to the top of

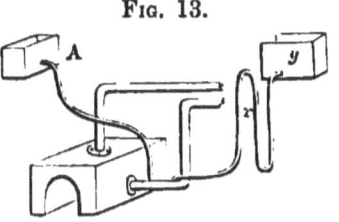

Fig. 13.

the boiler. But a still better plan is to bend the pipe, attached to the cistern, into the form shown by $x\,y$, which is a preventive to the escape of any heat or

E

vapour at that part, as the legs of the siphon x generally remain quite cold.

(41.) One very important part of the subject of expansion is the necessity which exists for allowing sufficient room for the elongation of the pipes when they become hot. Want of attention to this has caused several accidents; for the expansive power of iron, when heated, is so great, that scarcely anything can withstand it. The linear expansion of cast iron, by raising its temperature from 32° to 212°, is ·0011111, or about $\frac{1}{900}$ part of its length, which is nearly equal to $1\frac{3}{8}$ inches in 100 feet. Therefore it is necessary to leave the pipes unconfined, so that they shall have free motion lengthways, to this extent at least. Instead of confining them, as sometimes has been done, facilities should be provided for their free expansion, by laying small rollers under them at various points; for as the contraction on cooling is always equal to the expansion on heating, unless they can readily return to their original position when they become cool, the joints are very likely to get loose, and to become leaky. These rollers may be made simply of a piece of rod iron, about half inch or five-eighths of an inch diameter, which may be fixed in a frame to support the pipes, or they may lie loose on a stone or brick pier, the pipes being supported by this means at about every alternate joint. In very short ranges of pipes this provision is not necessary, as the longitudinal expansion is not sufficient to become a matter of importance.

CHAPTER III.

On the Resistance by Friction—Relative size of Main-Pipes and Branch-Pipes—Vertical and Horizontal Main-Pipes—Small connecting Pipes—Branch-Pipes at difference Levels —Stop-cocks and Valves—Their Use and proper Size—Their place supplied by Cisterns—Inconvenience of them —Remedies.

(42.) WHEN treating, in the preceding chapter, on the velocity of the circulation of water, it was observed that the theoretical velocity is always considerably reduced by friction. Although the calculation of the friction of water, in passing through pipes, is intricate,* the *relative* friction for different sizes of pipes is easily ascertained; and this appears to be nearly all that is necessary to be acquainted with for the purpose of the present inquiry.

The friction occasioned by water passing through small pipes is very much greater than in those which are larger. This arises from two causes—the increased surface with which a given quantity of water comes into contact by passing through a small pipe, and the greater velocity with which the water circulates, in consequence of losing more heat per minute.†

(43.) The relative friction for different sizes of pipes, when the velocity with which the water

---

* See Art. 22.

† See Chap. IV., Art. 61. This latter remark of course only applies to water circulating in a hot-water apparatus: the former applies to all cases of hydraulics.

passes is the same in all, may be seen in the following table :—

Diameter of Pipes ½, 1, 2, 3, 4 inches.
Friction   .   . 8, 4, 2, 1·3, 1.

Taking the friction, in pipes of four inches diameter,
as unity, that of a pipe two inches diameter is
twice as much, and a one-inch pipe four times as
much as the pipe of four inches ; the friction being
as the surface *directly*, and the whole quantity of
water *inversely.**

(44.) The friction which arises from increased
velocity is nearly *as the square of the velocity* ; but
this calculation it is unnecessary to enter into
here, because the velocity of circulation of the
water, in a hot-water apparatus, is constantly subject to fluctuation. For as the friction increases
with the velocity of circulation, so the velocity is
checked by the increased friction ; and it finally
assumes a mean rate, proportioned to the friction
on the one hand, and the theoretical velocity on
the other, calculated according to the rule (Art. 21)
in the preceding chapter.

(45.) Closely connected with the subject of
friction is the question of the proper size for
leading or main pipes. It has been supposed by
many persons that, where two or more circulating
pipes are attached to one main-pipe, the area or
section of the main-pipe ought to be equal to the
sum of the areas of all the branch-pipes. This has
led to most inconvenient arrangements having been
resorted to in particular cases. In some instances,
pipes as large as nine inches diameter have been
used for the main-pipes, where those of four inches
would have answered the purpose infinitely better ;
and other proportions equally erroneous have
frequently been adopted.

* Dr. Young's Hydraulics, *Nicholson's Journal*, vol. iii., p. 31.

(46.) It has been already explained (Art. 23), that the motion of water is more rapid in an upright than in a horizontal pipe. If four branch-pipes be supplied by one upright main-pipe, this latter needs be very little, if any, larger than the circulating pipes; but if only two, or even three, branches are to be supplied by one main-pipe, it will be quite unnecessary in ordinary cases that the upright main-pipe should be any larger than the branches, unless the length of the horizontal pipe be unusually great. If the branches exceed this number, it may be desirable to increase the diameter of the upright main-pipe, in a moderate degree; but the motion of the water through it, however, will be just so much the more rapid in proportion as there are more branches for it to discharge the water into. It is evident that, if the outlet from the boiler be by a pipe four inches diameter, the flow of water will be more impeded, than if a pipe of six inches diameter were used; and the water will therefore become specifically lighter in the boiler than in the descending pipe, in a greater degree in the former case than in the latter; and this will consequently cause a more rapid circulation through the apparatus. But, though the friction of the water will be greater in the ascending pipe by this arrangement, yet it will not be of importance, except when very small pipes are used. For the friction is extremely small in a vertical pipe to an ascending current of water. As a general rule, it may be observed, that all *horizontal* leading pipes require to be much larger in proportion to the branches than is requisite with *vertical* leading pipes, in consequence of the friction being so exceedingly small in the latter case.

(47.) Another advantage will arise from this arrangement, in consequence of a small pipe, *under* *these circumstances,* losing less of its heat than a large one. For, suppose four branch-pipes, four inches diameter, are to be supplied by one main-pipe; one pipe of eight inches diameter would have the same sectional area as the four pipes of four inches diameter; but if, instead of being eight inches diameter, the main-pipe be made only four inches diameter, then the water must travel four times faster through this pipe than it would do through the one of eight inches diameter, in order to supply the same quantity of heat to the branch-pipes. This it will do very nearly; and it may easily be deduced that, under these circumstances, the water will only lose one-half as much heat by passing through the small pipe as it would in passing through the larger one; for the loss of heat which the water sustains is *directly* as the time and the surface conjointly.*

(48.) On the same principle, it will frequently be found exceedingly convenient, when two rooms or buildings somewhat distant from each other are required to be warmed by one boiler, to make the connecting pipe between them smaller than the pipe used for radiating the heat to warm the buildings. For, on the principle already men-tioned, there will be a saving as well of heat as in the cost of the apparatus, by reducing the size of the pipe in that part which is not required to give off heat, but which is merely used to connect different parts together.† In this manner, a pipe of one inch diameter may frequently be made to supply a pipe of four inches diameter; and it will sometimes be found convenient to connect pipes

* See Chap. xiii., Art. 270.

† As all alterations in the size of the pipe, either by enlarg-ing or contracting its diameter, materially alter the velocity of the circulation of the water, care should be taken that these alterations be not made capriciously, and without some decided advantage appears to be attainable by so doing. Venturi

of large diameter with the boiler by this means, when the total length of pipe is not great. But it must always be borne in mind, that the size of the leading or main pipes ought to bear some proportion to the total quantity of water that is required to pass through them. And while it might be extremely convenient, in particular instances, to connect pipes of four inches diameter to the boiler by the intervention of small pipes of one-inch bore, when the total quantity of pipe is small, it would be a fatal mistake to adopt the same arrangement if three or four hundred feet of large pipe were connected to a boiler by means of so small a pipe as is here stated. The connecting pipe therefore ought to bear some proportion to the total quantity of water that has to pass through it. It should be larger in diameter when the quantity of water contained in the apparatus is greater, and smaller when the quantity of water is less.

(49.) Very important results frequently arise from errors with respect to main-pipes which are placed vertically, in consequence of the great velocity with which the water circulates through such pipes. It frequently happens that the pipes which branch out from an upright main-pipe are required to circulate at very different heights; as, for instance, in warming the several floors of a warehouse or manufactory. In this case either one of two methods may be adopted: the pipe may either rise to the highest floor or level first, and, after passing round such uppermost room, descend and circulate round that which is below it; then pro-

found by experiments that enlargements in a pipe reduce the velocity of discharge as follows:—When a given quantity of water was discharged through

A straight pipe in . . . . 109 seconds,
A pipe with one enlargement required 147 „
    „   three . . . . 192 „
    „   five . . . . 240 „

cecd to the next lowest, and so on, till it finally
returns again to the boiler; or each floor may
have a separate range of pipes branching off from
the main-pipe, and finally returning, either together or separately, into the boiler. In the first
of these cases it is obvious that the temperature
of the water and of the pipes will be much greater
in the higher floors than in the lower. For the
water, by having passed through a great length
of pipe before it reaches the lower rooms, will be
much reduced in temperature, and the upper rooms
will be heated long before the others become at all
warm; and at all times the temperature will be
very unequal. To obviate this, each floor should
be warmed separately by its own range of pipes.
But this requires particular management. For, if
the several pipes merely branch out sideways from
a vertical main-pipe, the whole of the hot water
will rise to the upper pipe in consequence of the
extremely rapid circulation in a vertical pipe, and
leave all the lower lateral branches without any
circulation of hot water through them. In order
to avoid this, it is necessary either to have a
separate pipe rising directly from the boiler for
each floor, or some means must be adopted to
check the water in its upward course through the
vertical main-pipe when only one is used for supplying the several different floors. In fig. 14 it
will be perceived that the water which passes up
the pipe B receives a check at B, and thereby, of
course, facilitates the flow of water through the
first horizontal pipe, which would otherwise have
been left stagnant. The same occurs also at c for
the same reason; and by this means a nearly
equal flow of hot water may be obtained; while,
had the supplying main from the boiler passed
directly upwards in a perfectly straight line, all
the hot water would have passed at once into the

upper pipe, and the lower ones would have been left comparatively cold. But even with this arrangement, it is extremely difficult sometimes to prevent all the hot water passing up to the upper floor, and leaving the lower floors unheated. It is, therefore, oftentimes very necessary to make a distinct rising pipe for the lower floors, in order to equalize the flow of hot water to the different levels. The vertical main pipes should likewise diminish in size as they rise upwards. If one main only be used, it is very desirable considerably to reduce

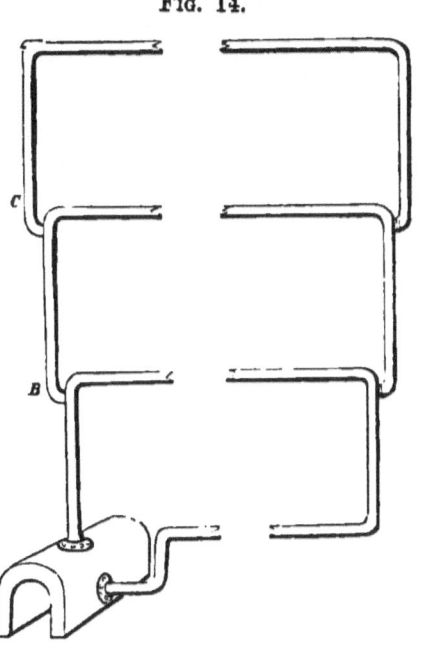

Fig. 14.

the size as it passes upwards, commencing perhaps with 4-inch pipe for the first floor, reducing it to 3 inches as it passes upwards to the second floor, and then reducing it still further in size as it leads upwards to the third floor. And unless some such contrivance be used, it is almost impossible to pre vent the whole of the hot water passing to the upper floor first, and making a most defective circulation in the lower ranges of pipes.

Whenever lateral pipes are taken off at right angles from an upright or vertical pipe which passes in a straight direction to a still higher level, the water will *always* pass by such lateral pipes unless checked in its upward course as here described. But the same effect very often occurs also in horizontal pipes, particularly where the branch pipes are smaller than the mains or leading pipes. One mode of preventing this in the case of vertical pipes has already been shown. But another mode, applicable both to vertical and horizontal main pipes, when these main pipes are sufficiently large in diameter, is the following: Suppose A to be a main pipe, either vertical or horizontal, and B C D branch pipes leading out of this main pipe. The water is supposed to flow in the direction shown by the arrows. Now in a vertical pipe *always*, and in a horizontal pipe frequently—particularly when the branch pipes are much smaller than the main pipe—the hot water will pass by the branches B C and D, and leave them quite cold and without any circulation. To prevent this a small strip of thin sheet copper, extending rather less than half way across the main pipe as shown *b c d*,

Fig. 14B.

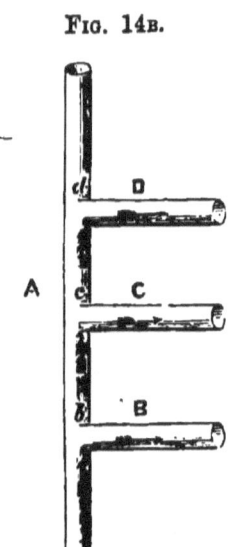

may be fixed on to the upper side of the branch pipe when the main pipe is vertical; or to the opposite side to the direction of the current, when all the pipes lie horizontal. This slight check to the direction of the current is sufficient to cause a circulation in the lateral or branch pipes under almost any possible circumstances, and when done with judgment it causes no sensible diminution of the circulation through the main pipe.

(50.) It is perfectly immaterial how many pipes lead out of or into the boiler; but it will generally much simplify the apparatus if branch-pipes be used, as in fig. 14, instead of making several separate outlet pipes, and various inlet pipes to the boiler.

Very frequently it happens that several branch-pipes are required from the boiler, to circulate nearly at the same level, particularly in horti-cultural buildings, where two or three hothouses are 'required to be warmed by one boiler. This seldom presents much difficulty, unless it be re-quired occasionally to stop off certain of these houses, while the others are heated. In these cases, a complicated and expensive arrangement of cocks or valves become necessary. But here the rule, which has already been given (Art. 46 and 48), for connecting pipes may likewise be followed, where stop-cocks are required occasionally to shut off the communication between different parts of an apparatus, so as only to warm one particular room or part of a building. The cocks used for this purpose need not always be so large as the bore of the pipes. Some judgment, however, must be exercised in all such cases; for, both with connecting pipes and cocks, if the size be very disproportionate, the free circulation of the water will of course be impeded. In many cases a cock of two inches diameter will be sufficiently large to use with pipes of four inches diameter; and a cock of one and a half inch diameter, with pipes of three inches diameter: but, for very small pipes, the relative proportions should perhaps be more nearly equal to the size of the pipes, on account of the increased friction. It should also be ob-served, that when an apparatus has but an in-, different circulation, this alteration in the bore of the water-way will be very objectionable, and

likely still further to impede it, though the exact
result will depend upon a variety of causes, for
which it is not easy to lay down a general rule.
In such an arrangement of pipes as fig. 14, it
is frequently desirable to use small-sized cocks or
valves, for the purpose of checking the flow of
water in particular directions; while in an apparatus like fig. 9 the same proportionate sizes of
cocks might be very injudicious, and greatly impede the circulation.

(51.) When cocks or valves are used to stop off
the circulation of a particular part of the apparatus, it is not sufficient merely to stop the upper or
flow-pipe; but the corresponding pipe which returns the water to the boiler must also be stopped,
otherwise the hot water will circulate backwards
through the return-pipe, and pass into the flow-
pipe, and thus the whole will become heated.
This more particularly applies to those cases
where the boiler is placed at any considerable
depth below the circulating pipes; for then, as
already stated, the circulating power will be much
increased. But it may be laid down as a general
rule, that where the circulation is tolerably good,
it is not enough to place a cock or valve on the
flow-pipe alone; but the return-pipe requires to be
stopped also. Cases undoubtedly sometimes occur
in which a valve or cock placed in the flow-pipe
will effectually shut off the circulation, without
also putting one in the return-pipe. But it will
generally be found that in these instances the
boiler is either too small for its work, or that it
is not worked at its full power, or that there is
something peculiar in the arrangement, which renders the portion of the circulation thus stopped
off naturally more sluggish and inert than the
rest of the apparatus; and therefore a very small
obstruction will be sufficient to stop entirely the

circulation through it. But in almost all the apparatus which have a good circulation and a boiler sufficiently powerful, it will be found necessary to stop the return-pipe as well as the flow-pipe, in order effectually to shut off the circulation.

(52.) In order to avoid the expense of cocks or valves in these cases, an open cistern, as in fig. 15, has sometimes been used. From this cistern all the several flow-pipes are made to branch out; and then, by placing a wood plug into any one or more of these pipes,

Fig. 15.

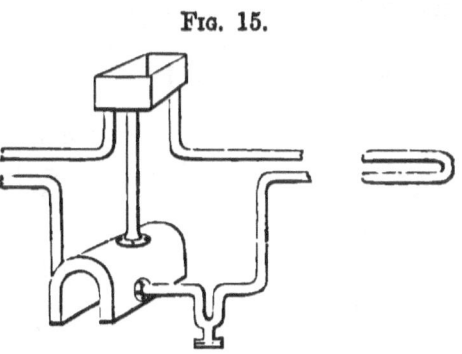

the circulation will be stopped in those particular pipes until the water throughout the whole apparatus becomes heated, when it will generally flow back through the return-pipe, as above mentioned. This inconvenience, however, may be prevented by such a contrivance as shown in the return-pipe of fig. 15, which is simply an inverted siphon of a few inches in depth. This will not prevent the circulation when the flow-pipe remains open; but if that be closed by a plug in the cistern, then the hot water will not return back through the lower pipe. This inverted siphon, however will, in process of time, be liable to be choked up with dirt, which will accumulate in the lower part of the bend; and for the purpose of removing this it will be necessary to make a cap or covering to the lower part of it, which can be removed at pleasure, for the purpose of clearing away any sediment that may accumulate. These cisterns, however, when thus used, are at best but a clumsy way of supplying the place of cocks or valves. The latter are sometimes made to stop off both the flow and return-pipe at one operation; but, in whatever way they are arranged, they are generally one of the most troublesome parts of the apparatus, as cocks and valves of all kinds are liable to get out of repair unless they are in the hands of those who perfectly understand them, and will keep them in proper order.

(53.) Though some of the statements respecting the relative sizes of connecting pipes, main-pipes, and cocks, may appear to be at variance with the laws of hydraulics, they will nevertheless be found correct; because several of the effects are to be referred either entirely to hydrostatic laws, or to a complicated result of hydrodynamics, and therefore they are not to be judged of by simple hydraulic principles. They rest not on mere hypothesis, but may be relied on as practical results obtained under almost every variety of circumstances, and too fully tested to admit of any doubt of their accuracy.

CHAPTER IV.

Permanence of Temperature—Rates of Cooling for different sized Bodies—Proper sizes for Pipes—Relative Size of Pipes and Boiler—Various Forms of Boilers, and their Peculiarities—Boilers heated by Gas—Objections against contracted Water-way in Boilers—Proper Size of Boilers for any given lengths of Pipe—What constitutes a good and efficient Boiler—Durability of different Materials for Boilers—Effect of Impure Fuel.

(54.) ONE of the greatest advantages which the plan of heating by the circulation of hot water possesses over all other inventions for distributing artificial heat is, that a greater permanence of temperature can be obtained by it than by any other method. The difference between an apparatus heated by hot water and one where steam is made the medium of communicating heat, is not less remarkable in this particular than in its superior economy of fuel.

(55.) It seldom happens that the pipes of a hot-water apparatus can be raised to so high a temperature as 212°; and, in fact, it is not desirable to do so, because steam would then be formed, and would escape from the air-vent or safety pipe, without affording any useful heat. Steam-pipes, on the contrary, must always be at 212° at the least, because, at a lower temperature, the steam will condense. A given length of steam pipe will therefore afford more heat than the same quantity of hot-water pipe: but if we consider the relative permanence of temperature of the two methods, we shall find a very remarkable difference in favour of pipes heated with hot water.

(56.) The weight of steam at the temperature of 212°, compared with the weight of water at 212°, is about as 1 to 1694; so that a pipe which is filled with water at 212° contains 1694 times as much *matter* as one of equal size filled with steam. If the source of heat be cut off from the steam-pipes, the temperature will soon fall below 212°, and the steam immediately in contact with the pipes will condense: but in condensing the steam parts with its *latent heat*; and this heat, in passing from the latent to the sensible state, will again raise the temperature of the pipes. But as soon as they are a second time cooled down below 212°, a further portion of steam will condense, and a further quantity of latent heat will pass into a state of heat of temperature; * and so on, until the whole quantity of latent heat has been abstracted, and the whole of the steam condensed, in which state it will possess just as much heating power as a similar bulk of water at the like temperature; that is, the same as a quantity of water occupying $\frac{1}{1694}$ part of the space which the steam originally did.

(57.) The specific heat of uncondensed steam compared with water is, for *equal weights*, as ·8470 to 1: but the latent heat† of steam being estimated at 1000°, we shall find the relative heat

---

* The heat of temperature is that which is appreciable by the thermometer; and the term is used in contradistinction to *latent* heat, which is not capable of being measured in a direct manner by any instrument whatever.

† The results of different experiments on the subject of the *latent* heat of steam, although somewhat various, are yet sufficiently near for all practical purposes. Watt's experiments gave 900° to 950°; Lavoisier and Laplace, 1000°; Mr. Southern, 945°; Dr. Ure, 967° to 1000°; and Count Rumford, 1000°.

obtainable from *equal weights* of condensed steam and of water by reducing both from the temperature of 212° to 60°, to be as 7·425 to 1; but for *equal bulks* it will be as 1 to 228; that is, bulk for bulk, water will give out 228 times as much heat as steam, on reducing both from the temperature of 212° to 60°. A given bulk of steam will therefore lose as much of its heat in one minute as the same bulk of water will lose in three hours and three quarters.

(58.) When the water and the steam are both contained in iron pipes, the rate of cooling will, however, be very different from this ratio; in consequence of the much larger quantity of heat which is contained in the metal itself than in the steam with which the pipe is filled.

The specific heat of cast iron being nearly the same as water (see Table V., Appendix), if we take two similar pipes, four inches in diameter, and one quarter of an inch thick, one filled with water, and the other with steam, each at the temperature of 212°, the one which is filled with water will contain 4·68 times as much heat as that which is filled with steam: therefore if the steam-pipe cools down to the temperature of 60° in one hour, the pipe containing water would require four hours and a half, under the same circumstances, before it reached the like temperature. But this is merely reckoning the effect of the pipe and of the fluid contained in it. In a steam apparatus this is all that is effective in giving out heat; but in a hot-water apparatus there is likewise the heat from the water contained in the boiler, and even the heat from the brickwork around the boiler, which all tends to increase the effect of the pipes, in consequence of the circulation of the water continuing long after the fire is extinguished; in fact, so long as the

F

water is of a higher temperature than the surrounding air of the room. From these causes,
the difference in the rate of cooling of the two
kinds of apparatus will be nearly double what is
here stated; so that a building warmed by hot
water will maintain its temperature, after the fire
is extinguished, about six or eight times as long
as it would do if it were heated with steam.

This is an important consideration, wherever
permanence of temperature is desirable; as, for
instance, in hothouses, conservatories, and other
buildings of a similar description. And even in
the application of this invention to the warming of
dwelling-houses, manufactories, &c., this property,
which water possesses, of retaining its temperature
for so long a time, and the very great amount of
its specific heat, prevents the necessity for that
constant attention to the fire which has always
been found so serious an objection to the general
use of steam apparatus.

(59.) The velocity with which a pipe or any
other vessel cools, when filled with a heated fluid,
depends principally upon two circumstances—the
quantity of fluid that it contains, relatively to its
surface and the temperature of the air by which it
is surrounded; or, in other words, the excess of
temperature of the heated body above that of the
surrounding medium. The subject of the radiation
of heat, and the rate at which a heated body cools,
under various circumstances, will be fully considered
in another chapter (see Chapter XII.). But for
temperatures below the boiling point of water, and
under such circumstances as we are now considering
with regard to hot-water pipes, the velocity of
cooling may be estimated simply in the ratio of the
excess of heat, which the heated body possesses
above the temperature of the surrounding air. The
variation in the rate of cooling, arising from a

difference in proportion of the superficies to the mass, is, for bodies of all shapes, *inversely, as the mass divided by the superficies.* Therefore, the relative ratio of cooling, for any two bodies of different shapes and temperatures, is the inverse numbers obtained by dividing the mass by the superficies, multiplied by the direct excess of heat above the surrounding air; provided the temperature of the heated bodies be below 212°. Thus, suppose the relative ratio of cooling be required, for two cisterns filled with hot water, one a cube of 18 inches, at the temperature of 200°; the other a parallelopiped, 24 inches long, 15 inches wide, and 3 inches deep, at the temperature of 170°; the surrounding air in both cases being 60°. Then, as,

	INCHES.		INCHES.		
The cube contains	5832,	divided by	1944,	the superficies	= 3·0
The parallelopiped contains.	1080,	do.	954,	do.	= 1·13

The *inverse* of these numbers is, to call the cube 1·13, and the parallelopiped 3·0. Then multiply 1·13 by 140 (the direct excess of temperature of the cube), and the answer is 158·2: and multiply 3·0 by 110 (the direct excess of temperature of the parallelopiped), and the answer is 330·0; therefore the parallelopiped will cool, in comparison with the cube, in the proportion of 330 to 158, or as 2·08 to 1. So that if it required two hours to cool the cube any proportional part of its excess of heat, the other vessel will lose the same proportional part of its excess of heat in one hour.

(60.) It is evident that these different velocities of cooling are quite independent of the total effect that the respective bodies will produce in warming a given space. For as the cube contains six times as much water as the other vessel, so it would warm six times as much air, if both vessels were of the same temperature. But if six of the oblong vessels

F 2

were used, they would heat just the same quantity
of air as the cube; but the latter would require
rather more than two hours and a half to do what
the oblong vessels would accomplish in one hour,
supposing the temperature to be the same in both
cases. In the previous example, the temperatures
are supposed to be different; otherwise the relative ratio of cooling of the two vessels would have
been as two and a half to one, instead of two to
one, as stated.

(61.) In estimating the cooling of round pipes,
the relative ratio is very easily found, because the
inverse number of *the mass divided by the superficies*,
which gives the relative cooling for *all* bodies,
is exactly equal to the *inverse of the diameters*.
Therefore, supposing the temperature to be alike
in all,

If the diameter of the pipes be 1,   2,   3,   4 inches.
The ratio of cooling will be........ 4,   2,   1·3, 1.

That is, a pipe of one inch diameter will cool four
times as fast as a pipe of four inches diameter;
and so on with the other sizes. These ratios,
multiplied by the excess of heat which the pipes
possess above that of the air, will give the relative
rate of cooling when their temperatures are different, supposing they are under 212° of Fahrenheit. But if the temperatures are alike in all, the
simple ratios given above will show their relative
rate of cooling, without multiplying by the temperatures. When the pipes are much above 212°,
as, for instance, with the high-pressure system of
heating, the ratio of cooling must be calculated by
the rules given in Chapter XII.

(62.) The unequal rate of cooling of pipes of
various sizes, renders it necessary to consider the
purpose to which any building is to be applied
that is required to be heated on this plan. If it

be desired that the heat shall be retained for a
great many hours after the fire is extinguished,
then large pipes will be indispensable; but if the
retention of heat be unimportant, then small pipes
may be advantageously used. It may be taken as
an invariable rule, that in no case should pipes of
greater diameter than four inches be used in any
ordinary building, because, when they are of a
larger size than this, the quantity of water they
contain is so considerable, that it makes a great
difference in the cost of fuel, in consequence of the
increased length of time required to heat them. .
(See Art. 108.) For hothouses, greenhouses,
conservatories, and such like buildings, pipes of
four inches diameter will generally be found the
best; though, occasionally, pipes of three inches
diameter may be used for such purposes, but rarely
any of a smaller size. In churches and manufactories, &c., pipes of either four or three inches
diameter will generally be found most convenient,
in consequence of the difficulty which almost
always occurs of placing a sufficient quantity of
pipes of smaller diameter in these buildings. But
in dwelling-houses pipes of two inches diameter
will generally be preferable, for they will retain
their heat sufficiently long for ordinary purposes,
and their temperature can be sooner raised than
larger pipes, and, on this account, a somewhat less
number of *superficial feet* will suffice to warm a
given space.

(63.) In adapting the boiler to a hot-water
apparatus, it is not necessary, as is the case with
a steam boiler, to have its *capacity* accurately proportional to that of the total quantity of pipe which
is attached to it :* on the contrary, it is sometimes
desirable even to invert this order, and to attach
a boiler of small *capacity* to pipes of large size. It

* See Chapter X. on heating by steam.

is not, however, meant, in recommending a boiler
of small capacity, to propose also that it shall be
of small superficies; for it is indispensable that it
should present a surface to the fire proportional
to the quantity of pipe it is required to heat; and
in every case, the larger the surface on which the
fire acts, the greater will be the economy in fuel,
and the greater also will be the effect of the
apparatus.

(64.) The sketches of the boilers, figs. 16 to 30,
are several different forms which present various
extents of surface in proportion to their capacity.

All except the first two, however, have but a
small capacity, relatively to their superficies, compared with boilers which are used for steam.
There is no advantage whatever gained by using
a boiler which contains a large quantity of water.
For, as the lower pipe brings in a fresh supply of
water, as rapidly as the top pipe carries the hot
water off, the boiler is always kept absolutely full.
The only plausible reason which can be assigned
for using a boiler of large capacity is, that as the
apparatus then contains more water, it will retain
its heat a proportionably longer time. This,
though true in fact, is not a sufficient reason for
using such boilers: for the same end can be
accomplished, either by using larger pipes, or by
having a tank connected with the apparatus,
which can be so contrived, by being enclosed in
brick or wood, or some other non-conductor, as to
give off very little of its heat by radiation, and yet
be a reservoir of heat for the pipes after the fire
has been extinguished. If this tank communicate
with the rest of the apparatus by a stop-cock, the
pipes can be made to produce their maximum
effect in a much shorter time than if this additional quantity of water had been contained in the
boiler; and a more economical and efficient appa-

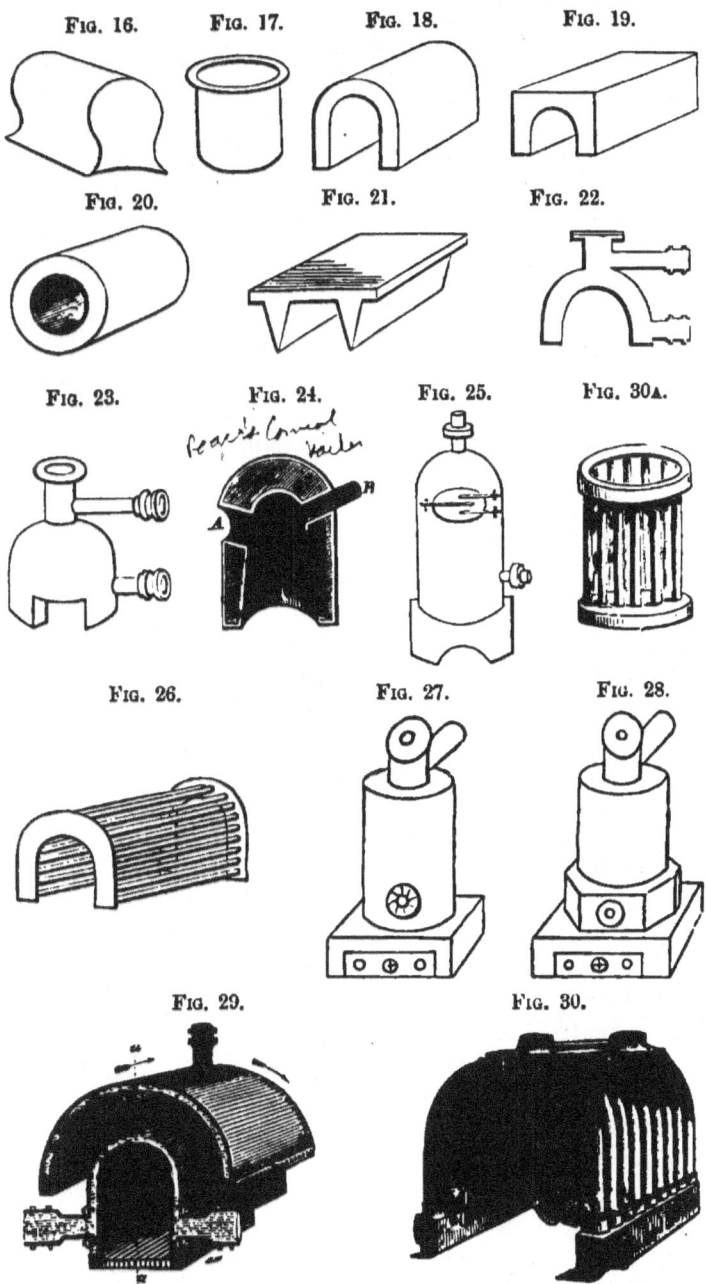

FIG. 16. FIG. 17. FIG. 18. FIG. 19.

FIG. 20. FIG. 21. FIG. 22.

FIG. 23. FIG. 24. FIG. 25. FIG. 30A.

FIG. 26. FIG. 27. FIG. 28.

FIG. 29. FIG. 30.

ratus will be obtained. The circulation will
likewise be more rapid from a boiler which contains but a small quantity of water, because the
fire will have greater effect upon it, and will
render the water which is contained in it relatively
lighter than that which is in the descending or
return-pipe.

(65.) The boilers, figs. 16 and 17, are but
seldom used for hot-water apparatus. Fig. 18 is
an excellent form of boiler; it is, in fact, the very
best boiler for general use that has ever been
made, and has been far more extensively used
than any other. It is generally made of wrought-
iron. Fig. 19 is something similar, though decidedly inferior, on account of the inconvenience of
a flat top; which not only prevents the easy flow
of the hot-water to the ascending pipe (which
ought always to be placed on the top), but also
the flues do not act so efficiently on the flat top of
this boiler. Fig. 20 is a good boiler, but is best
for either very small or very large apparatus (and
not for intermediate sizes), depending on the mode
of setting; which subject will be described in the
following chapter. Fig. 21 is only suitable for
a very small apparatus. Figs. 22 and 23 (the
former of which is a section and the latter an
elevation) represent a cast-iron circular boiler of
very efficient construction, and suitable for either
large or small apparatus. Fig. 24 is a section of
the boiler known as Roger's conical boiler; which
is a circular boiler, externally resembling the fig. 25.
This boiler has undergone much alteration of form
since its first invention. It was first open at the
top, and the fuel supplied there; this, however, is
now supplied at A, and B is the smoke flue. Fig. 25
is a boiler nearly similar to the last, but contrived
so as to be used without any brickwork. The
radiation of heat from the surface of this boiler is

of course considerable, and is generally entirely wasted; though when the boiler is placed inside the room or building to be warmed, this loss may be avoided. Fig. 26 is a boiler consisting of a double row of pipes (of which the external row alone is shown), connected at each end by an arch, by which the water is supplied to the pipes forming the body of the boiler. This boiler heats rapidly, but is necessarily very wasteful of fuel, as no flues can be formed in setting it. Fig. 27 is also a boiler to be used without brickwork, and for many purposes is very useful, particularly where the heat from the external surface of the boiler can be beneficially employed. Fig. 28 is another boiler very similar to the last, but can be adapted to a much smaller size, as the fire is contained in a fire-clay receptacle, by which means a smaller fire may be made to keep alight for many hours without attention; and by this contrivance a very much smaller boiler may be made to act efficiently than it otherwise could do. Both these boilers answer extremely well. The boilers, figs. 29 and 30, are more remarkable for their ingenuity than for their practical utility, and it is probable that the complication of their construction renders them peculiarly liable to accidents. The boiler, fig. 30A, is a useful boiler in some cases when used of small size, but it becomes wasteful when made of large dimensions.

(66.) Boilers heated by gas are occasionally used, and they often possess advantages when coal or coke cannot be conveniently employed. The cost of burning gas in this way is however considerable. In the earlier editions of this work the cost of burning gas was estimated at six times the cost of coal, when the products of combustion were *not* allowed to escape into the open air; and at twelve times the cost of coal when the products of combustion passed away directly into the open air. These estimates were perfectly correct when they were given ; but since that time improved methods of burning gas have been invented, by mixing the gas with about twelve times its volume of atmospheric air, previous to the gas reaching the place of its actual combustion. This invention is popularly known by the term "atmospheric gas." By means of this invention and also by considerable reductions in the price of gas relatively to the price of coal, the cost of burning gas in this manner may now be estimated at about three times the cost of coal. In experiments made with this method of burning Carburetted Hydrogen Gas, it has been found that 720 cubic feet of gas gave the same amount of heat that could be obtained by 75 lbs. of good average coal, and this at the price of three shillings and sixpence per thousand cubic feet for the gas, and twenty-four shillings per ton for the coal, would be nearly in the proportion of about three to one. When gas is to be burned in this way, the boiler ought to be made of copper ; and perhaps the best form of boiler for the purpose is like the upper part of that shown in fig. 25, which would then represent a shallow dome-topped boiler, well suited to receive the burners now generally used for the "atmospheric gas." Or another form of boiler for this purpose consists of a copper drum with several vertical copper tubes, approaching nearly to a locomotive boiler standing vertically instead of horizontally.*

(67.) There are many other forms of boilers which have been proposed for the hot-water apparatus, and, in fact, the multiplication of them appears almost without limit. It is, perhaps, scarcely any exaggeration to say they amount to

* See also Art. 306 for remarks on improved method of burning Carburetted Hydrogen Gas.

hundreds. When strictly considered, however, there is scarcely one that presents any real novelty, and generally they are mere colourable adaptations of some one of those which have been described; and the wonderful effects which sometimes are attributed to them arise either from the parties being deceived in the results, or from their being unacquainted with what has previously been accomplished by others. The principles on which a boiler must be constructed in order to become efficient, are as fixed and immutable as the laws of nature; and the modes by which these principles are to be applied are all determinable by experience, and can be correctly judged of by certain rules, beyond the possibility of error. The mode of doing this may, perhaps, in some degree tend to prevent the erroneous notions which frequently prevail upon this subject.

(68.) The adoption of boilers of small capacity having been recommended (Art. 64), it is necessary to accompany the recommendation with a caution against running into extremes; for this error has been the cause of failure, and of the inefficiency of the apparatus in many instances. The boiler, fig. 21, is an instance of this sort, in which an absurd extreme has occasionally been adopted. The contents of a boiler of this shape sometimes do not exceed a couple of gallons, even when applied to a very large furnace; and though this boiler presents a large surface to the fire, the space allowed for the water is so small that the neutral salts and alkaline earths, deposited by the water which evaporates from the apparatus, contract the water-way, already far too small, and effectually impede the circulation, and also prevent the full force of the fire from acting on the water. In a very small apparatus, however, this form of boiler has occasionally been used with advantage, the fire being less intense.

(69.). But perhaps the more immediate cause of failure of this shaped boiler arises from a different and very singular circumstance. The quantity of water which it contains being so small, and the heat of the fire, therefore, when the furnace is large, being very intense upon it, a repulsion is caused between the iron and the water, and the latter does not receive the full quantity of heat. This extraordinary effect is not hypothetical; it has been proved to exist by the most satisfactory experiments; particularly some which were made by the Members of the Franklin Institution of Pennsylvania. The repulsion between heated metals and water they ascertained to exist, to a certain extent, even at very moderate degrees of heat; being appreciably different at various temperatures below the boiling point of water. But, as the temperature rises, the repulsion increases with great rapidity; so that iron, when red hot, completely repels water, scarcely communicating to it any heat, except, perhaps, when under considerable pressure.*

The boiler in question, however, seldom or never reaches the temperature of luminosity, though it is still sufficiently high to make a con-

---

* Mr. Jacob Perkins brought this curious fact prominently forward during his ingenious experiments on high-pressure steam. It has, however, long been known as a philosophical fact, and was first observed in 1756, by M. Leindenfrost. M. Klaproth subsequently investigated it, and published some experiments on the subject (*Nicholson's Journal*, vol. iv., p. 208). In the "Parliamentary Report and evidence on the Scotch Distilleries for 1798 and 1799" (p. 610), there is a quotation from "Chaptal's Chemistry," showing that he was well acquainted with the fact; and also some experiments by M. Zeigler, by which he ascertained that a drop of water took 89 seconds to evaporate from metal heated to 520° Fahrenheit,

siderable difference in the heating of the water. Added to this, the form of it prevents the full effect of the heat being communicated to the pipes ; for the extreme smallness of the water-way prevents the rapid communication between the various parts, and therefore the upright or flow-pipe receives its principal supply of heat from that portion of the boiler which is imme-diately beneath where it is fixed, instead of that equable communication of heat from all parts, which is the ordinary process in boilers of good proportions. There is likewise a probability that steam would form in this boiler, which would still farther interfere with the circulation of the water. But were the water-way to be enlarged, all these inconveniences and probable causes of failure would proportionably decrease. All these causes of inefficient action may not exist simultaneously, yet they may act at different stages of the working of the apparatus. But they all apply equally to every boiler in which the rational limits of the surface, relatively to the capacity or contents of the boiler, have given place to wild chimeras and fanciful no-tions, not based on sound principles of philosophy.

These remarks are exemplified in a boiler which has received the name of the Trentham boiler, and was at one time much lauded for its extraordinary economy. In shape it is exactly like the Cornish

but that it only required one second when the metal was at 300° Fahrenheit.

In the recent experiments "On the Explosion of Steam Boilers," by the Franklin Institution of Pennsylvania, a very thick iron ladle was perforated with a number of small holes, and then made red hot. When water was poured into this ladle, none of it escaped through the holes, until the ladle cooled down below redness ; and the quantity which after-wards passed through increased with every reduction of the temperature, the difference being quite appreciable even be-tween the temperature of 60° and 80° Fahrenheit.

boiler, fig. 20. But the theory was propounded that
the less water it contained the greater the effect
would be; and in large boilers of eight or ten feet
long, the water-way was reduced all round the boiler
to about an inch and a quarter. Practical working
proved this to be a fallacy: and the much lauded
Trentham boiler has now resolved itself into the
ordinary Cornish boiler, which has been known
and extensively used for a century and a half.

(70.) It is obvious that the extent of surface
which a boiler ought to expose to the fire should
be proportional to the quantity of pipe that is
required to be heated by it; and it is not difficult
to estimate these relative proportions with sufficient accuracy, notwithstanding the various circumstances which modify the effect.

(71.) It has been proved by experiments that
four square feet of surface of an iron boiler will
evaporate one cubic foot of water per hour, when
exposed to the *direct action* of a tolerably strong
fire. This, however, requires free exposure to
the radiant heat of the fire; for the heat communicated to the flue surfaces is only equal to
one-third of that which is derived by the direct
action of the fire, acting upon the bottom or sides
of the boiler.* And it can be ascertained by

* Mr. Robert Stephenson's experiments on this subject
clearly prove this proportion between the relative heating
of flue surface and boiler, surface to be correct. In his experiments the flues consisted of tubes passing through the
water; and he found that while six square feet of boiler surface evaporated six gallons of water in 38 minutes, the flue
surface, consisting of 24½ square feet, had, in the same time,
evaporated eight gallons. This will be found equal to evaporating one cubic foot of water per hour, from 3·7 square feet
of the boiler surface exposed to the direct action of the fire;
and the same quantity of water evaporated by 11·9 square
feet of the flue surface; being in the proportion of 1 to 3.
(See *Wood's Treatise on Railroads*, 3rd edition, p. 524.) In the
best locomotive engines, the power of the boiler is equal to

calculation, that the same extent of heating surface which will evaporate one cubic foot of water per hour from the mean temperature of 52°, will be sufficient to supply the requisite heat to 232 feet of pipe four inches diameter, when the temperature of this pipe is to be kept at 140° above

one cubic foot of water evaporated per hour by 1·7 square foot of boiler surface exposed to the direct action of the fire. This appears to be almost the greatest effect that can be produced at present. (*Experiments on Great Western Railway*, 1838.) In 1834, the Chevalier Pambour found the average of the engines on the Liverpool and Manchester Railway to be one cubic foot of water evaporated per hour, from 2·5 square feet of surface exposed to the direct action of the fire. The flue surface in all these experiments was calculated as equal to one-third that of the boiler surface. This high evaporating power can only be maintained when a very powerful draught is produced by mechanical means; and in all these cases there is a very great waste of fuel. In the original experiments of Watt, on steam boilers, he found that the average of eight square feet of boiler surface was required to evaporate one cubic foot of water per hour. This proportion is still very generally used; and by employing a large heating surface, economy of fuel is always produced. This is strikingly exemplified in the Cornish engines, the boilers of which have a larger surface than any others; and the consumption of fuel (per horse's power of the engine) throughout Cornwall only averages one-fourth of that of the manufacturing districts of England. The whole of this saving, however, is not due to increased surface of the boilers; a large proportion of it is owing to the mode of using the steam expansively, which is there carried to the extreme. (See Art 95.)

Pambour has questioned the accuracy of the estimate for the proportionate effect of flue and boiler surfaces, which he considers do not differ so much in heating power as is generally supposed; but the doubt only applies to the boilers of locomotive engines where a very powerful blast is applied, and he agrees that, in other cases, the proportionate heating power of 3 to 1 for the boiler surface and flue surface is nearly correct. (*Pambour's Treatise on Locomotive Engines*, 2nd edition, p. 269.)

Tredgold also gives the proportions as about three feet of boiler surface exposed to the direct action of the fire, to evaporate one cubic foot of water per hour, when the furnace bars are one-fourth the area of the boiler surface.

that of the surrounding air.* From this, then, it appears that one square foot of boiler surface *exposed to the direct action of the fire*, or three square feet of flue surface, will be sufficient in a hot-water apparatus to supply the necessary heat to about 58 superficial feet of pipe; or in round numbers, the proportion may be stated as one foot of boiler to 50 feet of pipe. As this, however, is almost the maximum quantity of pipe which can be heated, or, in other words, the maximum effect which can be produced without mechanical means of producing draught, it is very desirable in all cases to allow an increased surface of the boiler; bearing in mind that not only will economy of fuel be thereby produced, but the apparatus will be much easier managed, and thus become more effective and certain in its operation. The following Table gives the *maximum* quantity of pipe which a boiler will heat, calculated by the above rule, and supposing the best coals alone to be used :—

TABLE II.

Surface of Boiler exposed to the direct action of the Fire.	4-in. Pipe.	3-in. Pipe.	2-in. Pipe.
4 square feet will heat	200 feet,	or 266 feet,	or 400 feet.
6	300	400	600
8	400	533	800
10	500	666	1000
14	700	933	1400
20	1000	1333	2000

* It appears by calculation (Art. 99), that a four-inch pipe will lose ·851 of a degree of heat per minute, when the excess of its temperature above the circumambient air is 125°. If, therefore, this excess were 140°, the loss per minute would be ·953 of a degree of heat. Calculating, therefore, this loss to be 57·18 degrees per hour, and estimating also (the latent heat of steam being 1000°) that the cubic foot, or 1728 cubic inches of water evaporated, has received 1160° of heat, and that one foot in length of a four-inch pipe contains 150·7 cubic inches of water, we shall obtain $\dfrac{1728 \times 1160}{150 \cdot 7 \times 57 \cdot 18} = 232$ feet, as stated in the text.

A small apparatus ought always to have more surface of boiler, in proportion to the length of pipe, than a larger one; as the fire is less intense, and burns to less advantage in a small than in a larger furnace.   The effect also depends greatly upon the quality of the coal, the height of the chimney, the rapidity of draught, the construction of the furnace, and many other particulars; and it will always be found more economical, as regards the consumption of fuel, to work with a larger surface of boiler at a moderate heat, than to keep the boiler at its maximum temperature.*

(72.) But beside all these causes that modify the effect, there is another, that will alter the proportions which may be employed.   The data from which the calculation of the boiler surface is made, assume the difference to be 140° between the temperature of the pipe and the air of the room which is heated; the pipe being 200°, and the air 60°.   But if this difference of temperature be reduced, either by the air in the room being higher, or by the apparatus being worked below its maximum temperature, then, in either case, a given surface of boiler will suffice for a greater length of pipe.   For if the difference of temperature between the water and the air be only 120° instead of 140°, the same surface of boiler will supply the requisite degree of heat to one-sixth more pipe; and if the difference be only 100°, the same boiler will supply above one-third more pipe

* A useful formula for calculating the effect of boilers and furnaces in most common cases is as follows :—

One square foot of boiler surface exposed to the *direct action* of the fire, will boil 11 gallons of water, from 52° to 212° Fahrenheit per hour; or, in other words, it will add 160° of heat to 11 gallons of water.   Or the same surface will *evaporate* 1½ gallon of water per hour.

The area of the furnace bars should be about one-sixth that of the boiler surface (see note to Art. 79).

than the quantity before stated. It will, therefore, sometimes occur in practice (where economy in construction is the primary object), that the quantity of pipe in proportion to a given surface of boiler may be even increased beyond the amount which is given in the preceding Table; because, in forcing-houses, for instance, the temperature of the air will always be above $60°$; and in the warming of churches, warehouses, or other large buildings, the temperature of the water will generally be considerably below $200°$—the pipe not being required to be worked at its greatest intensity,— and, therefore, in both these instances, a larger proportion of pipe may be applied to a given sized boiler. It therefore follows, that although a smaller boiler surface would really supply a sufficient quantity of heat, under strict management and constant attention, it will generally be better not to reduce the size of the boiler below what has here been stated; for not only will the apparatus need less attention, but also the required temperature of the building can be thus much sooner attained, as well as more easily continued. A very good proportion, suitable for nearly every purpose, is to allow about one foot of boiler surface (calculated, as already described, Art. 71) to about 40 superficial feet of pipe, or other radiating surface, or about one-fifth more boiler surface than the preceding Table states.

(73.) It may be desirable here to state what are the peculiar characteristics of a good boiler for this purpose, and how the qualifications of each particular shape are to be judged of. A minute detail of the peculiarities of each of the various forms would scarcely be worth the space such a description would require. The principal recommendations of a boiler are, that it shall expose the largest surface to the fire in the smallest space;

that it shall effectually absorb the heat given out from the fuel, so that as little heat as possible shall escape up the chimney; that it shall allow free circulation of the water throughout its entire extent; and that it shall not be liable to get out of order, nor rapidly deteriorate by continued use. The first of these qualifications is of itself a compound question. We have seen (Art. 71) that any surface exposed to the direct action of the fire, or, in other words, to the *radiant heat*, receives *three times* as much heat as a similar surface exposed merely to the conducted heat, or that which is afforded by the products of combustion after they are thrown off from the burning fuel. Here, then, is a very important distinction in boilers; for as radiant heat passes in straight lines in every direction, it follows that the largest possible surface ought to be exposed immediately over the burning fuel, and that, too, at the least possible distance; because the effect of radiant heat decreases as the square of the distance between the radiating and the recipient bodies (Art. 235). It is no recommendation of a boiler, therefore, to say that it contains a certain number of square feet of heating surface in a given space; for unless this surface can be acted upon by the *radiant* heat of the furnace, a boiler of less than one-half the superficial measurement, if judiciously contrived for this object, may greatly exceed it in power.*

* The most remarkable illustration of the effect of exposing a large surface to the direct action of the radiant heat is afforded by the evidence given before the Committee of the House of Commons, in 1798, on the Distillers of Scotland. Owing to the mode of levying the duty at that time, it became an object to work off the liquor from the still as rapidly as possible, irrespective of the cost of the apparatus or the expenditure of fuel. To such an extent was this carried, that the stills were actually charged, the wash distilled, and the refuse discharged about 520 times in 24 hours, or $2\frac{3}{4}$ minutes for each charge of

It is by not attending to this distinction that so many people deceive themselves in the construction of boilers. In order to increase the surface of the boiler as much as possible, they overlook this important distinction—whether the surface so added is or is not exposed to the *radiant heat* of the fire; and they frequently contract the surface exposed to the radiant heat in order to add a rather larger surface somewhere else; overlooking the fact that, unless they can add *three feet* of surface in place of each foot they subtract from the former, the boiler will have less power than it had before. In this way complicated forms of boilers have been constructed, expensive in making, and inferior in power, as well as in economy of fuel, to other more simple forms. There are no boilers which possess these advantages in a greater degree than the boilers shown in figs. 18, 22, and 23; the former being the arched boiler, which appears to accomplish all that can be desired as an efficient useful boiler, and the latter is the bell-shaped boiler, now comparatively little used, but still possessing great merit as a most efficient form of construction. The comparative disuse of this latter boiler probably arises from the great difficulty of setting it properly, and from being not so easily worked and kept clean in the flues as some others. It is almost neces-

---

16 gallons. There is no other instance known in the least approaching this extraordinary result, in which a small vessel of the measurement of 40 gallons could distil a charge of 16 gallons of wash in 1¾ minutes, half a minute more being required for charging the still, and the like time for discharging the refuse. This was accomplished by having the still exceedingly flat, so that the largest possible surface was exposed to the direct action of the radiant heat, and the flame acted intensely upon the whole bottom surface of the still, and then passed off at once into the chimney. The waste of fuel was of course immense, but the rapidity of action was fully accomplished.—*Report on Scotch Distilleries*, 1799, pp. 517-731.

sarily made of cast iron, as the form is difficult to make of wrought iron. The part exposed to the fire is covered with a series of ribs two inches deep, and about one-fourth or three-eighths of an inch thick, radiating from the crown of the arch, at an average distance of two inches from each other. These ribs, it is evident, must increase the surface exposed to the fire to an enormous extent, and that, too, precisely where the effect is by far the greatest, being immediately over the burning fuel, and receiving the whole of the radiant heat from the fire.*

(74.) The second qualification, that the boiler shall absorb the greatest quantity of heat from the fuel, is partly dependent on the cause already explained, and partly on the conducting power of the metal itself. In this respect the boiler (fig. 18) possesses an advantage over the other, in consequence of being made of wrought iron, and therefore very much thinner. Were it made of copper, its effect would be still further increased (see Art. 245); but the greater expense of copper is an objection.

---

* As early as the year 1828, the author adopted the plan of increasing the heating surface of these boilers by means of a great number of protuberances cast on the bottom, which protuberances were one inch long and seven-eighths of an inch diameter, and placed two inches apart from each other. Subsequently these pins, or protuberances, were still further extended, so as to form continuous bars or ribs, radiating from the centre of the boiler; and they were made one inch and a half deep, and three-eighths of an inch thick, and they were then placed as well on the surface exposed to the water as that exposed to the fire. This plan of increasing the heating surface was, in the year 1835, patented by Mr. Sylvester, both for cast and wrought-iron boilers; and in 1841, Mr. C. W. Williams patented the same plan for wrought-iron steam boilers, the pins being, by his plan, screwed into the substance of the plate, instead of being formed by rolling, as proposed by Mr. Sylvester.

It has been generally supposed that all sharp protuberances

(75.) The third recommendation of a boiler, that it shall allow of a free circulation of the water, is entirely dependent on its form; and on this subject some remarks have already been made (Art. 68). And the last test of a good boiler, that it shall not be liable to get out of order, nor rapidly deteriorate, is one that depends partly on the goodness of the materials and workmanship, and partly on the mode of producing the combustion of the fuel. Some very important chemical effects appear to result occasionally, both from the fuel employed as well as from the method of combustion. The effects on copper are the most destructive; and instances have occurred where the bottoms of copper boilers have separated entirely from the sides, as though cut through with a chisel, just at the part where the principal action of the ¡fire occurs; and others have become entirely riddled throughout the surface exposed to the fire. These are very rare occurrences, and the cause appears somewhat obscure; but the subject will be better explained in a subsequent chapter (Chapter VI., Part II.). Generally speaking, a wrought-iron

---

inside a boiler caused a more rapid ebullition of the water than a flat surface; and the author in 1828 originally adopted this mode of increasing the water-surface as well as the fire-surface; and the plan was afterwards followed both by Mr. Sylvester and Mr. Williams in their patents. Subsequent experiments, however, have convinced the author that it is unnecessary to increase the water-surface by these means. In particular Mr. Josiah Parkes's experiments in 1840 (*vide* his published Report), on Mr. M. A. Perkins' patent steam-boiler, proved that, owing to the great conducting power of water, the whole of the heat generated by 117 superficial feet of iron exposed to the fire was abstracted by 44 superficial feet exposed to the water; or that the water will absorb the heat at least 2·6 times as fast from the iron as the iron can receive it from the fire; and therefore it appears the internal protuberances on the *water-surface* of a boiler are unnecessary. See also Art. 251.

boiler will heat quicker, and with less expenditure of fuel, than a cast-iron boiler : it will also be less liable to accidents, as cast-iron boilers occasionally crack, particularly if they contain sharp angles in their construction, or if they are so formed that the fire acts unequally on them, and expands one part more than another. From these defects wrought-iron boilers are exempt. But, on the contrary, wrought iron will corrode much more rapidly than cast iron ; and in very damp situations, where the stoke-hole or boiler-house is sometimes left for half the year some inches deep in water, a wrought-iron boiler would be rapidly corroded, when a cast-iron one would be comparatively uninjured.

Very destructive effects have sometimes been produced by using fuel strongly impregnated with sulphur. The effect produced by such fuel on the durability of boilers is very rapid. Boilers heated with such fuel will oftentimes be destroyed in one-fourth the time that they would have lasted, had a better fuel, free from sulphur, been employed. With a sulphurous coal the plates are very rapidly destroyed, and still more the rivet heads, which are sometimes entirely eaten away, leaving the boiler in a leaky and dangerous state.

CHAPTER V.

On the Construction of Furnaces—Combustion dependent on size of Furnace-bars—Furnace-doors, and other Parts of Furnace—Proportionate Area of Furnace-bars to the Fuel consumed—Confining the Heat within the Furnace—Directions for building the Furnace for different Boilers—Advantage of large Furnaces—Modes of Firing—Size of Chimneys.

(76.) THE construction of the furnace for a hot-water apparatus is a matter which requires considerable care; for although, from the small size of the boilers generally used, the furnaces are by no means difficult to construct, it is a very common fault in building them to allow of such a very easy exit for the flame and heated gaseous matter, that a large portion of the heat passes up the chimney, instead of being received by the water in the boiler. This arises principally from the shortness of the flues in these boilers, in comparison with those of steam-engine boilers; and, in setting boilers for hot-water apparatus, it therefore requires great caution to prevent an unnecessary waste of fuel by erroneous principles in constructing the furnaces.

In giving some general instructions on the subject of furnaces for hot-water apparatus, it is not intended minutely to describe the proper furnace for each different form of boiler; but the plan of building the furnaces for three or four different forms of boilers will be given, and the application of the principles to other forms must be left to the discretion of those who erect them.

(77.) The rate of combustion of the fuel in a furnace depends very little upon the total size of the furnace, but chiefly on the proportionate size of the furnace-bars. A furnace which possesses, for instance, an area of 12 square feet would not necessarily burn a much larger quantity of fuel per hour than one that had only an area of eight square feet, provided the area of the furnace-bars was the same in both cases, and that no more air was admitted to the former than to the latter. But, by building the furnace of considerable dimensions, and with a moderately small area of fire-bars, the fuel can be made to burn for a much longer period without attention or renewal; and this is a very important object for this description of apparatus. For, so intense a fire is not required as is the case with a steam-boiler. A very small degree of attention is necessary with a furnace of a hot-water apparatus, which, when well constructed, ought to burn for ten or twelve hours without replenishing the fuel.*

(78.) In all cases, a good and perfectly tight furnace-door is requisite; for, if the door does not fit accurately, a large quantity of cold air enters, and passes between the fuel and the bottom of the boiler, and cools the boiler to a considerable extent.† The furnace-door should always be double; ‡ and also a door to the ash-pit should be used, in order to shut off the excess

---

* In some steam boilers, particularly in the Cornish boilers, the fuel is burned with slow combustion ; but the furnaces and boilers are very large in proportion to the work done by them, and great economy of fuel results from this plan of heating them (see Chapter VI., Part II.).

† In a subsequent chapter, the combustion of smoke will be discussed, and it will then be shown that the admission of air at or near the furnace-door is sometimes desirable, but only in particular stages of the combustion.

‡ Count Rumford first introduced these double furnace-doors, of which many modifications have been since adopted.

of air below the furnace-bars when the fire is required to burn slowly for a great length of time. Immediately within the furnace-door there should be a dumb-plate; and the larger this is the better, provided it does not project the furnace-bars too far back, so as to cause the most active part of the combustion to take place at the posterior part of the furnace, instead of immediately under the boiler. The use of a large dumb-plate in front of the furnace-bars is to allow the fuel to be gradually coked, by placing it first on this dumb-plate, and then, when well heated, pushing it backward upon the furnace-bars, where it enters into active combustion, and then a fresh charge of fuel is to be again laid on the dumb-plate, in order to undergo the same operation. By this plan of coking the coals on the dumb-plate, nearly all the smoke from the furnace may be consumed; by which a considerable saving of fuel will be effected,* and a great nuisance prevented.

(79.) The size of the fire-grate, or furnace-bars, must be regulated by the quantity of pipe or other heating surface which the apparatus contains. The quantity of heat given off by a certain extent of iron pipe, or other heated surface, can be exactly ascertained, and will be shown in the next chapter. From the data there given, we learn the quantity of coals required to be burned per hour in order to maintain the required temperature. Having already given (Art.

---

* See Chapter VI., Part II., on the "Combustion of Smoke," where it is shown that nearly 40 per cent. is saved by the combustion of the smoke. Those who wish to learn the endless forms which may be given to furnaces may see *several hundred* different forms and arrangements described in the various volumes of the following periodicals, namely :— *The Technical Repository,—The Repertory of Arts and Patent Inventions,—The Mechanics' Magazine,—The Quarterly Journal of Science,—The Reports of the British Scientific Association, &c.*

71) the extent of boiler surface required to heat a given quantity of pipe, it will be desirable now to show the area of the furnace-bars which will be required. It has already been stated (Art. 72) that the extent of boiler surface exposed to the fire may with advantage be increased beyond the dimensions already given; and that economy of fuel will generally result from this increased surface. But the quantity of fuel that is burned ought not to be also increased in the same way; and therefore the size of the furnace-bars, which alone regulates the quantity of fuel consumed, should be proportioned rather to the quantity of surface which radiates heat into the building, instead of bearing an exact ratio to the surface of the boiler.

With ordinary furnace-bars, the spaces for the admission of air will generally vary from one-fourth to one-third of the total area of the space occupied by the furnace-bars. In such cases one square foot of furnace-bars will be sufficient to burn about 10 lbs. or 11 lbs. of coal per hour, under ordinary circumstances;* and on this cal-

* The consumption of fuel, on any given area of furnace-bars must depend upon the rapidity of the draught. In locomotive engines, with an artificial blast from the steam, the consumption of fuel is about 80 lbs. from each square foot of fire-grate per hour. In some furnaces the consumption is not more than 6 lbs. per square foot, or about $\frac{1}{13}$ of the locomotive-engine furnace; but the quantity given in the text is a mean rate. Mr. Andrew Murray (*Minutes of Institution of Civil Engineers*, June, 1844) estimates the average consumption in steam-engine furnaces at 13 lbs. of coal per square foot of furnace-bars, and that 150 cubic feet of air should pass through the furnace for each pound of coal consumed in order to produce perfect combustion. Taking Dr. Ure's estimated velocity of 36 feet per second as the average rate at which the air passes through a furnace, Mr. Murray estimates the size of the opening over the bridge of the furnace ought to be 26 square inches for each square foot of furnace bars. The temperature of an ordinary furnace is assumed to be about 1000° of Fahren-

culation the following Table has been constructed :—

TABLE III.

Area of Bars.			4-in. Pipe.	3-in. Pipe.	2-in. Pipe.
75	square inches will supply	150 feet, or	200 feet, or	300 feet.
100	„	„	200 „	266 „	400 „
150	„	„	300 . „	400 „	600 „
200	„	„	400 „	533 „	800 „
250	„	„	500 „	666 „	1000 „
300	„	„	600 „	800 „	1200 „
400	„	„	800 „	1066 „	1600 „
500	„	„	1000 „	1333 „	2000 „

Thus, suppose there are 600 feet of pipe, four
inches in diameter, in an apparatus; then the
area of the bars should be 300 square inches, so
that 14 inches in width and 22 inches in length
will give the requisite quantity of surface.*

heit, and at this temperature the gases passing through it
would be expanded to three times their original bulk; and
these proportions will allow sufficient air to pass through the
furnace to compensate for that portion which is not completely
consumed. Tredgold, in his calculations, assumes the temperature of an average furnace at 800° Fahrenheit. See also
Art. 397 and the notes appended thereto.

* The proportions deducible from the above Table, and
those given Art. 71, for ascertaining the boiler surface, are
very different from those generally used in steam-engine boilers.
It will be observed that, by the rules here given, the area of
the furnace-bars is about one-sixth of the area of the boiler-
surface exposed to *the direct* action of the fire, whereas in steam
boilers the flue and boiler surfaces conjointly are usually in
proportion to the surface of the furnace-bars as 11 to 1, and
sometimes even as 18 to 1 (*British Scientific Reports* for 1842,
p. 107). When this latter calculation, however, is reduced to
the same standard as the other, viz. three feet of flue surface
being equal to one foot of boiler surface exposed to the radiant
heat, the difference will not be near so great as it here appears.
And it must also be remarked that in large boilers the proportions must necessarily differ from those of the very small
boilers required for a hot-water apparatus; for the effect of
radiant heat decreases as the square of the distance between
the recipient surface and the hot body; and therefore it is very
easy to see how a considerable difference may arise between
surfaces placed so differently in this respect as they necessarily
must be in large and in small boilers.

When it is required to obtain the greatest heat in the shortest time, the area of the bars should be proportionably increased, so that a larger fire may be produced; and, on the contrary, when the object is to obtain slow combustion of the fuel, and when the rapidity with which the apparatus becomes heated is of little or no consequence, then the area of the bars may be reduced. The best method, however, will generally be found in using a sufficiently large surface of fire-bars for the maximum effect required, and to regulate the draught by means of an ash-pit door and a damper in the chimney; by these means almost any required rate of combustion can be obtained, with any common degree of care.*

(80.) When the size of the furnace will allow of it, a portion of brickwork should extend at the

---

* When a rapid draught and quick combustion are required, the furnace-bars may very advantageously be made very narrow and deep, so as to allow a larger proportionate space for the entrance of the air. Instead, therefore, of using furnace-bars one and a half or one and three-quarters of an inch wide, with half an inch air-space between the bars, they may be made about three-eighths or half an inch in width, and about four and a half or five inches deep, tapering at the lower edge to about one-eighth of an inch, and made with shoulders, as usual, to allow about half-inch air-spaces. Bars of this kind will have many advantages in particular cases. They will allow more than twice the quantity of air to pass through that the other bars will do, and therefore twice the quantity of coal can be burned on each square foot of the bars; and they will last longer than bars of the ordinary construction. The author, at the latter end of 1842, suggested the use of these bars in loco-motive engines, which is the most severe test they could be put to, and the result proved completely successful. Owing to the extreme thinness of the bars, the air passing between them keeps them always cool, which is impossible if the bars much exceed this thickness. The great depth of the bar gives the necessary stiffness; and the result of nearly twelve months' trial, and with nearly twenty locomotive engines, was a very great increase in the durability of the furnace-bars, in addition to the obvious advantage of admitting much more air into the

back of the furnace-bars, and level with the bars, so as to make a dead-plate behind the bar as well as in the front, which makes the furnace hold more fuel without actually increasing the consumption.

(81.) It is a matter of very great importance, that the heat should be confined within the furnace as much as possible, by contracting the farther end of it, at the part called the throat, so as to allow only a small space for the smoke and inflamed gases to pass out. The neglect of this causes an enormous waste of fuel; for, in consequence of the shortness of the flues of these boilers, the heated gaseous matter passes too readily from the boiler, and escapes through the chimney at a very high temperature. The only entrance for the air should be through the bars of the grate,* and the heated gaseous matter will then pass directly upwards to the bottom of the boiler, and should be there detained as long as possible by the contraction at the throat of the

furnace. Some of these bars, after having been used for ten months, and with which the engines ran nearly 20,000 miles, were still perfectly good, after having done nearly four times the work of ordinary bars. The best size for this purpose is five and a half inches deep by half an inch thick, tapered to a quarter of an inch. In furnaces which have less intense heat than a locomotive engine, bars of four or four and a half inches deep are quite sufficient.

When the old form of furnace-bars is used, and they are required to bear a very intense heat, their durability is increased by making a longitudinal groove in the upper surface about three-eighths of an inch deep. This groove becomes filled with ashes, which, being a slow conductor of heat, preserves the bars from the intense heat of the fire.

* These observations apply exclusively to the small furnaces and boilers used for hot-water apparatus, and not to large furnaces for steam-boilers or for other purposes. In the latter, air may very advantageously be introduced at or near the furnace-door, or in many other ways, as will be shown in the chapter on the " Combustion of Smoke." Even in these

furnace ;* and if this part of the furnace be properly constructed (by not making the throat too near the crown of the boiler, and making it sufficiently small in proportion to the total quantity of gaseous matter required to pass through it), a reverberatory action of the flame and heated gases will take place, by which a far greater effect will be produced than if too easy an exit were allowed into the flues and chimney.

(82.) A furnace constructed on this plan is shown in fig. 31, 32, and 33, which is the mode in which many thousands of furnaces have been constructed and applied to the boiler shown in fig. 18; this plan having been recommended by the author many years ago, and used with the most uniform success.†

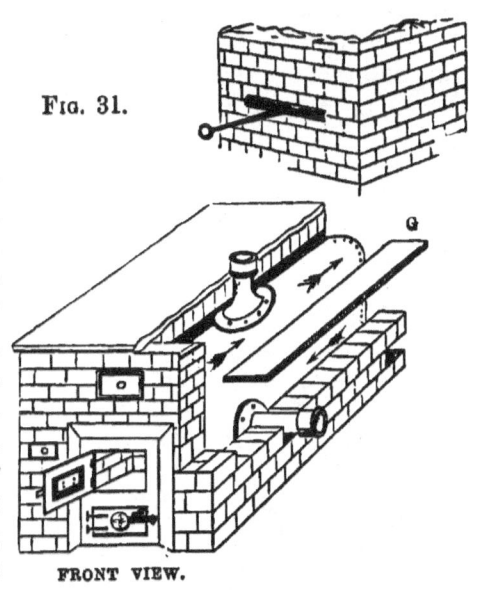

Fig. 31.

FRONT VIEW.

The way to construct this furnace is as

small furnaces, a limited quantity of air might *in certain stages of the combustion* be advantageously introduced ; but this would require so much more attention than is usually given, or indeed required, for an apparatus of this kind, that the rule given in the text will be found most advisable for ordinary practice.

* See also note to Article 79, *ante.*

† This plan of setting boilers was first used by the author in 1830, and was published subsequently, by consent, in the *Horticultural Transactions.*

follows. After building up the foundation of the boiler to the proper height, the bars are to be placed so that the front of the bars shall lie even with the *front* of the boiler, and the upper surface of the bars shall be level with the *bottom* of the boiler. The bars will generally be about two-thirds the length of the boiler; the remainder of the length of the furnace being made up with brickwork beyond the bars towards the back end of the boiler. . In front of the bars should come the dead-plate, about 9 inches wide; and immediately in front of this dead-plate should come the furnace-door: the bottom of the furnace-door, the dead-plate, the furnace-bars, and the brickwork beyond the furnace-bars being *all exactly on the same level.* The distance between the front of the boiler and

Fig. 32.   Fig. 33.

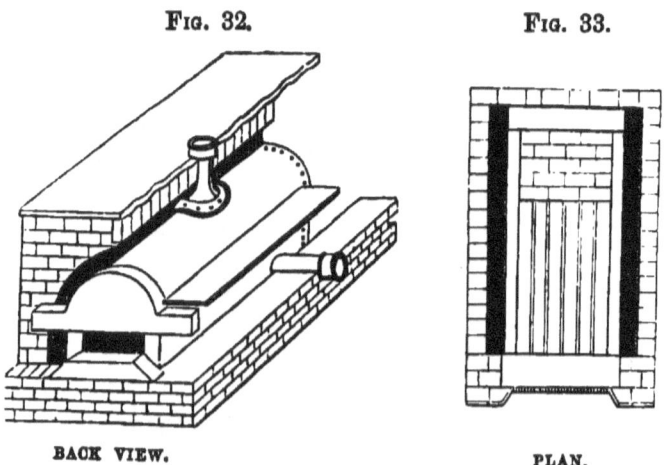

BACK VIEW.   PLAN.

the furnace-door will, by this arrangement, be just nine inches; and this space is made up with the brickwork, which forms the general front of the whole furnace. In this brickwork are placed three soot-hole doors, to clean out the three flues. Two of these doors are shown in fig. 31; the other one is supposed to be removed, to show the inside of the flue. At the back of the boiler are placed two fire-lumps, shown in fig. 32, and which are made to fit the size of the boiler. When these two fire-lumps are placed as shown in fig. 32, they leave an opening between them of from $3\frac{1}{2}$ to $4\frac{1}{2}$ inches, according to the size of the boiler. This opening should come, as nearly as may be, to *the centre* of the arch of the boiler; and it is the only passage for the flame and smoke to escape into the flues, which here divide to the right and left hand, passing along the sides of the boiler, and following the direction of the arrows, fig. 31. Two cast-iron plates are built into the brickwork, in the position shown by o, fig. 31, to divide the lower from the upper flue. These flue-plates do not come to the front of the boiler by from four to five inches, thus leaving a passage from the lower into the upper flue, and from the latter into the chimney. The whole of these flues, of which there is one on each side of the boiler and one on the top, are to be made as *deep* as ever the boiler will allow, and the brickwork should stand off from the boiler about $4\frac{1}{2}$ inches, forming the *width* of the flues. By this arrangement it will be perceived, that nearly the power of a reverbera-tory furnace is obtained. The upper fire-lump, shown in fig. 32, partly stops the flame, and retains it as long as possible in contact with the inner arch of the boiler; and on this mainly depends the economy and efficiency of this plan of setting these boilers. The flame and smoke then pass between the two fire-lumps, dividing right and left into the two lower flues, and then in the direction of the arrows, fig. 31, into the two upper flues, and from thence passing into the top flue of the boiler, which latter should be nearly as large as the other two together. A damper must be placed

H

in the chimney in such a position as to be easily got at.

(83.) This plan of setting arched boilers is perfectly available for all boilers, from eighteen inches long up to about six feet long; for boilers beyond that length, it is advisable somewhat to modify the plan, by placing a fire-lump *within the inner arch*, so as to form a bridge, which bridge should be formed to fit the arch of the boiler, but leaving a space of about five inches in *depth*, and the width of the arch, between the crown of the arch and the top of the fire-lump. When this bridge is used, the bottom fire-lump, already described, is not required. This latter mode of setting is only suitable for such boilers as are six feet long and upwards. In this case the bridge is best placed at the distance of about five feet from the front end of the boiler.

(84.) The boiler, figs. 22 and 23, requires nearly the same arrangement; but in this boiler the aperture for the escape of the flame and smoke is generally made a part of the boiler itself. This opening is also somewhat lower down towards the level of the furnace-bars, and the boiler being circular, the flue generally winds round the boiler, instead of passing separately on the right hand and on the left. The boiler, fig. 21, may be set in the same kind of furnace as the boiler, fig. 18. If the two legs or protuberances at the bottom be very short and close together, the fire may be made to act upon the whole under-side of the boiler (the bars being fixed at some distance below), and the flame returned through a flue along the top.

(85.) The boiler, fig. 20, may be set in two different ways. When the inside tube is sufficiently large, it is best to place the fire inside this tube, the furnace-bars being placed at about one-third the diameter of the tube from the bottom. In this case the action of the furnace becomes very similar to that already described for the boiler, fig. 18; except that the water-way is continued below as well as above the fire. The throat of the furnace must be contracted, as already described for fig. 18; but in this case the flues must first pass directly under the boiler, and then pass along the two sides and top.

When this boiler, fig. 20, is very small, the fire must be made entirely below the boiler; and the boiler is then best made of an oval or flattened shape, both externally and in the tube. The flame, in this case, passes from the surface below, first through the tube, and then returns over the top of the boiler, and from thence the heated gases escape into the chimney. Or still another plan may be used, by making the fire act first on the bottom of the boiler, then return to the front through the central flue, and then divide right and left on the outside of the boiler, then so on to the chimney as in fig. 33A.

Fig. 33A.

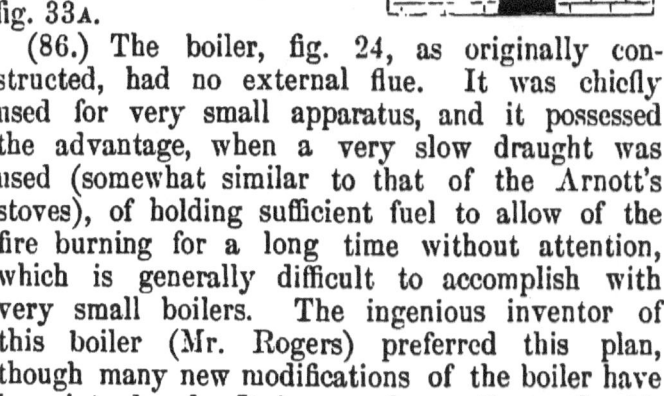

(86.) The boiler, fig. 24, as originally constructed, had no external flue. It was chiefly used for very small apparatus, and it possessed the advantage, when a very slow draught was used (somewhat similar to that of the Arnott's stoves), of holding sufficient fuel to allow of the fire burning for a long time without attention, which is generally difficult to accomplish with very small boilers. The ingenious inventor of this boiler (Mr. Rogers) preferred this plan, though many new modifications of the boiler have been introduced. It is now frequently used with

an external flue. The temperature of this boiler
is somewhat more difficult to regulate than that
of the arched boiler; as the more the fuel burns
away, the greater the heat becomes, in consequence
of a larger surface of the boiler being then exposed to the radiant heat, and also because the
fuel burns quicker, in consequence of the air meeting less obstruction in passing through it. In this
case, the greatest heat is produced when about
two-thirds of the fuel has burnt away. When
the boiler has an external flue, the best mode of
setting is to make the flue proceed from openings
of about three inches by six inches, left at the
bottom of the boiler, and leaving a free space for
the flue, around the boiler, of two and a half or
three inches, or thereabouts. The draught of air
meeting less obstruction in passing through the
external flue than by passing through the large
body of fuel contained in the body of the boiler,
the whole exter-
nal surface be-
comes available
for receiving heat
from the fire, in-
stead of being
entirely useless,
as in the other
mode of setting.
Of course, the
same sized boiler
will by this ar-
rangement heat a

Fig. 34.   Fig. 35.

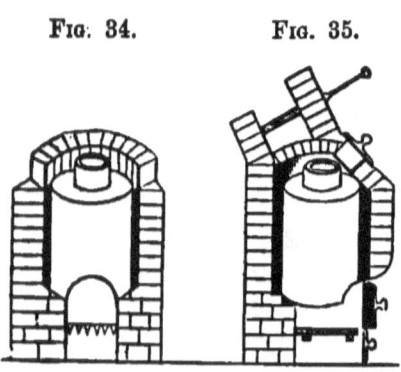

larger quantity of pipe. Figs. 34 and 35 show
this mode of setting these boilers. These boilers,
however, will, under any form, expose but little
surface to the radiant heat of the fire, and the
external surface will scarcely exceed the flue surface of the arch-boiler, in its power of absorbing

heat. This flue surface, we have already seen, only possesses one-third the absorbent power which those surfaces have that are exposed to the direct action of the radiant heat. The fire of this boiler, however, is not difficult to manage, and burns with but little attention.

(87.) In the boiler, fig. 26, there is necessarily a considerable waste of fuel, in consequence of the flame escaping immediately into the chimney without passing through any flues,—this form of boiler not admitting of any kind of flues being used. The flame passes between the several pipes which form the boiler, and of course can only act upon their under side. If the draught be rapid, a partial vacuum must be formed on the upper sides of the pipes, the flame passing in straight lines upwards; and, therefore, a loss of heat by radiation would take place from the upper side of the pipes which form this boiler. The boiler, however, heats rapidly, as the consumption of fuel in the furnace, owing to the rapid draught, is very considerable.

(88.) The advantage of making the furnace to contain a large quantity of fuel has already been mentioned. But, independent of the smaller degree of attention required, when sufficient fuel to last for many hours is supplied at once, it is found practically that great economy results from this plan. From experiments made on this subject with steam-engine furnaces, it appears that the increased consumption of fuel always bears a direct proportion to the frequency with which it is supplied to the furnace; and that in the experiments in question the greatest economy resulted when the fuel was supplied only once a day.*

* Mr. Josiah Parkes on "The Evaporation of Water from Steam Boilers," in the *Transactions of the Institution of Civil Engineers* for 1838. This result, however, was obtained by a peculiar kind of furnace, in which air was admitted at the bridge as well as through the fire-bars.

When this plan is followed, the combustion is less intense than with more frequent firing; and, therefore, a larger boiler surface is always required. Care also should be taken to prevent the ingress of an undue quantity of air through the ash-pit, when the fuel burns away and the furnace-bars thus become unequally covered; for, in this case, a large quantity of cold air will rush in and cool the boiler.

(89.) The rate of combustion materially depends upon the thickness of fuel on the furnace-bars, and on its compact or open state, as illustrated in the two cases of small coal and of large well-burned coke. The quantity of air passing through the fire-grate or bars must be very different in these two cases, and the combustion wholly depends upon the quantity of air admitted to the fuel. For unless a sufficient quantity of air be admitted to convert the whole of the carbon into carbonic acid gas, it will escape in the form of carbonic oxide, and a loss of effect will thereby arise (see Chap. VI., Part II., on the Combustion of Smoke).

(90.) The greatest economy of fuel is produced when the fires are kept thin and bright; the coal well coked, by means of a large dumb-plate in the front of the furnace, and the damper kept as close as possible consistent with allowing a sufficient draught. The Cornish engines, so celebrated for their economy of fuel, are thus worked. The thinner the fire, the less is the probability of the formation of carbonic oxide, which always causes a loss of heat. When thick fires are used, this loss is frequently very considerable, unless (as in Mr. Parkes's experiments already mentioned) air is supplied above the fuel as well as through the furnace-bars.* In the small fur-

---

* In locomotive engines, the fires are frequently as much as 17 inches thick; and the quantity of carbonic oxide formed in

nace of a hot-water apparatus, it is frequently difficult, if not impossible, to adopt this plan of using a dumb-plate sufficiently large to coke the whole of the fuel which is used; but the principle should be borne in mind in all cases, and applied as far as circumstances will permit. The theory of combustion will be given in the chapter on the Combustion of Smoke.

(91.) It may, perhaps, not be amiss here to give some rules for the proper size of chimneys. Very elaborate rules have been given for this purpose by different authors, and the most extraordinary differences exist between them; their calculations giving results totally at variance with each other. But the practical rules are very simple. Mr. Murray* estimates the area of the chimney should be about 18 square inches for a boiler consuming 12 lbs. of coal per hour. Mr. Armstrong estimates the area at 20 square inches for the same consumption of coal.† Tredgold's‡ calculations give an area of about 14 square inches for the same quantity of coal consumed per hour, when the boiler is worked at a low temperature, and very considerably less than this when the temperature is high. Some of these calculations, however, are

consequence of this great thickness is very considerable, and the loss of heat enormous. The thinner the fire, the more perfect must be the combustion. Carbonic oxide is formed by the carbonic acid (which is the result of perfect combustion) passing through the red-hot coke, by which it imbibes an additional quantity of carbon, and is converted into carbonic oxide. On all the carbonic acid that undergoes this change there arises a loss of one-half the heat derived from its original conversion. The various methods of admitting air at the bridge, and at other places above the fuel, are all intended to obviate the loss; reconverting the carbonic oxide into carbonic acid, by supplying it with an additional dose of oxygen.

* *Minutes of Institution of Civil Engineers*, June, 1844.
† Armstrong on "Steam-boilers," p. 80.
‡ Tredgold on "Warming and Ventilating," &c., p. 114.

for much higher chimneys than are ever used for the purpose that we are here considering, and the lower the chimney the larger the area ought to be. But, from what has been stated, sufficient may be gathered to estimate the size of chimneys for such common purposes as are here supposed to be required.*

* Those who wish to investigate this subject further may refer to Péclet's *Traité de la Chaleur,* p. 79 *et seq.* ; Wyman " On Ventilation," &c., p. 392 *et seq.* ; Sylvester " On Chimneys " ; Rees' " Encyclopædia, and Annals of Philosophy," &c. ; Gilbert " On Ventilation," &c.; *Quarterly Journal of Science* for 1822 ; and Tredgold " On Warming and Ventilating," p. 114 *et seq.* ; and Dr. Ure, *Phil. Transactions,* 1836.

CHAPTER VI.

ESTIMATE OF THE HEATING SURFACES REQUIRED
TO WARM ANY DESCRIPTION OF BUILDINGS.

Heat by Combustion—Quantity of Heat from Coal—Specific
Heat of Air and Water—Measure of Effect for Heated Iron
Pipe—Cooling Power of Glass—Effect of Vapour—Quantity
of Pipe required to warm a given Space—Time required to
heat a Building—Facile Mode of calculating the Quantity
of Pipe required in any Building—Quantity of Coal consumed.

(92.) Having in the preceding chapters investigated the fundamental principles of the hot-water
apparatus, we proceed to consider some particulars
which are necessary to be known in order to apply
the preceding remarks, and correctly to apportion
the various parts of the apparatus, and calculate
the effects which will be produced under various
circumstances.

Very erroneous notions are entertained by many
persons as to the absolute quantity of heat contained in different substances. This subject has
already been mentioned; and in the present chapter
we shall have occasion to apply this law of specific
heat in several important calculations.

(93.) It will, however, be desirable first to ascertain the quantity of heat which can be obtained
by the decomposition of combustible materials by
fire; for in this also, it may be observed, very
erroneous notions prevail. The quantity of heat
obtainable by the combustion of any substance is

not, as many persons appear to consider, illimitable, but it is as fixed and determinate as any other of the laws of heat. The amount of heat by combustion depends on the chemical composition of the particular substance; but although this heat may be either wasted or advantageously applied, according as the apparatus used for its combustion is imperfect or otherwise, still it must be remembered there is a maximum effect, which has been accurately ascertained, and which cannot be exceeded in any form of apparatus; though in no apparatus yet invented has it been possible absolutely to render available the whole of this heat.

Although every kind of fuel differs in the quantity of heat that it affords, it is unnecessary here to inquire into any other than the ordinary descriptions used for purposes similar to that we are now considering. The calculations, therefore, will be made with reference only to coal and coke of ordinary and average qualities.

(94.) It is stated by Watt that one pound of coal will raise the temperature of 45 lbs. of water from 55° to 212°. Rumford states the same quantity of coal will raise $36\frac{2}{10}$ lbs. of water from 32° to 212°; and Dr. Black has estimated that one pound of coal will make 48 lbs. of water boil, supposing it previously to be at a mean temperature. These quantities, when reduced to a common standard, vary but little from each other. Watt's experiment, of 45 lbs. of water being heated from 55° to 212°, is equal to $39\frac{1}{4}$ lbs. only, if heated from 32° to 212°; and this nearly agrees with Count Rumford's calculation; at least, the variation is not more than might be expected from a slight difference in the quality of the coal. Dr. Black's estimate is as much in excess over the experiments of Watt as Rumford's is in defect; we may, therefore, take the average of these three experiments, which will give us a result, that 39 lbs. of water may be heated from 32° to 212° by one pound of coal.

(95.) The results of later experiments show that, as an average effect, the above calculations are very accurate, when practically applied on a large scale. Mr. Parkes * found that the greatest effect he could produce, by his improved mode of firing, was 10·3 lbs. of water at the temperature of 212° evaporated by one pound of coal; and that, by the ordinary methods of firing, the average obtained is only 7·5 lbs. of water of the like temperature evaporated by one pound of coal. The first of these is equal to 57·2 lbs. of water heated from 32° to 212° by one pound of coal; and the latter is equal to 41·6 lbs. of water heated to the like extent, and which very nearly agrees with the experiments of Watt, Rumford, and Black. In the Cornish engines, however, a much higher result is obtained. Mr. Parkes has given the results obtained in the "United Mines," during eight months, from which it appears the greatest evaporation is 15·3 lbs., and the average quantity 11·8 lbs. of water evaporated from the temperature of 212° by one pound of coal. The former of these gives 85 lbs., and the latter 65½ lbs. of water, raised from 32° to 212° by one pound of coal; which results appear to be the highest that are practically attainable, and are very much greater than can be produced with any other boilers, or qualities of coal, than those with which the experiments were made. In all the subsequent calculations, therefore, the average of

* Mr. Parkes "On the Evaporation of Water from Steam-Boilers," in the *Transactions of the Institution of Civil Engineers*, 1838; and these results very nearly agree with the results arrived at by Mr. Herepath from purely physical calculations. (Herepath's *Mathematical Physics*, vol. i., p. 351.), See also note to Art. 71, *ante*.

the experiments of Watt, Rumford, and Black
will be adhered to, as being the most correct for
ordinary practice; and we shall shortly have
occasion to apply them, in elucidating that branch
of the subject which is included in the present
chapter.

(96.) In order to ascertain the effect that a
certain qnantity of hot water will produce in
warming the air of a room, there appears to be no
better method than that of computing from the
specific heat of gases compared with water.

(97.) Every substance, it is well known, has its
own peculiar specific heat: that is, a given weight,
or volume, of any particular substance at a certain
temperature, contains a definite amount of heat,
which, if imparted to any other substance, will
produce upon this last a certain known effect,
though it will be different for every different body
or substance. Now, it is ascertained that one
cubic foot of water, by losing one degree of its
heat, will raise the temperature of 2990 cubic feet
of air the like extent of one degree; and by
losing 10° of its heat, it will raise the temperature
of 2990 cubic feet of air 10° or 29,900 cubic feet
one degree, and so on.*

(98.) But this calculation regards only the
ultimate effect which will be produced, without

---

* The specific heat of *equal weights* of water and air, by the
experiments of Berard and Delaroche, is found to be as 1 to
·26669 : but as the volume or bulk of an equal weight of atmospheric air is to water as 827·437 to 1, we shall have ·26669 :
1 : : 827·437 = 3102, which is the number of cubic feet of air
that has the same specific heat as one cubic foot of water.
This, however, appears to be rather too high a calculation ; for
Dr. Apjohn, in a memoir recently published (*Rept. Brit. Sci.
Assoc.*, vol. iv.), gives the result of a new mode of determining
the specific heat of permanently elastic fluids, by which he
makes the specific heat of atmospheric air ·2767 when that of
water is represented by unity.   Therefore ·2767 : 1 : : 827·437
= 2990, which is the number given in the text.

reference to the time which will be required to obtain the result. To ascertain the time that is required to heat the air, which is a most essential element in every calculation connected with the subject under consideration, recourse must be had to direct experiments; for the rate at which a given quantity of hot water will impart its heat to the surrounding air depends upon the nature and extent of surface of the body which contains it, as well as upon the degree of motion which the air possesses. The effect of the velocity of the air, however, is not necessary here to be considered, as it is only to a still atmosphere in a building that these calculations are to be applied. But as the radiating and conducting powers of different substances vary considerably, it is necessary to make experiments with the same substance or material as the pipes for which we wish to estimate the effect, before we can arrive at any conclusions as to the quantity of heated surface that will be required to produce any desired temperature in a building.

(99.) From experiments made to determine this question, it appears that the water contained in an iron pipe of four inches diameter internally, and four and a half inches externally, loses ·851 of a degree of heat per minute, when the excess of its temperature is 125° above that of the circumambient air. Therefore (by Art. 97) one foot in length of pipe four inches diameter will heat 222 cubic feet of air one degree per minute, when the difference between the temperature of the pipe and the air is 125°.*

---

* From the data given in Art. 270, Chap. XIII., it appears that 171·875 cubic inches of water, exposed to the cooling influence of the air by 287·177 square inches of surface of cast iron, loses ·8 of a degree of Fahrenheit per minute when the air is 79° colder than the pipe : therefore $\dfrac{125 \times \cdot 8}{79} = 1 \cdot 265$

This calculation will serve as the basis by which we may estimate the quantity of heating surface for any building. But before we can apply it practically, we must know what quantity of heat the building will lose per minute, by the cooling power of the glass, by ventilation, radiation, and all other causes which may tend to lower its temperature; for on these several causes must obviously depend the quantity of heat that is required to be added to it by the warming apparatus.

(100.) The quantity of air required for ventilation, and the method of ventilating buildings, are considered in subsequent chapters (Chapters III. and IV., Part II.). It is unnecessary, therefore, in this place to pursue the subject further than to state that, in large buildings and rooms of dwelling-houses, a quantity of air equal to from three and a half to five cubic feet for each individual the room contains must be changed per minute, in order to preserve the wholesomeness and purity of the atmosphere.*

will be the loss of heat per minute when the temperature of the pipe is 125° above that of the air. But this quantity of water, if exposed in a pipe of four inches diameter inside, and four and a half inches outside, will only be surrounded by 193·435 square inches of radiating surface; therefore $\frac{193\cdot435 \times 1\cdot265}{287\cdot177} = \cdot851°$, will be the loss of heat per minute by a four-inch pipe, when the excess of temperature is 125° above the circumambient air. As all pipes are technically known by their internal diameter, this mode of measuring is here used, although the external measurement would be a more correct definition for these calculations.

* This estimate is given for ordinary buildings. In the Houses of Parliament, and in some other buildings, where expense is of no consideration, a much larger quantity of air has been introduced for ventilation very beneficially; and when the cost of the apparatus is unimportant, it may be assumed that the larger the quantity of fresh air introduced, the greater will be the comfort and salubrity. In Chap. iv.,

(101.) The loss of heat in all buildings having any great extent of glass we shall find to be very considerable. It appears by experiment* that one square foot of glass will cool 1·279 cubic feet of air as many degrees per minute as the internal temperature of the room exceeds the temperature of the external air; that is, if the difference between the internal and the external temperature of the room be 30°, then 1·279 cubic feet of air will be cooled 30° by each square foot of glass, or, more correctly, as much heat as is equal to this will be given off by each square foot of glass; for, in reality, a very much larger quantity of air will be affected by the glass, but it will be cooled to a less extent. The real loss of heat from the room will therefore be what is here stated.

(102.) But though this amount is only calculated for a still atmosphere, as intense cold is seldom or never accompanied with high winds,† no additional allowance needs be made for this cause, provided we estimate sufficiently low for the external temperature. For the highest winds are generally about March and September, and the average temperature of the former month is

Part 2, where the ventilation of public buildings is considered, it will be observed that twelve feet per minute is assumed as the minimum quantity for each person; and as a general rule it may be stated that the more people who are crowded into a given space the larger should be the supply of air to each individual.

* "Experiments on Cooling," Art. 271.

† That intense cold is rarely accompanied by high winds is matter of common experience. The obliquity of the sun's rays on the higher latitudes of the northern hemisphere, when near the time of the winter solstice, prevents the atmosphere of those places which are distant from the tropics from receiving any considerable quantity of heat; and, therefore, the air being all of nearly equal density, there is but little tendency to aerial currents in the lower strata.

46°, and the latter 59½°. The greatest diurnal variation of the thermometer is 20° in March, and 18° in September; so that the average temperature of the nights will be 36° in March, and 50° in September.* But we shall presently find (Art. 106) that when the external atmosphere is at 36°, the quantity of pipe required to warm a building to 65° is only about one-half of what would be necessary were the external air at 10°. Therefore, in calculating the quantity of pipe to warm buildings used during the night, we should estimate that the *external* temperature may fall as low as 10° Fahrenheit. Or if the building is required to be warmed during the daytime only, we may estimate the external temperature may fall as low as 25° Fahrenheit. If we adopt these as the *external* temperatures we shall find that generally no further allowance needs be made for the effect of high winds; because such high winds only occur when the external air is much above these limits, and therefore the quantity of pipe calculated by these rules will be correct both for cases of extreme cold and also for cases of very high winds; which two conditions, as already stated, never occur simultaneously.†

---

* These temperatures are for the neighbourhood of London. In March, 1837, the night temperature, obtained by a register thermometer, only averaged 31·1°, which is nearly 5° lower than has been known for many years. Mr. L. Howard, in his "Climate of London," states that the average temperature, ascertained by observation for ten years, is as follows : —

		In London.	In the Country.
March . .	Mean highest temperature .	47·31°	48·46°
	Mean lowest temperature .	37·32°	34·57°
September	Mean highest temperature.	65·91°	65·52°
	Mean lowest temperature .	52·45°	47·08°

† By reckoning the external air at the above temperatures, the wind may have a velocity of from twenty to thirty miles an hour, without producing any diminution of the internal

But in such situations as are very much exposed to high winds, it will perhaps be prudent to calculate the external temperature from *zero*, to compensate for the increased cooling power of the wind ; and, in very warm and sheltered situations, a less range in the temperature will be sufficient. Local knowledge of the situation will therefore be necessary to guide the judgment in particular cases.

(103.) The difference between the cooling effect of glass which is glazed in squares and that which is lapped is very trifling in those buildings where the air contains much moisture. This is the case in hothouses, where the plants are constantly steamed ; and therefore, for such buildings, no farther allowance should be made on this account for loss of heat.*   But in skylights of dwelling-houses, in consequence of the greater dryness of

---

temperature; for it is probable that the cooling effect of wind on ordinary window glass is not above one-half so much as appears by the experiments, Art. 271, in which the glass was so much thinner than ordinary window glass.

* The calculations of the specific heat of air, given in the note, Art. 97, are only for dry air. If the temperature be at 60° and the air saturated with moisture, then the same quantity of heat will only raise the temperature of 2,967 cubic feet of this saturated air any given number of degrees, which would have raised 2,990 cubic feet of dry air to the like temperature. This 2,967 cubic feet of saturated air will contain 68 cubic inches of water; and this quantity of water will absorb as much heat during its conversion into vapour as would raise the temperature of 117,507 cubic feet of air one degree. This is equal to the entire heat that 46 feet of pipe, four inches diameter, will give off in ten minutes, when the temperature is 140° above that of the air. The glass will, however, cool much less of this saturated air than of dry air, for the mixture of air and vapour has greater *specific heat* than dry air. With lapped glass the loss of heat will be less with saturated than with dry air, because the vapour, when condensed upon the glass, will run down and nearly fill up the crevices between the laps, and effectually prevent the escape of the air, and thereby avoid the loss of heat.

I

the atmosphere, the heated air will escape through the laps of the glass in greater quantity, in proportion as less vapour is condensed on the surface. The height of the skylight will also make a considerable difference in the velocity of the escape of air through the laps, as it depends upon the same principles which have been explained (Art. 21) as governing the motion of water, the increased velocity being relatively as the height and the difference of temperature between the internal and external air.*

(104.) In making an estimate of the quantity of glass contained in any particular building, the extent of surface of the woodwork must be carefully excluded from the calculation. This is particularly necessary in buildings used for horticultural purposes, where, from the smallness of the panes, the wood-work occupies a considerable space. The readiest way of calculating, and sufficiently accurate for ordinary purposes, is to take the square surface of the sashes, and then deduct one-eighth of the amount for the woodwork. In the generality of horticultural buildings, the wood-work fully amounts to this quantity; but in some expensively finished conservatories, &c., it is considerably less, and therefore the allowance must be made accordingly. When the frames and sashes are made of metal, the radiation of heat will be quite as great from the frame as from the glass; therefore, in such cases, no deduction must be made.†

* See also Chapter V., Part II.

† Some persons have imagined that the loss of heat from a glass roof will vary greatly with the angle which the roof forms with the horizon. But this variation in the effect cannot be very considerable. It can only be that portion of the heat lost by conduction of the air that can vary in this manner; and calculating the ordinary excess by which the

Some loss of heat will likewise arise from the imperfect fitting of doors and windows. In these cases

temperature of the hothouse exceeds the temperature of the external air, this portion of the heat is only about three-sevenths of the whole (see Art. 227, &c.). But a small part only of this quantity will be affected by the angle of the glass. For the cooling effect of wind will be in proportion to the number of particles of air brought into contact in a given time; and with a horizontal wind this will be directly as the sine of the angle which the roof forms with the horizon. Supposing, then, the roof to be at an angle of 34°, as recommended by Mr. Knight (*Horticul. Trans.* vol. i. p. 99), we shall find that the sine of the angle multiplied by the above-mentioned portion of the heat affected by the conducting power of the air will give as a result that, at the angle of 34°, there will be about two-ninths more of the heat carried off by the conducting power of the air than would be the case if the glass were placed horizontally. But, practically, this loss will be very materially lessened by an effect which is not capable of being reduced to any exact calculation. When a stream of air strikes an upright surface of glass, it is not reflected back again upon itself, but glides along the surface, and by the increased heat will be directed upwards in a vertical line. ("Quetelet's Philosophy," note 5, Appendix.) Passing, then, in this direction, it meets another stream of air proceeding in a line parallel to the original line of its motion; and it is by this again driven more closely in contact with the glass. But, having been warmed by the contact in the first instance, it will abstract less heat from the glass, and will thus prevent, to a considerable extent, the further loss of heat, until by the upward motion of the air it finally escapes into space. The same effect will be produced by a glass roof lying at any angle; but it is clear the heated particles will escape upwards more easily in proportion as the angle of the roof is smaller. Now, these effects are exactly the opposite of each other. The cooling effect of the wind *increases* with the angle of the roof, and is greatest on a vertical surface; while the counteracting influence, by the interference of the particles of air with each other, also increases in nearly the same proportion; and therefore the variation in the angle of the roof makes far less difference than might at first be expected in the cooling effect. The difference, however, between the cooling of a vertical pane of glass and a perfectly horizontal one is not inconsiderable in high winds; but the angle of a roof must be very small indeed before it can escape the influences above described, and be brought to assimilate with a horizontal roof.

the circumstances vary very considerably; but, in the majority of instances, no allowance is necessary for these sources of loss of heat, the external temperature of the air having been reckoned (Art. 102) sufficiently low to supersede the necessity of any farther deduction.

(105.) From the preceding calculations, the following corollary may be drawn : The quantity of air to be warmed *per minute* in habitable rooms and in public buildings must be from three and a half to five cubic feet for each person the room contains, and one and a quarter cubic feet for each square foot of glass.* This air has to be heated from the external temperature, to the temperature at which the room or building is required to be kept. For conservatories, forcing-houses, and other buildings of this description, the quantity of air to be warmed *per minute* must be one and a quarter cubic feet for each square foot of glass which the building contains ; and this air also will have to be heated from the external temperature to the proposed temperature of the building. When the quantity of air *to be heated per minute* has been thus ascertained, the quantity of pipe that will be necessary to heat the building may be found by the following rule :—

RULE :—Multiply 125 by the *difference* between the temperature at which the room is purposed to be kept, when at its maximum, and the temperature of the external air; and divide this product by the *difference* between the temperature of the pipes and the proposed temperature of the room ; then the quotient thus obtained, when multiplied

* As corrugated sheet-iron is coming much into use, it may be proper to observe that the loss of heat from this kind of material is exactly the same as from the like extent of glass (see " Experiments," Chapter XIII.), and must be allowed for accordingly, whenever it is used.

by the number of cubic feet of air to be warmed *per minute*, and this product divided by 222, will give the number of feet in length of pipe, four inches diameter, which will produce the desired effect.*

When the pipes which are to be used are three inches diameter, then the number of feet of four-inch pipe, obtained by this rule, must be multiplied by 1·33, which will give the length of three-inch pipe ; or, to obtain the quantity of two-inch pipe, the length of pipe four inches diameter, obtained by the rule, must be multiplied by two ; the length required of three-inch pipe being one-third more than four-inch, and the length of two-inch pipe being double that of the four-inch, when the temperatures are the same in all.

(106.) By the following Table, however, even the simple calculations given in this rule may be dispensed with. The Table shows the quantity of pipe four inches in diameter which is required to heat 1,000 cubic feet of air *per minute* any number of degrees. The temperature of the pipes is assumed to be 200° of Fahrenheit, this being the most usual temperature at which they can be easily maintained. But, according to the length of pipe

---

* Let p be the temperature of the pipe, and t the tempera-ture the room is required to be kept at, then $\dfrac{125}{p-t} = x$, which will represent the number of feet of pipe that will warm 222 cubic feet of air one degree per minute, when $p-t$ is different to the proportions given in Art. 98. If d represents the dif-ference between the internal and the external temperature of the room, and c the number of cubic feet of air which are to be warmed per minute, then $x \cdot \dfrac{d.c}{222} = F$ will be the number of feet of pipe, four inches diameter, which will warm any quantity of air per minute, according to the calculations, Art. 98.

The rule given in the text has been arranged in such a manner that it may be worked without decimals.

which is heated by one boiler, the temperature
will sometimes be greater and sometimes less than
this estimate, the temperature of the water being
generally higher when only a small quantity of
pipe is used. When the quantity of air to be
warmed *per minute* is greater or less than 1,000
cubic feet, the proper quantity of pipe will be
found by multiplying the length given in the
Table by the actual number of cubic feet of air to
be warmed *per minute*, and dividing that product
by 1,000.

(107.) If the building which it is designed to
warm is required to be used only during the day,
the air, in this part of the country at least, is
scarcely likely to be below 25°; but if—as for a
forcing-house, for instance—it is required to be
heated both by day and by night, then, perhaps,
10° will not be too low to calculate from, or 22°
below the freezing point. Suppose, now, we want
to calculate the quantity of pipe required to heat
a forcing-house to 75° in the coldest weather—
which we will assume to be 10° of Fahrenheit's
scale, or 22° below freezing. We have already
seen (Art. 101—105) that the quantity of heat
required for horticultural buildings is merely so
much as is necessary to replace the heat given off;
or, in other words, to compensate for the loss
sustained by the glass. The actual cubic measurement of the house signifies nothing in this case.
It is the glass alone which gives off any appreciable heat; and therefore whatever quantity of
pipe will compensate for this loss of heat by the
glass will also warm the house in the first instance
and maintain it at the required temperature afterwards; because, until the air of the house is heated
to its maximum temperature, the glass will cool
proportionally less air, the cooling power of the
glass being obviously exactly proportional to the

TABLE IV.

Table showing the Quantity of Pipe, four inches diameter which will heat 1,000 Cubic Feet of Air per Minute, any required number of Degrees: the Temperature of the Pipe being 200° Fahrenheit.

Temperature of external Air. Fahrenheit's Scale.	Temperature at which the Room is required to be kept.									
	45°	50°	55°	60°	65°	70°	75°	80°	85°	90°
10°	126	150	174	200	229	259	292	328	367	409
12°	119	142	166	192	220	251	283	318	357	399
14°	112	135	159	184	212	242	274	309	347	388
16°	105	127	151	176	204	233	265	300	337	378
18°	98	120	143	168	195	225	256	290	328	368
20°	91	112	135	160	187	216	247	281	318	358
22°	83	105	128	152	179	207	238	271	308	347
24°	76	97	120	144	170	199	229	262	298	337
26°	69	90	112	136	162	190	220	253	288	327
28°	61	82	104	128	154	181	211	243	279	317
30°	54	75	97	120	145	173	202	234	269	307
Freezing Point. 32°	47	67	89	112	137	164	193	225	259	296
34°	40	60	81	104	129	155	184	215	249	286
36°	32	52	73	96	120	147	175	206	239	276
38°	25	45	66	88	112	138	166	196	230	266
40°	18	37	58	80	104	129	157	187	220	255
42°	10	30	50	72	95	121	148	178	210	245
44°	3	22	42	64	87	112	139	168	200	235
46°	..	15	34	56	79	103	130	159	190	225
48°	..	7	27	48	70	95	121	150	181	214
50°	..	..	19	40	62	86	112	140	171	204
52°	..	..	11	32	54	77	103	131	161	194

⁎ To ascertain by the above Table the quantity of Pipe which will heat 1,000 cubic feet of air per minute, find, in the first column, the temperature corresponding to that of the external air; and at the top of one of the other columns find the temperature at which the room is to be maintained; then, in this latter column, and on the line which corresponds with the external temperature, the required number of feet of pipe will be found.

difference between the internal and external temperature. The pipe, therefore, gives off more heat to the air in the earlier stages of the operation than the glass transmits by radiation to the external atmosphere. This difference in the effect is actually the rate at which the building becomes heated; and the increase of the temperature of the building continues until the radiating power of the glass exactly balances the heat given off by the pipes. But the heat given off by the pipes, it may be observed, constantly *decreases*, while the cooling power of the glass *increases* with every addition to the temperature of the internal air. We see, then, that when we have estimated the surface of glass in such a building, we can calculate the quantity of pipe that will heat it. For suppose the house has 800 square feet of glass: we find (Art. 105) that every square foot of glass cools *one and a quarter cubic feet of air per minute* as many degrees as the internal exceeds the external temperature. If therefore the external temperature be 10°, and the internal temperature is required to be 75°, then 800 square feet of glass will (as above stated) cool 1,000 cubic feet of air *per minute* from 75° down to 10°. By the Table, then, in the *column* marked 75°, and on the *line* marked 10° for external temperature, we find the quantity 292; which is the number of feet in length of pipe, four inches diameter, that are required to heat this 1,000 cubic feet of air per minute the required number of degrees. This quantity of pipe, therefore, will heat a building having 800 square feet of glass, whatever the actual size of the building may be. And whenever the quantity of air to be heated per minute is either greater or less than 1,000 cubic feet (or, in other words, when the quantity of glass is greater or less than 800 square feet), then the proper quantity of pipe will be obtained by the rule of proportion, as already stated (Art. 106).

(108.) The above rule will not, however, give the length of time required to heat any particular building. This will, of course, depend upon many circumstances; nevertheless, some approximation may be made to the average time required. Suppose the maximum temperature of the pipe to be 200°, the water being at 40° before lighting the fire; then the maximum temperature in horticultural buildings will be attained with

Four-inch pipes, in about four and a half hours.
Three-inch pipes, in about three and a quarter hours.
Two-inch pipes, in about two and a quarter hours.

But if a larger quantity of coal than that given by the Table (Art. 114) be used—if the surface of the boiler be much increased in proportion to the length of pipe—if the quantity of pipe used be excessive—or the temperature of the external air be higher than the estimated amount, then, in each of these cases, the time required for heating will be less. If, on the contrary, the required temperature be not attained in the time given above, then either too small a quantity of pipe, too small a surface of boiler, or too small a quantity of coal has been used.

It should, however, be observed, that although the *maximum* temperature will not be reached, on an average, in less time than is above stated, still the required temperature will very often not take longer than half or two-thirds of this time, to be attained; because the quantity of pipe being always apportioned to meet the case of extreme cold, when the external temperature is above that extreme limit, the pipe, by being superabundant, will warm the same space in a shorter time.

(109.) These calculations of the time required to heat buildings will only apply to those cases where the cooling surfaces are very large, and the proportion of the pipe very considerable, relatively to the actual dimensions of the building. Wherever, on the contrary, the cubical content of the building is large in proportion to its cooling surfaces, the time required to raise its temperature will be greater than is above stated, in consequence of the very small proportionate quantity of pipe; and this will be found to vary greatly in different descriptions of buildings. Churches and other large buildings (which only require a small heating surface relatively ,to their cubic contents) will generally require the longest time to heat; dwelling-houses will require a shorter time, and horticultural buildings the shortest of all. The length of time above estimated only applies to the latter description of buildings. For the other kinds of buildings the period will be very variable, and can scarcely be determined, except for each individual case.

(110.) Various circumstances may, however, interfere to diminish the effect of the apparatus; such, for instance, as damp walls—particularly if the building is new—excess of ventilation, &c. The effect of damp walls in reducing the apparent power of an apparatus is very considerable, in consequence of the great quantity of heat which is necessary to evaporate the moisture. It will require as much heat to vaporise one gallon of water from the walls of a building as would raise the temperature of 47,840 cubic feet of air 10°. The true power of an apparatus can, therefore, never be ascertained unless the building be perfectly dry. The same cause, though in a much less degree, becomes operative in buildings which are only occasionally warmed; and a longer time will always be necessary to heat such places than those that are in constant use.

(111.) For estimating the quantity of pipe which is required to warm any building, rules of a much more facile character, though at the same time more loose and inaccurate than those that have been already given, may be deduced.* They are the results of experience, and they have been found so generally useful in practice, and in most cases so nearly accurate in their results, that they are here given at considerably greater length than in the earlier editions of this work.

CHURCHES AND LARGE PUBLIC ROOMS.—To heat these when they have an average number of doors and windows, and only moderate ventilation, divide the cubic measurement of the building by 200, and the quotient will be *the number of feet in length of pipe four inches diameter*, that will be required to produce a temperature of about 55° in very cold weather.† This is equivalent to allow-

---

* The following rules must not be confounded with that already given. The former rule gave the result entirely from an estimate of the quantity of glass, *without any reference to the cubic contents of the building;* the present rules, on the contrary, are founded entirely on the cubic contents of the building without direct reference to the quantity of glass. The results, however, of the two rules will be found to agree with sufficient accuracy for most practical purposes.

† Churches and other buildings for containing large assemblages of people ought never to be heated to a very high temperature, on account of the great quantity of animal heat given off in crowded congregations. It has been ascertained, by calculations founded on the amount of oxygen consumed, that a man generates a quantity of heat in 24 hours sufficient to raise 63 lbs. of water from the freezing to the boiling point. Of this quantity, as much heat is expended in forming the vapour that passes off by perspiration and by transpiration from the lungs as would heat about 36½ lbs. of water 180°; and the remainder of the heat, which is equal to raising the temperature of 26½ lbs. of water 180°, passes off by radiation from the body. (*Quetelet's Philosophy*, Art. "Heat.") Now these

ing five feet of four-inch pipe for every thousand cubic feet of space which the building contains. If the apparatus is so contrived that the warming of the air is effected before it actually circulates in the room, and that the same portions of air are not returned to be heated a second time, but fresh portions of external air are brought successively in contact with the heating apparatus, it will require from 50 to 70 per cent. more pipe to produce the same effect; but the air will, of course, be more pure and fresh.*

results, if reduced to the same standard that has been adopted in the preceding calculations, will lead to the conclusion that as much heat is given off per minute from the body of an adult man as would be produced by an iron pipe four inches diameter and three-and-a-half feet long, filled with water at 200°. This estimate, however, in practice, would be found too high; for where there is no muscular exertion less heat is produced; and the increased temperature of the surrounding medium would also prevent its free radiation. For, as all bodies only give off heat in proportion to their excess of temperature, the human body being constantly at the temperature of 98° nearly twice as much heat would be given off (if the body were freely exposed) when the surrounding medium is at 50° as would be the case if the latter were raised to 70°. It is found also that, on an average, women only consume about half as much oxygen as men (*Combe's Principles of Physiology*, 4th edition, p. 222), and therefore they can only produce half as much heat; the consumption of oxygen always being proportional to the heat generated. From these facts it will appear that not only should buildings such as those we are now considering, not be too highly heated, but that the pipes should be moderately small in diameter, in order to allow of the temperature being more easily lowered when the building is filled with people. Some experiments on the heat thus given off from the human body are given in Wyman "On Ventilation," p. 185, Boston, 1846.

* This mode of calculating the quantity of pipe will differ from the previous mode of calculating by the surface of glass and the allowance necessary for ventilation, to a much greater extent in the case of churches than almost any other kind of building. The reasons for this are twofold: not only does the proportion of glass to the area of the building differ to a

DWELLING-ROOMS.—These will generally require about 12 feet of four-inch pipe to every thousand cubic feet of space contained in them, to give a temperature of about 65°. To raise the temperature to 70° will require about 14 feet of four-inch pipe.

HALLS, SHOPS, WAITING-ROOMS, ETC., will require about ten feet of four-inch pipe to every thousand cubic feet of space, to raise the temperature to about 55°. For a temperature of 60°, about twelve feet of four-inch pipe will be required.

WORK-ROOMS, MANUFACTORIES, ETC., where a temperature of about 50° to 55° only is required, will generally be sufficiently heated by six feet of four-inch pipe for every thousand cubic feet of space they contain. For a temperature of 60° about eight feet of pipe will be required.

very great extent in different churches; but, besides this, if the quantity of pipe were only just sufficient to compensate for the loss by the cooling power of the glass, it would require far too long a time to heat the church to a required temperature, in consequence of its very great area in proportion to the total heating surface. In college chapels, and some other ecclesiastical buildings, where the quantity of glass is particularly small, this is remarkably the case; and the former rule would hardly give one-fourth the quantity of pipe found by the latter. Now, in these cases, when the quantity of glass is so very small, a less quantity of pipe would certainly suffice than that obtained by the latter rule of allowing five feet of four-inch pipe for each thousand cubic feet of space: and it must be decided by experience which rule shall be adopted in such cases. In those cases where the warming apparatus is kept constantly in operation, the smaller quantity of pipe obtained by the first rule would suffice; but in cases where the apparatus is only heated once or twice a week, it requires to be much more powerful, in order to produce the required effect in a sufficiently short time. In a very large majority of churches, the last rule given will be correct; but when any doubt may exist as to its applicability to any particular case, a very safe plan will be to calculate the quantity of pipe by both rules; add the results together, and divide the resulting quantity by two. This plan will give a result which will safely meet almost any case that can arise.

SCHOOLS, AND LECTURE-ROOMS, requiring a temperature of 55° to 58°, will require from six to seven feet of four-inch pipe to every thousand cubic feet of space.

DRYING-ROOMS, or closets, for drying wet linen and other substances, require from 150 to 200 feet of four-inch pipe to every thousand cubic feet of space to raise the temperature to 120° when empty, or about 80° when the room is filled with wet linen.*

DRYING-ROOMS, for curing bacon, or for drying paper, or leather, or damp hides, will require twenty feet of four-inch pipe to every thousand cubic feet of space to give a temperature of about 70°.

GREENHOUSES AND CONSERVATORIES, requiring a temperature of about 55° in the coldest weather, must have 35 feet of four-inch pipe for each thousand cubic feet of space they contain.

GRAPERIES AND STOVE-HOUSES, requiring a temperature of 65° to 70° in the coldest weather, will require 45 feet of four-inch pipe for each thousand cubic feet of space; and if a temperature of 70° to 75° is required, 50 feet of four-inch pipe must be allowed for each thousand cubic feet of space.

PINERIES, HOTHOUSES, and CUCUMBER-PITS, requiring a temperature of 80°, must have about 55 feet of four-inch pipe for every thousand cubic feet of space the house contains.

Modern refinements have introduced heating by hot water in many other forms of buildings than originally contemplated. Thus STABLES are often thus heated, and they usually require about five feet of four-inch pipe per thousand cubic feet of space. COACHHOUSES require about the same quantity, and so also do DOG HOUSES and FOWL HOUSES. DAIRIES are also now frequently warmed

* See also Arts. 176 and 177.

in this manner. They require about 16 to 18 feet of four-inch pipe per 1000 cubic feet of space to give a temperature of about 56° Fahrenheit.

The quantity of pipe estimated in this way will only suit for such buildings, whether horticultural or otherwise, as are built quite upon the usual plan and of the ordinary proportions; for, if they vary much from the most ordinary construction, these rules will not be accurate, and the method given in the former part of this chapter should then be employed.

(112.) Although these calculations are all made on the supposition of using pipe of four inches diameter as the heating surface, it is by no means intended to recommend that as the best size for all purposes. For all horticultural purposes it is the best, where it can be used; but for most other purposes, smaller pipes, or even other forms of heating surfaces, may generally be more advantageously employed. If the pipes used are only three inches diameter, we must add one-third to the quantities here given; and if pipes of two inches diameter are used, double the quantity will be required.

(113.) It should here be mentioned, that the calculations for the quantity of pipe required for horticultural buildings have been made with a view to the most economical mode of effecting the desired object. Some of the most successful horticulturists, however, have adopted the plan of using a much stronger heat in their forcing-houses, and allowing, at the same time, a much greater degree of ventilation than usual. This plan is stated to produce a finer fruitage; but it will only be obtained at an increased cost in the apparatus, and by a larger expenditure of fuel. Where economy is not required, it may perhaps be desirable to adopt this plan; and then the quantity of pipe which is used must be proportionally increased above the estimates which are given in this chapter.

(114.) The quantity of coal necessary to supply any determinate length of pipe is easily ascertained, from the data given in Art. 270. After the water in the pipes is heated to its maximum, the quantity of coal consumed is, obviously, just what is required to supply the heat given off from the pipes. Now, by Art. 99 we find that when pipes four inches diameter are 146·8° hotter than the air of the room, the water contained in them loses exactly 1° per minute of its heat. By Art. 94, we find that 1 lb. of coal will raise the temperature of 39 lbs. of water 180°; and as 100 feet in length of four-inch pipe contains 544 lbs. of water, it will require 13·9 lbs. of coal to raise the temperature of this quantity of water 180°. If, therefore, the water loses 1° of heat per minute, or 60° per hour, this quantity of coal will supply 100 feet in length of pipe for three hours, if its temperature continue constant with regard to the air of the room. On this principle the following Table has been constructed. The temperature of the pipe is assumed to be 200°: then, knowing the temperature of the room, if we take the *difference* between the temperature of the pipe and that of the room, by looking in the Table for the corresponding temperature, we shall find under it the number of pounds weight of coal which will be required per hour for every 100 feet in length of pipe, in order to maintain the stated temperature. Thus, suppose the pipe to be four inches diameter, and its temperature 200°, while the room is at 75°, then, under the column headed 125° (which is the difference between these two temperatures), we find 3·9 lbs. as the quantity required per hour for every 100 feet of pipe. The quantities stated in the Table are given in pounds and tenths of a pound.

TABLE. V.

Table of the Quantity of Coal used per Hour to heat 100 Feet in length of Pipe of different Sizes.

Diameter of Pipe, in Inches.	Difference between the Temperature of the Pipe and the Room in Degrees of Fahrenheit.														
	150	145	140	135	130	125	120	115	110	105	100	95	90	85	80
4	4·7	4·5	4·4	4·2	4·1	3·9	3·7	3·6	3·4	3·2	3·1	2·9	2·8	2·6	2·5
3	3·5	3·4	3·3	3·1	3·0	2·9	2·8	2·7	2·5	2·4	2·3	2·2	2·1	2·0	1·8
2	2·3	2·2	2·2	2·1	2·0	1·9	1·8	1·8	1·7	1·6	1·5	1·4	1·4	1·3	1·2
1	1·1	1·1	1·1	1·0	1·0	·9	·9	·9	·8	·8	·7	·7	·7	·6	·6

(115.) It should, however, be borne in mind that an apparatus will not always consume the same quantity of coal; in fact, it will seldom require near so much as the Table shows, because that is the calculation for the maximum effect. Suppose the quantity of pipe in a room has been accurately calculated, in order to maintain the temperature at 75° when the external air is at 30°, the consumption of coal for pipes of four inches diameter will then be 3·9 lbs. per hour for every 100 feet of pipe. But should the external temperature now rise to 40°, 77 feet of pipe would produce the same effect as 100 feet would in the former case ; therefore the pipe must be heated to a lower temperature ; and it will be found by calculation, that only 3 lbs. of coal would be used, instead of 3·9 lbs. As much coal, therefore, as would supply 77 feet of pipe at the maximum temperature would suffice for 100 feet at this reduced temperature. The quantity of

K

fuel which is consumed will, therefore, be continually subject to variation, as it will alter with
the temperature of the external atmosphere ; and,
in general, the average quantity of coal required
will be fully one-third less than the amount given
in the Table.

It is almost unnecessary to observe that, in calculating this Table, it has been assumed that the
boiler and furnace are of good construction; for
on no other basis could an estimate be formed.
Very great differences, however, exist in this
respect; and for such cases no estimate whatever
can possibly be made.

CHAPTER VII.

Various Modifications of the Hot-Water Apparatus—Kewley's Siphon Principle—The High-pressure system—Holmes and Coffey's modifications—Eckstein and Busby's Rotary Float Circulator—Fowler's Thermo-siphon—Price's improved Hot-water Boxes—Rendle's Tank System—Corbett's Trough System of Evaporation—Theory of Evaporation.

(116.) Under the common and generic term of "hot-water apparatus" various plans have been brought forward by different inventors, which, though essentially different in some of their features from those that have been already described, are, nevertheless, merely modifications of the general principles that have been explained. In the present chapter some of these peculiar modifications of the invention will be investigated; and it will appear that the original principles of all are the same, but that other of the fundamental laws of Nature are here brought into action conjointly with those that we have already examined, and give rise to an apparent diversity of operation.

(117.) The first notable invention of this sort which shall be mentioned, is Kewley's siphon principle. The sketch, fig. 36, shows this apparatus in its simplest form. The boiler is open at the top, and the two pipes dip into the water; the pipe A descending only a very short distance below the surface, and the pipe B reaching nearly to the bottom of the boiler. A small flexible metal pipe, x, is attached to the highest part of the pipes. To

this an air-pump is connected, and the air in the
pipes being exhausted by this means, the atmospheric pressure forces the water up the pipes and
fills them completely.

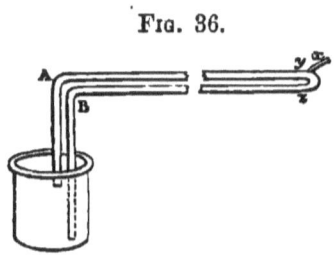

Fig. 36.

This avoids the ne-
cessity of having a
reservoir of water
higher than the top
of the boiler; for it
is well known, that
the usual atmospheric
pressure is capable of
raising a column of water in a vacuum to about
30 feet in height, varying, however, with the degree
of pressure shown by the barometer.

The water in the longer pipe B will acquire a
preponderance of weight over that in the pipe A
even if it be at first of an equal temperature and
density; because the pipe B only receives the particles of hot water which rise immediately under
its base, while the other receives the heat from all
parts of the bottom as well as the sides of the
boiler; the water on the top being hotter than
that at the bottom. But as soon as the water circulates through the pipes, it parts with its heat,
and the whole length of the pipe B will then be
colder than the pipe A, and the water will descend
through B with greater force.

In consequence of the long pipe B being surrounded by the hot water in the boiler, the water,
while descending through it, receives a small portion
of heat, which lessens the difference of temperature
between the two pipes, and reduces the velocity
of the circulation. It appears probable, therefore,
that additional velocity of circulation would be
gained by placing the descending pipe B outside
the boiler, and attaching it to the side in the same
manner as the return-pipe in fig. 5. The prin-

cipal inconvenience attending this would be the difficulty of stopping the ends of the two pipes A and B, which is now done by the simple contrivance of a plate screwed moveably to the base of each of the pipes, by means of an external rod passing over the pipes A and B, with a screw attached on the top; and by turning this the plate is drawn up into close contact with the end of the pipe. This completely stops the water when necessary, the ends of the pipes being turned true to the plates, to make them water-tight; and by reversing the action of the pump, attached to the pipe x, and thus making it into a force-pump, the soundness of the joints can then be ascertained. A leaky joint is difficult of detection by any other means, as there is no emission of water from it in the usual way. The only immediate consequence of a leaky joint is the immission of air, and it is not observable except by its stopping the circulation of the water, which occurs by the air accumulating and cutting off the connexion of the water between the two pipes.

If this plan of having the return-pipe placed outside the boiler were found to increase the motive power of the apparatus, an advantage would be gained in all those cases where the pipes are required to pass under a doorway, because, in all such cases, the boiler for this apparatus must be set much further below the level of the floor than is required for the common hot-water apparatus. But by increasing the motive power, a less height would be sufficient; and it would therefore prevent the inconvenience sometimes found to attend this particular form of the apparatus, arising from the great depth the furnace is required to be sunk beneath the level of the pipes, in consequence of the very large size of the boiler which is generally used.

(118.) A singular fact is connected with this invention, which deserves notice, because it arises from a philosophical principle, which, in some other instances, has been applied in a most useful manner;[*] though, with this particular invention, it is rather disadvantageous than otherwise. It has already been stated, that the height to which the water will rise in a vertical column, by the atmospheric pressure, is about 30 feet above the boiler. Supposing this to be the extreme limit to which the water will ascend in the pipes, the slightest elevation above this will cause a vacuum to be formed, similar to that at the top of a barometer, and the water at the top of the pipe will, in this case, be *without any pressure*. But if, instead of 30 feet, the pipe be continued upwards only 15 feet, then the pressure on the water, in the upper part of the pipe, will be 7½ lbs. on the square inch, or half the usual atmospheric pressure; and so on for other heights. Now, the boiling-point of all liquids varies with the pressure. Water boils at 212°, under the mean pressure of 15 lbs. per square inch; but by reducing the pressure, it boils at a lower temperature; so that at half the mean pressure of the atmosphere it boils at about 186°. Suppose now that the pipes just described rise 30 feet above the boiler, the water at the top will boil at the temperature of 161°, and will form steam in the upper part of the pipe; and this, by its great expansion, will force the water down and overflow the boiler or the supply cistern. For, at the ordinary pressure of the atmosphere, steam occupies about

---

[*] The boiling of liquids *in vacuo* is well known, and has been most extensively applied in many cases. The boiling of sugar in vacuum-pans is one of the most successful applications of science to the arts which modern times has produced.

1,700 times as much space as the water from which it is formed, and still more at a diminished pressure, its expansion being inversely as the pressure. When the pipes rise to other heights above the boiler than that described above, the boiling-points will be as follows :—

At 5 feet high, the boiling point will be 203°
10 . . . . . . 195°
15 . . . . . . 186°
20 . . . . . . 178°
25 . . . . . . 169°
30 . . . . . . 161°

Therefore the water in the boiler must always be kept below these temperatures, according to the height to which the pipes ascend.*

This peculiarity, which applies only to pipes on the siphon principle, is more a philosophical fact than a practical difficulty; for the water can generally be kept at a temperature sufficiently low for any ordinary height that is required. And, in fact, the boiling-point will generally be higher than the temperatures here stated because a small portion of air always remains in the pipes, which increases the pressure on the water, and

* These calculations are made by Wollaston's rule for his thermometric barometer. But this rule, although accurate at moderately small differences of pressure, becomes erroneous at considerable reductions of pressure. Professor Robinson estimates the boiling-point of water *in vacuo* at only 88°, instead of 161°, which the above calculation shows; and it is probable that the relative proportion between the pressure and the boiling-point is in a logarithmic ratio, instead of the common arithmetical proportion of Wollaston's rule. This, in fact, is found to be the case at temperatures above 212°. But it is probable that, in the present case, Wollaston's rule will give a more accurate result than the other, because, as the vacuum in the pipes cannot be at all perfect, the boiling-points will be much higher than the calculated temperature; perhaps even higher than stated in the text. See *Robinson's Mechanical Philosophy*, vol. ii., pp. 22–37 ; and Wollaston " On the Thermometric Barometer," *Philosophical Transactions*, 1817, p. 183.

makes the boiling-point higher than the calculated amount. This form of the apparatus answers the intended purpose when worked with care, and is a very ingenious application of scientific principles ; but it requires more care in working than the ordinary apparatus, and is now comparatively but little used.

(119.) The next invention which we shall consider is the High-Pressure hot-water apparatus.* This apparatus consists of a coil of small iron pipe, built into a furnace, the pipe being continued from the upper part of the coil, and passes round the room or building which is to be warmed, forming a continuous pipe when again joined to the bottom of the coil. The diameter of this pipe is one inch externally, and half an inch internally. A large pipe, of about two and a half inches diameter, is connected in some part of the circulation, either horizontally or vertically, with the small pipe, and is placed at the highest point of the apparatus. This large pipe, which is called "the expansion pipe," has an opening near to its lower extremity, by which the apparatus is filled with water, the aperture being afterwards secured by a strong screw ; but the expansion pipe itself cannot be filled higher than the opening just named. After the water is introduced, the screws are all securely fastened, and the apparatus becomes then hermetically sealed. The expansion pipe, which is thus left empty, is calculated to hold about $\frac{1}{12}$ as much water as the whole of the small pipes ; this being necessary in order to allow for the expansion that takes place in the volume of the water when heated, and which otherwise would inevitably burst the pipes, however strong they may be. For the expansive force of water is almost irrepressible, in consequence of its possess-

* *Repertory of Arts, &c.*, vol. xiii. (1832), p. 129.

ing but a very small degree of elasticity; and the increase which takes place in its volume, by raising the temperature from 39° (the point of greatest condensation) to 212°, is equal to about $\frac{1}{23}$ part of its bulk, and at higher temperatures the expansion proceeds still more rapidly.*

The temperature of these pipes, when thus arranged, can be raised to a very great extent; for, being completely closed, and all communication cut off from the atmosphere, the heat is not limited, as usual, to the point of 212°, because the steam which is formed is prevented from escaping, as it does in the common form of hot-water apparatus. The most important consideration respecting it, however, is the question as to its safety; for most persons are aware that steam, when confined beyond a certain point of tension, becomes extremely dangerous; and in this apparatus the bounds of what hitherto has been used in other cases are very far exceeded.

(120.) On the first introduction of this plan, it was usual to make the coil consist of one-fourth part of the total quantity of pipe which was used in the apparatus; and it was considered that, when this proportion was observed, the heat of the pipes could not be raised so high as to endanger them by bursting. But in practice this has not always proved a preventive to accident, even when the proportion which the coil bears to the radiating surface is much smaller than is here mentioned.†

---

* See Table IV., Appendix. The force which would be exerted on the pipes by this expansion of $\frac{1}{23}$ of the volume of the water would be equal to 14,121 lbs. per square inch, according to the experiments of Professor Œrstead. *Report, British Scientific Association*, vol. ii., p. 353.

† The specification to the patent for this invention states, that when the radiating surface is three times that of the coil, the pipes cannot burst. It has, however, been found neces-

The *average* temperature of these pipes is stated to be generally about 350° of Fahrenheit. But a most material difference of temperature occurs in the several parts of the apparatus, the difference amounting sometimes to · as much as 200° or 300°. This arises from the great resistance which the water meets with, in consequence of the extremely small size of the pipes, and also from the great number of bends, or angles, that of necessity occur, in order to accumulate a sufficient quantity of pipe. In these angles, the bore of the pipe, already extremely small, is still further reduced, which causes the water to flow so very slowly, that a great portion of its heat is given out long . before it has circulated round the building which is to be warmed. The temperature of the coil, however, is what we must ascertain, if we wish to know the pressure this apparatus has to sustain, and thence to judge of its safety: for, by the fundamental law of the equal pressure of fluids, whatever is the greatest amount of pressure on any part of the apparatus must also be the pressure on every other part.

(121.) Now the temperature of this apparatus is found to vary, not only with the intensity of the heat of the furnace, but also with the proportion which the surface of the coil bears to the surface of the pipe which radiates the heat. In some apparatus, if that part of the pipe which · is immediately above the furnace be filed bright, the iron will become of a straw colour, which proves the temperature to be about 450°.* In other instances it will become purple, which

sary greatly to increase the proportion of radiating surface, in order to prevent the bursting by excessive pressure ; and the radiating surface is now frequently made *ten* times that of the coil in the furnace, in order to secure its safety.

* See Table VI., Appendix.

shows the temperature to be about 530°; while, in some cases, it will become of a full blue colour, which proves that the temperature is then 560°. By this means the pressure on the pipes may be known; for, as there is always steam in some part of the apparatus, the pressure may be calculated so soon as the temperature is ascertained. By referring to Table I. in the Appendix, we shall find that a temperature of 450° produces a pressure of 420 lbs. per square inch, while a temperature of 530° makes the pressure 900 lbs.; and when it reaches 560°, the pressure is then 1,150 lbs. per square inch.

(122.) Those who are acquainted with the working of steam-engines are aware that a pressure of three or four atmospheres is considered as the maximum for high-pressure boilers: but we see that in this apparatus the pressure varies from ten times to twenty times that amount. And it will also be borne in mind that, in consequence of the extremely small quantity of water used in these pipes, the slightest increase in the heat of the furnace will cause an immediate increase in the pressure on the whole apparatus. For it appears by a reference to the Table last mentioned, that if the temperature of the pipes be increased 50° above the amount before stated, the pressure will be raised to 1,800 lbs. per square inch; and by increasing the temperature 40° more, the pressure will be immediately raised to 2,500 lbs. per square inch; so that any accidental circumstance, which causes the furnace to burn more briskly than usual, may, at any moment, increase the pressure to an immense amount.*

* This increased pressure is also extremely likely to occur in this apparatus when a portion of the pipe is occasionally shut off by means of cocks or valves. In this case the coil in the furnace becomes too powerful for the apparatus, and an explo-

(123.) The pipes which are used for this appa
ratus are stated to be proved with a pressure of
2,800 lbs. per square inch.*   This is very probable :
for as wrought iron of the best quality requires a
longitudinal strain of 55,419 lbs. to break a bar one
inch square, so the force necessary to break a
wrought-iron pipe of one inch diameter externally,
and half an inch diameter internally, would be
13,852 lbs., which is equal to 8,822 lbs. per square
inch on the internal diameter.   But, on account of
the strain on these pipes being transverse to the
grain of the iron, and also in consequence of the
welded joint of the pipe not being so strong as the
solid metal, these pipes will not bear anything
like the calculated amount of pressure.   It is
evident, however, that no ordinary force can burst
them ; but, as this casualty does sometimes occur,
this great strength of the materials proves the im-
possibility of regulating the temperature in hermetically sealed pipes, so as to keep the expansive
force of the steam within even this immense
limit.

(124.) Although this description of apparatus
has been erected by many different individuals,
possessing various degrees of mechanical knowledge, and severally performing their work with
different degrees of excellence, much uniformity
appears in the result, in those cases where failure
has occurred.   From a comparison of a number of
cases where accidents have happened to apparatus
erected on this system, more than one-half have

sion is then very likely to occur, unless the utmost caution be
observed in regulating the fire.   This source of danger is peculiar to the high-pressure system of heating, and does not at all
apply to any of the other plans which have been described.
   * As pipes are always proved when they are cold, this does
not at all show the strain they will bear when heated.   On
this subject see the following Note.

arisen from the bursting of the coil, notwithstanding the increased size of the expansion pipe renders
this latter apparently the weakest part of the apparatus; the relative strength of pipes, with the
same thickness of metal, being inversely as their
diameters.

(125.) The cause of the explosions occurring
principally in the coil is owing to the iron becoming weaker in proportion as its temperature is
raised; so that, as the pressure increases, the iron
decreases in strength to resist the strain.* Another
circumstance also tends to produce the same effects.
It is found, on breaking one of these pipes, after it
has been used for some time in or near the fire,
that the iron has lost its fibrous texture, and that
it presents a crystallized appearance, similar to
what is known as "cold short iron." This singular change in the texture of iron has been
noticed in other instances. Mr. Lowe (*Report,
British Scientific Association*, 1834) has found that
wrought iron at a red heat exposed to the steam
of water for a considerable time, becomes crystallized; and in many other instances also, even
without the presence of steam, the same effect has
been observed. The cause of this phenomenon
has not been clearly ascertained; but, whatever it

---

* The temperature of maximum strength for cast iron has
been estimated at about 300°; but the "Committee on the
Explosion of Steam-Boilers," appointed by the Franklin Institution, consider that the maximum for wrought iron is higher
than this, and that 572° may be considered as the temperature
of maximum strength. After the temperature of maximum
strength is once passed, the decrease in the strength of wrought
iron is considerable: at a red heat, or about 800°, it loses about
one-fifth of its strength. The maximum strength of copper,
on the contrary, is at a very low temperature; for the strength
increases with every reduction of temperature down to 32°,
which is the lowest that has been tried. See Chapter XII.,
Arts. 265 and 266.

may be, the effect undoubtedly is to weaken the tenacity and cohesive strength of the metal to a very great extent.*

(126.) But we shall find that, enormous as the pressure appears to be, with which these pipes are proved, it is not adequate to the working pressure which they sometimes have to resist. It has been ascertained that the strength of wrought iron decreases considerably at temperatures above 572°, and as it also loses a great deal of its strength when it assumes the crystallized state, varying with the circumstances, and sometimes amounting to three-fourths of its original strength, it will appear that the proof pressure, when cold, for pipes which are to be used in this kind of apparatus ought, in fact, to be much greater than the amount to which they are actually proved ; and hence the cause of these pipes bursting after they have been in use for some considerable time, if they happen accidentally to get heated to very high temperatures.

(127.) The question has sometimes been asked, What would be the effect on this apparatus if the

* The author, in a paper which was read before the Institution of Civil Engineers (*Minutes of the Institution*, June, 1842), endeavoured to trace the cause of the extraordinary change which iron undergoes in these and some other circumstances. Percussion at certain high temperatures produces an instantaneous change, and, at lower temperatures, longer-continued percussion produces the same effect. Heating and rapid cooling likewise produce crystallization ; and, in every case, magnetism appears to accompany the phenomena ; but whether as cause or effect is not easy to determine. The subject is altogether of great interest, both in a practical and in a scientific point of view; and experiments on a large scale are in progress in order to determine the question. Other metals besides iron are, probably to some extent, affected in a similar manner; and it is probable that, under certain circumstances, spontaneous change in the molecular structure of iron occurs, though far more slowly than by the action of percussion and heat.

expansion-pipe were to be filled with water, as well as the small circulatory pipe? The almost immediate consequence would be the bursting of the pipes; for scarcely anything can resist the expansive power of water. The force necessary to resist its expansion is equal to that which is required for its artificial condensation. Now, at the temperature of 386°, water expands rather more than $\frac{1}{12}$ of its bulk; and to condense water this extent (Note, Art. 12) requires a pressure of 27,104 lbs. per square inch; therefore, the bursting pressure at this temperature would be enormous. If the pipes were filled completely full of cold water, without allowing any room for expansion, and if they were then hermetically sealed, as before described, by increasing the temperature of the water only about 60°, the expansion of the water would cause a pressure of 2,000 lbs. per square inch on every part of the apparatus, reckoned by the internal measurement.

(128.) The assertion has often been made, that the heated fluid contained in an apparatus constructed on this plan will not scald, even if the pipes should chance to burst, because *high-pressure steam*, it is well known, is not injurious in this respect. But this is quite a mistaken notion; for high-pressure hot water will scald, though high-pressure steam will not; and the fluid which would issue through any fissure that might occur in these pipes could only be partially converted into steam, unless its temperature were at least 1,200°. This is obviously impossible; were it the case, the water would be all converted into steam the instant that it issued from the pipe. The reason that high-pressure steam does not scald is in consequence of its capacity for *latent* heat being greatly increased by the high state of rarefaction it instantaneously assumes when suddenly liberated;

this lowers its *sensible* temperature, and causes it
to abstract heat, from everything that it comes in
contact with. The scalding effect of high-pressure
hot water, on the contrary, when suddenly projected from a pipe or boiler by explosion, will
always be the same, whatever its temperature
may be while confined within the pipe; for, the
instant it is liberated, a portion of it is converted
into steam, and the remainder sinks to the temperature of about 212°.

(129.) Among the advantages which have been
supposed to arise from the use of this invention,
it has been imagined that, in consequence of the
quantity of water which the pipes contain being
so small, the consumption of coal would be less
with this than with any other description of hot-
water apparatus. We have seen, however (Art.
114), that the quantity of coal which is used is in
proportion to the heat that is given off in the room
that is warmed; and a reference to the Table
(Art. 114) will show that the size of the pipe
makes no difference in the consumption of coal
per hour, provided the same effect is required to
be produced, the only difference being in the
length of time required to warm the water in the
first instance. But there will, on the contrary,
be a greater expenditure of fuel in this apparatus,
in consequence of the coil affording less surface
for the fire to impinge against than would be
obtained by using a boiler. In addition to this,
the colder any surface may be when exposed to
the action of a fire, the more heat will it receive
in a given time; therefore, as the heat of these
pipes is nearly three times as great as that of a
boiler, there must be a waste of fuel from this
cause.

(130.) In consequence of the intense heat of
these pipes, it is sometimes found that rooms

which are heated by them have the same disagreeable and unwholesome smell which results from the use of hot-air stoves and flues. In reality, the cause is the same in both cases; for it arises partly from the decomposition of the particles of animal and vegetable matter that continually float in the air, and partly from a change which atmospheric air undergoes by passing over intensely heated metallic surfaces. The electric state of the air is likewise altered by highly heated metallic surfaces, and exerts an important effect on all animal bodies exposed to its influence. Some experiments on this subject are recorded in the Philosophical Transactions of the Royal Society (Vol. XXVII. p. 199). Also in the year 1874 the French Royal Academy appointed a commission to examine the effect produced on atmospheric air by cast iron surfaces, made red hot. Rabbits were made to breathe air passed over stoves of cast iron heated to redness; and afterwards chemical examination of the blood of the animals was made to ascertain the presence of carbonic oxide. The Commissioners reported that " the experiments made upon rabbits do not permit us to fix with any precision the proportions of carbonic oxide absorbed by their blood, nor that of the oxygen which has been expelled from it; but the results all agree to show that the use of stoves of cast iron heated to a red heat, causes in the blood (by the presence of carbonic oxide) a gas eminently poisonous, and causes changes whose repetition may become dangerous." Whether wrought iron produces the same effect is not clearly ascertained.

(131.) The high temperature of these pipes, and the intensity at which the heat is radiated from them, have sometimes been urged as an objection against this invention, when applied to horticultural purposes, because any plants which

L

are placed within a certain distance of them are
destroyed. Although, no doubt, this effect really
takes place, it can be easily avoided with proper
care; for, as radiated heat decreases in intensity
as the square of the distance, it only requires that
the plants should be placed farther off from these
pipes than from those which are of a lower temperature. In comparing the effect of two different pipes, if one be *four* times the heat of
the other (deducting the temperature of the air
in both cases), the plants must be placed *twice*
as far off from the one as from the other, in
order to receive the same intensity of heat from
each. The only inconvenience, therefore, is the
loss of room, which in some cases may not be of
much importance. But a more serious objection
by far appears to lie in the inequality of temperature which any building heated by these
pipes must have, in consequence of their being so
very much hotter in one part than in another.
This difference of temperature between various
parts of the same apparatus has already been
stated to amount, in some cases, to as much as
200° or 300°; varying, of course, with the length
of pipe through which the water passes. From
what has been stated in Chapter IV., it will also
be observed that, owing to the smallness of these
pipes, this kind of apparatus cools so rapidly when
the fire slackens in intensity, that the heat of a
building which is warmed in this manner will be
materially affected by the least alteration in the
force of the fire, instead of maintaining that permanence of temperature which is so peculiarly the
characteristic of the hot-water apparatus with large
pipes.

(132.) This invention undoubtedly exhibits great
ingenuity; and could it be rendered safe, and its
temperature be kept within a moderate limit, it

would be an acquisition in many cases, in consequence of its facile mode of adaptation. Its safety would perhaps be best accomplished by placing a valve in the expansion-pipe, which, from its large size, would be less likely to fail of performance than one inserted in the smaller pipe. If this valve were so contrived as to press with a weight of 135 lbs. per square inch, the temperature of the pipes would not exceed 350° in any part; the pressure would then be nine atmospheres, which is a limit more than sufficient for any working apparatus where safety is a matter of importance.

(133.) A modification of this apparatus was proposed, and a patent taken out in 1832, by Mr. Holmes, for using oil instead of water.* As fixed oils boil only at very high temperatures, it was supposed there would be no liability to bursting the apparatus, as the temperature could not be raised sufficiently high to produce any pressure similar to that from steam. The temperature proposed to be employed was about 400°, but the plan entirely failed, in consequence of oil, when exposed to very high temperatures for any considerable length of time, becoming thick and viscid, finally losing its fluidity, and becoming a gelatinous mass. Of course, under these circumstances, no circulation of the oil could be produced, so as to render the apparatus practically useful.

(134.) A patent was also taken out in December, 1866, by Mr. John A. Coffey, for using distilled *mineral* oil, very nearly in the same manner as above described. This mineral oil, when properly distilled, is not inflammable, and it may be heated to nearly 1000° Fahrenheit *without producing any pressure* in the pipes, and still retain its fluidity. The death of the inventor, and other circumstances, have hitherto prevented the general use of this

* *Repertory of Arts, &c.*, vol. xv. (1833), p. 79.

invention, beyond a few isolated experiments ; but it appears to possess points of great utility where intense heat is required. The author has heated the refined mineral oil up to 800° Fahrenheit, without producing any sensible pressure on the apparatus. The only inconvenience appeared to be that by long continued boiling at these intense temperatures the oil deposited slightly a black sediment like powdered charcoal. This might endanger the stopping up the pipes of very small bore, unless some special provision were made for its removal. The sediment appears to be heavier than the oil, and would probably settle in the lowest parts of the apparatus. There is also a mechanical difficulty in making the various joints of the apparatus sound at these extreme temperatures, notwithstanding there is *no pressure* exerted by the fluid.

(135.) An apparatus of a totally different character from the preceding follows next to be described. It is an invention which, at first, appears to be singularly at variance with the general principles that have been laid down in this treatise. But however its mode of action may at first appear to differ from the principles which have been explained, it is certain that if these principles are derived from the laws of Nature, they must act equally at all times and under all circumstances ; for the operation of the physical laws can never be suspended, though they may be occasionally neutralized by a superior antagonist force. In the case of two opposing forces, the resulting action is proportional to their difference of power ; but when the antagonist force is removed, each will act according to its own peculiar laws.

This is the case with the invention now to be described. By it hot water is made to *descend* to any required depth below the boiler—apparently in opposition to the law of gravity—while the cold water will *ascend*, though of greater specific weight.

(136.) Eckstein and Busby's Patent Circulator, or Rotary Float, is an invention by which *centrifugal force* is made to overcome the *force of gravity*, in the circulation of hot water.* The boiler, which is either open or closed at the top, has a pipe, *a*, attached to its circumference, which is carried in any direction, either downwards, or around the room to be warmed, and finally returns into the boiler, and ends exactly in its centre, as shown at *b* in the annexed figure.

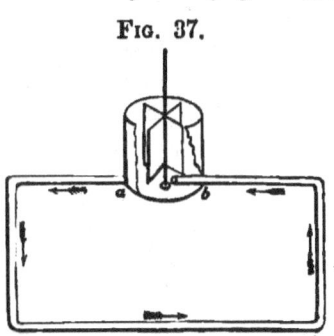

FIG. 37.

The float, or circulator, has motion given to it by means of a fly, similar to a smoke-jack, which is placed in the chimney, and is turned by the smoke of the fire that is used to heat the boiler, the float being fixed on centres, and revolving freely in the boiler. The centrifugal force imparted to the water by the rapid rotation of this float causes it to rise higher at the periphery than in the centre of the boiler; and the velocity with which the float moves determines the extent of this deviation from the level. The end of the pipe *b*, being *in the centre*, is then under a less pressure or head of water than the pipe *a*, the former being, by its position, removed from the greater pressure at the sides, which is caused by the centrifugal force imparted to the water by the float, which acts on the pipe *a* placed at the circumference.

* *Repertory of Arts, &c.,* vol. xiv. (1832), p. 137.

(137.) Suppose now the velocity of rotation to be such as to impart a centrifugal force sufficient to raise the water *one inch* higher at the circumference than in the centre, there will then be a pressure of 246½ grains per square inch upon the pipe *a* more than upon the pipe *b*, supposing the temperature of the water be about 180°. This additional pressure will allow the water in the pipe *a* to descend 42 feet below the boiler, if it does not lose more than 6° of heat before it returns back again to the boiler through the pipe *b* : if it lose 10°, then it will only descend 25½ feet, and so on for other temperatures. Now, as a pipe four inches diameter loses ·817 of a degree of heat per minute when its temperature is 120° above that of the room (Art. 270), this pipe may be of as great a length as the distance through which the water will flow in seven minutes and a half in the first case, or twelve minutes in the second.

(138.) The length of pipe through which the water will circulate in the above-mentioned times will depend upon the depth to which it descends below the boiler. In this apparatus, the shorter the distance through which the water flows, the greater is the rapidity of circulation—an effect which is the reverse of what occurs in the common form of hot-water apparatus. In general, the circulation is here very rapid; but the distance through which the water will travel is more limited than with the common plan of circulation. For, suppose the water to be raised by the centrifugal force one inch higher at the periphery than at the centre of the boiler, and that it descends 42 feet; if the water in the pipe lose six degrees of heat during its transit, the circulation will then be extremely slow; because, by the Table (Art. 15) we find that the difference of weight between two columns of water 42 feet high, and six degrees

difference of temperature, is 242 grains per square
inch on the area of the pipe; and this is within
four grains of the additional weight which the centrifugal force produces, under these circumstances,
at the side of the boiler. But if the difference
between the temperature of the two pipes be
only four degrees, then the difference between the
weight of the two columns will be 160 grains per
square inch of the area of the pipe; and (by Art. 21)
we shall find that this will give a velocity of
81 feet per minute, so that the pipe may in this
case be about 400 feet long. If the water only
lose three degrees of heat during its transit through
the pipes, then (by Art. 21) its velocity will be
100 feet per minute, provided it descends only
42 feet below the boiler; and therefore the pipe
may be about 350 feet in length. If the depth of
the descent below the boiler be only one half the
amount above mentioned, or 21 feet instead of 42,
than the length of pipe through which the water
will circulate will be just double the amount that
has been stated for the several differences of temperature.*

These calculations are all made for pipes of four
inches diameter; but if smaller pipes be used, the
distance through which the water will circulate
will be less, because, as the quantity of heat lost
in a given time by different sized pipes is *as the
inverse of their diameters,* so also will be the distance
that the water will flow, if the velocity of its
motion be the same.†

---

* This being exclusive of friction, the actual length of pipe
will be less than is here calculated.

† It will be observed, from what has been stated respecting
the common plan of circulation, that the whole of these effects
are exactly the reverse of what there occurs. In that, the
greater the difference of temperature between the pipes, the
more rapid the circulation: in this, the circulation is more

(139.) If greater velocity be given to the fly-wheel and float, the centrifugal force and the height of the water at the circumference of the boiler will both be increased; and the distances to which the pipes can be carried may then like-wise be extended.

(140.) By using a close boiler instead of an open one, a range of pipes may be taken upwards which will act on the common plan of circulation, while another range of pipes may proceed from the bottom, and act on the principle which has here been explained. In this case, the centrifugal force, of which the additional height at the cir-cumference of the boiler is merely the index or measure of effect, will still be of equal power, provided the velocity of the float continues the same; and the water will therefore descend to the same extent as before. The spindle of the float must, in this latter case, pass through a stuffing-box on the top of the boiler, or some other con-trivance to answer the same purpose must be adopted.

This invention, which is a happy application of dynamical principles to overcome one of the most constant of Nature's laws, by the develop-ment of an antagonist force, has hitherto been but little used. It is, however, clearly capable of being applied in cases where the same object cannot be accomplished by any of the more simple means which have been previously described.

(141.) A plan proposed by Mr. Edward Weeks for circulating water at different heights was some years since (1829) made the subject of a patent, but it is not necessary here to be described, as it

rapid in proportion as the pipes are nearer to the same tem-perature. In the former, the circulation is more rapid when the pipes are moderately small; in the latter, the larger the pipe the greater the velocity of circulation.

has long since been superseded by more efficient means. A boiler, also of his invention, is described in Art. 87.

(142.) Mr. Thomas Fowler, of Great Torrington, in 1828, took out a patent for an apparatus which he called a Thermo-siphon.* It was, however, wholly inapplicable to the purposes intended, in consequence of the complication of valves and cocks required to work it; the whole of which complication is overcome by the invention of Mr. Kewley, described in Art. 117.

(143.) An apparatus was invented by Mr. H. C. Price, and patented in 1829, which was designed principally to alter the form of the radiating surfaces.† These surfaces, instead of being composed of pipes, are formed of flat close vessels, about three feet square, and 2½ inches thick, placed edgeways (or vertically), so that when several of them are fixed together, a thin stratum of air passes upwards between them, and becomes heated in its passage. A main-pipe connects all these vessels together at the bottom, another similar pipe is fixed at the top; and these main-pipes lead to and from the boiler, thus keeping up the circulation of water in all these flat vessels at the same time. This is often a very convenient mode of applying a large extent of heating surface in a moderately small space; but the principle of the circulation is in no way different from the ordinary hot-water apparatus. The vessels containing the hot water are usually placed in a vault or chamber below the room or building to be warmed; and the air, when heated, ascends to the room above through ventilators in the floor, or other similar contrivances.

For large buildings this plan answers exceed-

* *Repertory of Arts, &c.*, vol. ix. (1830), p. 393.
† Ibid. vol. x. (1830), p. 65.

ingly well, and many very extensive apparatus have been erected with perfect success, as it is always combined with a system of ventilation, which is too often neglected in other methods of heating buildings.

(144.) A plan has lately been proposed by Mr. Rendle, of Plymouth, as a novel mode of heating horticultural buildings, and which he has denominated the "Tank System." The principal object proposed is to afford bottom-heat to plants, without the use of bark, tan, dung, or other expensive and troublesome materials. The plan recommended is to construct a tank, either of brick, wood, slate, stone, or metal, immediately below the beds of the hothouse, the tank being the full width of the bed, and about four or five inches deep. A partition divides this tank up the centre, except about two or three inches at the further end; and then, by connecting two pipes to the end of the tank nearest to the boiler—one leading from the flow-pipe of the boiler, and the other to the return-pipe—and placing these on the opposite sides of the before-mentioned partition, a circulation of water is produced, the hot water passing up one side of the tank and returning back along the other side, the partition merely separating the two currents. The tank is first covered with large and strong slates, and then a layer of loose stones is laid, and subsequently layers of sand and earth; and the sides of the tank are so constructed as to rise sufficiently high to make a convenient receptacle for these several layers of materials, through which the heat ascends to the plants placed in the earth on the top.

This form of apparatus is represented in fig. 38, which shows the tank without the cover, and also, without the sides which support the layers of stones and earth. The tank, if made of brick, requires to be well stuccoed, in order to make it

FIG. 38.

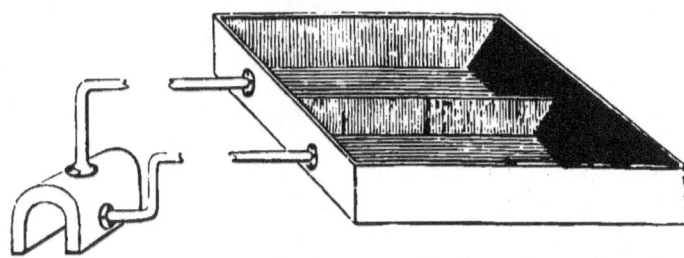

hold the water without percolation through the bricks; but the expansion caused by the heat renders it difficult to keep the tank perfect. Wood tanks are also liable to become leaky; and when used, require to be made in the most substantial manner.

(145.) The plan of affording bottom-heat in this manner possesses but small claims to novelty, except as regards the form of the vessel which contains the heated water. In 1788 steam was first used for giving a bottom-heat to plants by T. Wakefield, Esq., of Northwich. His experiments were continued for several years with considerable success, and in 1792 the plan was adopted at Lord Derby's gardens, at Knowsley, apparently on the model of Mr. Wakefield's, and was attended with perfect success.* From 1801 to 1806 the use of hot water, flowing through leaden pipes, placed about nine inches below the surface of the mould, was employed by T. N. Parker, Esq., of Sweeney Hall, Oswestry, for giving bottom-heat for melon-pits, the water circulating through a single pipe only, which was

* A long account of these experiments will be found in the *Repertory of Arts*, vol. xiv. (1st series), p. 235, *et seq.*; and also in the *Transactions of the Society of Arts and Manufactures*, &c.

attached to a small copper boiler. Mr. Weston, of Leicester, in 1800, employed for the same purpose leaden pipes three inches diameter, filled with hot water, which, from the slow conducting power of the metal, retained the heat for a great length of time.* In 1826, Mr. MacMurtie, gardener to Lord Anson, described a method of giving bottom-heat jointly by flues and steam, which he had successfully employed for a period of twelve years; † and in 1831 the author applied the circulation of hot water, on the common plan, through iron pipes of four inches diameter, for producing bottom-heat in forcing pits, without the use of dung.

All these plans have been successful to a considerable extent; but they all have one defect in common with Mr. Rendle's system. It is difficult to regulate the quantity of moisture; and by some of those plans—particularly those where steam was allowed to evaporate in large quantities—there was too much damp for the plants, and by others they were too dry. Mr. Rendle's tanks are liable to the same objections. It is probable that the most efficient way of applying hot water circulation for producing bottom-heat would be by passing large iron pipes of four inches diameter through troughs made water-tight, placed beneath the bed required to be heated, and filled with small loose stones. These stones, when once heated, will retain their temperature for a great length of time, and by pouring water into the trough, vapour can be raised to any extent that may be required, the quantity being much or little as circumstances may render desirable; or the heat may be continued without any vapour, whenever

* *Repertory of Arts*, vol. xiii., p. 238 (1st series).
† *Horticultural Transactions*, vol. vi.; and *Quarterly Journal of Science*, vol. xxii., p. 341.

a dry heat is required. The pipes need not be placed very close together; about 12 or 18 inches apart from each other would probably be a good distance when four-inch pipes are used, depending principally upon the quantity of vapour which is required to be raised.

(146.) Mr. Rendle proposes to make his tanks supply sufficient heat to warm the air of the house, as well as to produce a bottom-heat for the plants. Those who adopt this plan will soon find that it can only be used as an adjunct to any other method, and that it cannot supersede the ordinary modes of heating horticultural buildings. The chief merit of this plan consists in bringing into action a large reservoir of hot water, contained in slow-conducting materials; by which (when the tank is connected with the pipes that heat the air of the house in the usual way) greater equality of temperature can be maintained in a hothouse than by any other method which comprises only a smaller body of hot water. This, however, does not appear to form any part of Mr. Rendle's system. In a pamphlet, published by him, the apparatus is described as intended both to heat the air of the house, as well as to give bottom-heat to the plants; but when the tank is made of brick, according to his recommendation, the quantity of heat radiated from the exterior surface must necessarily be so small that sufficient heat to warm the air of the house in cold weather cannot possibly be obtained. Iron tanks, instead of brick, have been proposed by other parties; by which this latter objection would be overcome: but by the greater radiating power of the iron the advantages contemplated by Mr. Rendle would not be obtained. The brick tank, when used only for bottom-heat, and as an addition to the usual arrangement of pipes for warming the air, may

frequently form a very useful apparatus ; but it
will probably give way to some better arrangement,
by which the vapour can be regulated with more
certainty than this form of apparatus appears
capable of doing.

(147.) A patent was obtained in 1838, by Mr.
Corbett, of Plymouth, for heating buildings by
hot-water circulation in open troughs, instead of
pipes of the usual form.* The objection to this
plan is the difficulty of regulating the quantity of
vapour given off by the open troughs, to suit the
varying requirements of the plants. The inventor proposes to do this by means of moveable
covers to be placed on the troughs, which would
prevent the vapour rising so rapidly. But all
the advantages of this plan appear to have been
anticipated several years previous, by the use
of pipes with troughs of about three inches deep
cast on the top, and extending nearly the whole
length of the pipe. The trough, being formed by
the surface of the pipe itself, is always kept hot :
and any quantity of vapour can be obtained, up
to the actual saturation of the atmosphere of the
house, by filling these troughs partially or wholly,
as the case may require.

(148.) The last two inventions which have
been described appear to be founded on a very
imperfect knowledge of the physical laws which
they call into action. The tank system, when
used as the inventor of it recommends, must
necessarily fail to afford sufficient heat to the air,
as a very slight acquaintance with the laws of
radiant heat will show ; and the use of open
troughs, to supersede the previous method of
.pipes with troughs cast on them, is founded on a
misconception of the laws of vaporization.

(149.) Mr. Corbett states, in the specification

* *Repertory of Arts, &c.*, vol. xi. (1839), p. 346.

to his patent, that the quantity of vapour which is given off from the troughs depends upon the temperature of the air and the depth of the troughs. This last circumstance (inasmuch as by increasing the depth, the *radiating* surface must be enlarged and the temperature of the water thereby lowered) will certainly somewhat alter the rate of evaporation, though not necessarily its final quantity; but the temperature of the air in no way influences either the quantity or the rate of evaporation, unless it be so cold as to condense the vapour.

(150.) Dr. Dalton * has shown that the only circumstances which affect the vaporization of water in atmospheric air are—" First, the quantity evaporated is in direct proportion to the surface exposed, all other circumstances being alike. Second, an increase in the temperature of the liquid is attended with an increase of evaporation not directly proportionate. Third, evaporation is greater where there is a stream of air, than where the air is stagnant. Fourth, the evaporation from water is greater, the less the humidity previously existing in the atmosphere, all other circumstances being alike." The rate of evaporation, therefore, from water under 212° of Fahrenheit is directly as the surface of water exposed, and as *the evaporating force;* which latter is the name that has been given to the difference in the elasticity of the vapour which rises from the water as the given temperature, and that of the vapour already existing in the air. Dr. Dalton found that "the same quantity is evaporated with *the same evaporating force,* whatever be the temperature of the air."

(151.) This theory of evaporation is here more

* *Memoirs of the Philosophical Society of Manchester,* vol. v., p. 576, *et seq.*

particularly dwelt upon, because many persons
have adopted the notion that any quantity of
vapour, without limitation, can be produced by
means of open troughs, such as have been described. Nothing, however, can be more erroneous. When once the air is saturated, the *evaporating force* ceases,* until either a portion of the
vapour it contained be condensed, or the temperature of the evaporating fluid be increased,
when of course a new state of circumstances
obtains, and the condition of equilibrium ends.
Practically, however, evaporation never ceases
when the evaporating liquid is kept at a high
temperature; because the air being at a much
lower temperature, the vapour contained in it
must always have less elasticity than that given
off from the water. The result will be that the
water evaporated will be condensed by the air,
and will be precipitated; while, at the same time,
the latent heat of the vapour will be rendered
sensible, and will raise the temperature of the
air.† If the air were confined in a place where it
could not lose any portion of this acquired heat,
this process would ultimately stop the evaporation;
but in horticultural buildings the loss of heat
from the large surface of glass is so considerable,
that this result cannot follow.

* These remarks apply to the case of fluids under the temperature of 212°. A different state of things obtains when
the vapour formed exceeds the total atmospheric pressure, as
in the ordinary process of generating steam.

† The quantity of heat given out by the condensation of
vapour is very considerable. By the note to Art. 97, it will
be seen that a cubic foot of water will heat 2,990 cubic feet of
air as many degrees as the water itself loses. As the latent
heat of vapour is about 1,000°, it follows that every *cubic inch*
of water which is evaporated, and then again condensed from
vapour, as above mentioned, will communicate 5° of heat to
346 *cubic feet* of air.

(152.) As the quantity of vapour given off to the air depends, as we have here shown, on the surface and the temperature of the water exposed, it follows that no possible advantage can be obtained from a large quantity of water in an open gutter over the same surface of water obtained by the old method of troughs fixed or cast on the top of the pipes. The latter are more easily regulated to suit the requirements of different plants and different seasons, and are also free from several practical difficulties which attend the application of the other method.*

(153.) The preceding remarks are descriptive of the principal modifications which have hitherto been introduced into the hot-water apparatus, though many deviations of a minor character have been proposed, several of them, indeed, being introduced apparently for the sake of novelty. The advantage, however, which may be derived from these peculiar forms or modifications of the apparatus must depend upon the purpose for which it is required. Thus, in places where a long continuance of heat and uniformity of temperature are required, the form of the pipes, tanks, vessels, and other radiating surfaces should be such as to afford only a small surface, while they contain a large quantity of water. This may be obtained by using pipes of large diameter, or tanks of a large cubical content, and approaching in form either to the sphere or the cube; while, on the contrary, where the heat is required to be quickly raised, and permanence of temperature is unim-

---

* Those who wish to follow further the subject of spontaneous evaporation may with advantage consult the *Memoirs of the Manchester Philosophical Society*, already quoted; also *Nicholson's Journal*, vol. xxvii., p. 17; *Journal de Physique*, vol. lxv., p. 446; *Quarterly Journal of Science*, vol. xvii., p. 46; and *Philosophical Transactions*, vol. lxxxii., p. 400.

portant, the radiating surfaces may be greatly increased, in proportion to the actual quantity of water contained in the apparatus. In this case, therefore, pipes of small diameter, or tanks which are very flat in form, may advantageously be used; and many varieties of the apparatus will necessarily be adopted, amongst ingenious persons who practically apply this invention to the vast number of useful purposes to which it is applicable. For many of the purposes to which it is extremely suitable it has not hitherto been applied; and its advantage, in other cases, has not even yet been sufficiently appreciated. Such are the uses to which it may be adapted in various manufactories—in paper-making, calico-printing, dyeing, and starch-making; and also for druggists, seedsmen, and numerous other purposes of general utility; and for drying-rooms for every purpose where a mild and equable heat is desirable. For many of these purposes it is exceedingly convenient, as the form of the heating surface can be made to suit the peculiar object to which it is to be applied; and the inconveniences that arise from unequal degrees of heat, consequent on most other methods of warming, are by this means entirely avoided.

CHAPTER VIII.

General Summary of the Subject—Points requiring particular
attention—Abstraction of Air from the Pipes—Vertical
Alteration of Level—Effect of Elbows in compensating
unequal Expansions of the Pipe—Different Floors heated
by one main Pipe from the Boiler—Method of connecting
Coils to main Pipes—Reduced Effect from Pipes laid in
Trenches—Effect of Cold Currents of Air in neutralizing
heating Apparatus—Heating Apparatus placed in Vaults—
Cements for Joints—Sediment in Boilers—Use of Salt in
Pipes to prevent freezing—Deposition of Vapour in inhabited
Rooms, and necessity for Ventilation—Construction of
Drying-rooms.

(154.) Having, in the preceding chapters, arranged, under distinct heads, the various remarks
on the principles of warming by the circulation of
hot water, it may here be desirable to bring under
general review the principal facts which it has
been the object of this work to explain. There
are, besides, many minor points connected with
the invention that could not conveniently be
brought under notice in any of the foregoing
divisions under which the subject has been treated,
but which, nevertheless, may be found very useful
to those who are investigating its principles, or
adapting it to practice.

(155.) A correct knowledge of the cause of circulation of the water (it has already been observed)
is absolutely necessary to the successful application
of this invention in many of its more complicated
arrangements. Some estimate must be formed of
the amount of the motive power possessed by an

apparatus of this sort, otherwise it will be impossible to ascertain what will be the result of any particular position, or determinate length of the pipes, in many peculiar cases; as, for instance, in such forms of apparatus as figures 10, 11, and 28. It is also necessary, in order to make provision for the escape of the air from an apparatus of this kind, to have some knowledge of the laws which regulate the motion of fluids, in order to ascertain where the air will lodge, and why it should accumulate in one place rather than another. No circumstance connected with the subject requires greater caution than this. In almost every part of the apparatus where an alteration of the level occurs, a vent for the air must be provided; because, from the extreme levity of air compared with water (Art. 9), it is impossible that the air can ever descend, so as to pass an obstruction lower than the place where it is confined. Thus, in fig. 7, if the air accumulate in the pipe between A and e, it is evident that a vent at c, although it would take off the air from g h and from c d, could not receive any portion of that which is confined between A e, or between e f, because, in that case, it must *descend* through the pipe e f before it could escape. The principle is the same in all cases, however large or however small the descent may be; and the accidental misplacing of a pipe in the fixing, by which one end may be made a little higher than the other, will as effectually prevent the escape of air through a vent placed at the lower end as though the deviation from the level were as many *feet* as it may, perhaps, be *inches*. It is, however, impossible to give multiplied examples of this part of the subject, for probably no two instances precisely similar may occur; but it deserves the most serious attention in following out its practical consequences, for innumerable failures have arisen from its neglect, and no part of the subject is more puzzling to an experienced practitioner.

(156.) When any particular obstructions are required to be overcome, in consequence of numerous alterations in the level of the pipes; when the pipes are required to descend below the boiler; or, in short, when any other variation from what may be considered as the usual form and arrangement of the apparatus may be desirable,—it is essentially necessary to have some data on which to found a calculation as to what will be the practical result of the required deviation. For no partial experiment of a tentative character, nor even the effect shown by a miniature model, will give anything like an accurate idea of what will be the result, when the experiment is made on a large scale. The reason of this is obvious. It has been shown that the greater the distance through which the water flows, the greater does the motive power become, in consequence of the water being colder in the return-pipe relatively to the flow-pipe, while at the same time the friction is increased, though not always in an equal degree. This will, therefore, prevent partial experiments—that is, working models exhibiting only a particular portion of the whole apparatus—from being conclusive; and with a miniature model, although the decreased time and distance of transit are compensated by the reduced size of the pipe exerting a greater cooling power on the water, the friction being much greater in small than in large pipes, the velocity will be reduced in a very sensible degree, and the results rendered wholly inconclusive. In general, the successful working of a miniature model will be conclusive that the experiment on a larger scale will perform still better; but the failure of the model will be no proof that the larger apparatus will not be successful.

Calculations on this subject possessing any claims to accuracy are extremely difficult, and exact results, indeed, are perhaps impossible, in the present state of our knowledge on this difficult branch of hydraulics, notwithstanding the many eminent and learned men who have both written and experimented on the motion and resistance of water moving in pipes.* Notwithstanding this acknowledged difficulty, the remarks made in the preceding chapters will enable an approximate estimate to be formed as to the general effect that may be expected from any particular form of the apparatus, and whether the motive power will be increased or diminished by the arrangement proposed. And, by following out in detail the rules which have been given, a tolerably accurate judgment may be arrived at as to the result that may be expected, under almost every form of the apparatus that may be adopted.

(157.) In the diagrams which have been given in the first and second chapters, the simplest possible arrangements of the pipes have been shown. To give the various forms in which the pipes and heating surfaces may be laid would obviously be impossible; it must be left to the iugenuity of the adapters of the apparatus to deduce, from the general rules which have been given, the form of apparatus best suited to the particular case; and, while bearing in mind the causes which produce and increase the circulation, it will not be difficult to contrive an almost infinite variety of arrangements for accomplishing the desired result.

(158.) In the simple form of the apparatus as generally used in hothouses, there is seldom much difficulty, except when the pipes are required to dip below the doorways. When the boiler is

* See Chapter II., Art. 22.

placed sufficiently low (Art. 28), the difficulty of taking off the air from the pipes at the places where the dips occur is principally to be attended to. It has sometimes been the practice to place an *open* upright pedestal on the top of the pipe at *l*, fig. 9, for the supposed purpose of adding by the weight of its column of water to the downward pressure at that part, and by this means to increase the circulation. But the real effect of such a column of water pressing on the pipe must be to increase the pressure on every part of the apparatus alike, and not on this part in particular ; and therefore it is probable that any effect obtained by this means is attributable only to a more easy escape being provided for the air. Any air accumulating at this part must be particularly prejudicial, by occupying a portion of the space which ought to be filled with water, and thus diminishing the pressure on the descending column. A very small accumulation of air at this part of the apparatus might seriously retard, or even totally obstruct, the circulation, and therefore a ready escape for the air at this point is particularly desirable. The same effect is no doubt to some extent produced by the open cistern A, fig. 12 (Art. 32) ; for, although its effect is generally beneficial, it is difficult to account for it on the ground of increased pressure ; while it is obviously quite possible that benefit may arise from the air having so easy an escape,* and from the difference in the temperature of the two pipes.

---

* It has been supposed that, although this increased pressure by a vertical column of water undoubtedly extends every part of the apparatus, an advantage must arise in consequence of the bubbles of steam and hot water (which rise continually upwards from the bottom of the boiler, strongly acted on by the fire) being relatively lighter than the rest of the water, when the latter is thus condensed by the increased

(159.) It will be apparent from these remarks how essential it is that the pipes be free from any accumulation of air, when such difficult cases of circulation are attempted as those described in Arts. 30—34, or any of an analogous character. It is probable that many failures have arisen entirely from this cause, which a very slight alteration would have prevented; and attention to this point, in the construction of all hot-water apparatus, can hardly be urged too strongly.

(160.) When the pipes dip below the doorways of hothouses, it is sometimes very difficult to allow sufficient depth for two pipes of large diameter below the step of the door, when placed as shown in fig. 9. When this is the case, a smaller pipe may be used at this particular part; and just at the dip, the large iron pipe may be reduced to one of smaller diameter, and again increased when it rises from below the door. The circulation will by this means always be reduced to some extent (Art. 48), and it is by no means advisable to adopt it whenever it can be avoided. All unnecessary variations in the size of the pipe ought always to be avoided, and, when adopted, the alterations are better to be gradual than sudden, as there will be less obstruction to the water passing through a trumpet-shaped pipe, either from a large to a small pipe, or from a small pipe to a larger, than when these changes are made abruptly.

Inconvenient as these vertical dips in the pipe frequently are, they, as well as the horizontal bends, are sometimes useful in counteracting an

---

vertical height. This, however, must needs be erroneous, because steam produced under pressure is of greater density exactly in proportion to this pressure, and therefore the relative proportions between the densities of the steam-bubbles and the water must be preserved, however much the pressure may be increased.

inconvenience which otherwise might occur from the expansion of the pipe in its horizontal length. When there is a very long length of perfectly straight pipe passing along the side of a building and returning again in the same direction, it often happens that the expansion of one of these lengths of pipe is so much greater than the other, owing to its higher temperature (and this particularly happens when the circulation is slow), that some of the joints become loose, and a leakage occurs. This inconvenience is more likely to occur when three or four pipes are placed one above another; in which case the upper pipes always become heated first, and thence become more expanded in length. And so great is the force of this expansion, that unless there is some degree of elasticity given to the pipes, either by elbows intervening or by some other means, a leakage is almost certain to occur; and generally it happens at the extreme end of the pipes farthest from the boiler. When steam pipes were used for heating buildings, this cause of leakage was very inconvenient; and Count Rumford introduced an ingenious mode of counteracting it, by attaching both the upper and lower pipe to a copper drum, which, being more pliable than the cast-iron, corrected the evil. In all cases where there is a great length of pipe running in a straight line, it is necessary to bear in mind the certainty of expansion occurring, and to provide for its effects; for serious accidents have occurred to buildings by neglecting this precaution, as the expansion cannot be prevented under any circumstances, and its power is immense against anything that resists it.

(161.) In all cases where pipes are placed at various elevations above the boiler, for the purpose of warming different floors of a building, or

where from any other cause the pipes descend by
steps or gradations from a high elevation to a
lower before the water returns to the boiler, it is
desirable that the water should be made to ascend
at once from the boiler to the highest elevation.
By this means the best possible circulation is
always insured, as there will then be the greatest
difference between the weight of the ascending
and descending columns. It will, however, often
be found convenient, when different floors of a
building are required to be heated by one apparatus, to have a separate pipe leading from the
boiler to each floor; as it is difficult to make
horizontal branch-pipes effective when leading out
at different elevations from a general main-pipe
rising vertically from the boiler. This arises from
the extremely rapid motion of the water in vertical
pipes, by which means the whole of the heated
water passes directly to the highest level, without
delivering any to the lower horizontal branches.
This plan of having separate pipes for each floor
is better than having one pipe gradually *descending*
through them all, as, the shorter the circulation,
the more equal will be the effect, in consequence
of greater equality in the temperature of the pipes
(Art. 49). When it is inconvenient to have
separate flow-pipes from the boiler to each floor,
an equal circulation may be obtained by the
arrangements shown Art. 49.

When several coils are fixed in succession in
different places, being supplied by means of one
general main pipe with its return pipe passing
underneath the main or flow-pipe, it is not unusual
to connect the coils into both the flow and the
return-pipe, thus—

FIG. 38B.

This plan of connecting coils very often fails entirely, the circulation stopping at the second, third, or fourth coils, as the case may be. When the main-pipe is very large, say 3 or 4 inches diameter, and the connections from the main-pipe to the coils are about 1 inch diameter, this mode of connection succeeds tolerably well, and several coils in succession may be connected in this way; but when the main-pipe is small, such, for instance, as $1\frac{1}{2}$ inch or 2 inches diameter, it is very rarely that the circulation can be forced beyond the first or second coil, and sometimes it stops entirely after the first coil. The reason of this is obvious. The circulation returns at once to the boiler after passing through the first coil, the friction being too great and the quantity of heated water being too small, when passing through the very small main-pipe; and in this case the water returns to the boiler through the shortest route instead of flowing through the whole arrangement of pipes. This rarely takes place when the main-pipe is very large, unless the connections from the main-pipe to the coils are out of due proportion. Although it is a great mistake (but one that is often committed) to make the main-pipe too small, proportionately to the number and size of the coils it is expected to supply, a much smaller main pipe may be used when the coils in a range are connected in this manner—that

FIG. 38c.

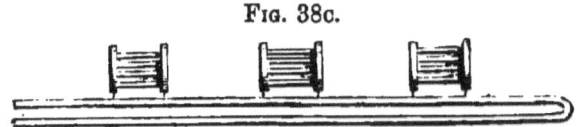

is, all the connections being made to the flow-pipe alone. By this means of connecting coils, a circulation can be secured through almost any number of coils without any inequality in their temperature.

When coils are connected by lateral pipes placed at some distance from the main-pipes, another difficulty frequently occurs. In this case the circulation is very likely to pass by the lateral or side connection, and refuses to be diverted from its straight course. Sometimes a *lateral* connection of 4 or 5 feet long, at right angles from the main pipe, will entirely stop the circulation into the coil, at other times it may be carried very much further. This difference probably arises from a variation in the velocity of the circulation ; for it has sometimes happened that the water in a coil fixed *laterally*, a few feet off from a main-pipe, will circulate when the apparatus is first set to work, and then gradually cease to act as the water in the general apparatus gets hotter and hotter. This effect may be prevented when the main-pipe is sufficiently large, by placing a stop half way across the main-pipe, as explained, Art. 49, with regard to vertical main-pipes.

(162.) In those cases where the radiating surfaces are not placed actually in the room or building which is to be heated, but which produce their effect by heating successive portions of air, which then pass into the room that is to be warmed, there will always be considerable loss of heat, which will require a correction to be made in the calculations already given. This applies also to those cases where the pipes are placed in drains or trenches below the floor, with trellis gratings or ventilators in the floor to cover the pipes and to emit the heat. The proportion of heat lost by this means will depend upon various circumstances. When the pipes freely radiate their heat into the room to be warmed, about one-fifth of the total effect is produced by *radiation*, and the rest by contact of the air with the heated surface (Art. 227, *et seq.*). But that portion of the heat obtained by *radiation* will be considerably reduced by placing the pipes in a trench. For the interior surface of the trench will become considerably heated ; and as the heat given off by *radiation* depends in a great degree upon the difference of temperature between the heated body and the medium by which it is surrounded, it follows that the hotter the trench becomes the less *radiant heat* will be given off by the pipes. The loss of effect by this means would be considerable were it not for the circumstance that the heated surface of the trench itself gives out heat by conduction to the air. The effect, therefore, becomes a very complicated one. But it may probably be assumed, as a tolerably correct result, that the loss of heat by placing the pipes in a trench which is well covered for nearly its whole length with trellis gratings, will not exceed five per cent. in ordinary circumstances. When the trench is only partially covered with trellis gratings, the loss of heat will be more ; and therefore, when this plan of placing the pipes is adopted, at least two-thirds of the entire trench ought to be covered with the trellis grating, and, if possible, the whole length should be so covered, in order that the freest possible escape be allowed for the heated air. The trench also should not be made too large. The best size will be, to have it just large enough to hold the pipes, and to allow the workmen easily to fix them ; and all the space beyond that will only tend to diminish the effect of the apparatus.*

---

* It is sometimes desirable to make the bottom and sides of these trenches full of holes, for the purpose of admitting fresh air into the building. Some remarks on this subject will be found under the head of "Ventilation," in a subsequent chapter (Art. 353). It requires great judgment in this case in order to prevent the cold air being discharged into the room instead of first warming it by proper contact with the pipes. A much smaller surface of trellis grating is required in this case, but a much larger surface of pipe is necessary.

The same remarks will, of course, apply to all cases where the pipes or other radiating surfaces are .placed in vaults or chambers below the rooms intended to be heated; and, in all cases of this kind, allowance must be made for the loss of effect arising from the causes here described. It should also be observed that when the heating surfaces are placed in vaults or chambers below the rooms to be heated, and are so contrived that the air from the rooms does not pass over them a second time, but fresh external air alone is made to pass over them and then to escape from the heated room by means of ventilators, from 50 to 70 per cent. more pipe is necessary to heat a given space than will be required if the air of the heated room be made to circulate repeatedly over the heating surfaces. The air will be more salubrious when it is not heated over again a second time, but the cost of the apparatus will be greatly increased.

Whenever the heating surfaces are placed in vaults or chambers, and not distributed generally in the rooms to be heated, it requires great judgment in placing the openings for the emission of heated air in a proper position: for if there is a current of cold air from one end of the room or building—as, for instance, in a church which has the entrance at one end—it is impossible to make the hot air travel to the extremity of the building; but, being met by the cold current from the door, it will be forced back and prevented from reaching the colder end. This circumstance has entirely destroyed the beneficial action of innumerable apparatus; and so strong frequently is the action of the cold current of air in such buildings, when rushing into a somewhat rarefied atmosphere, that it has not unfrequently happened that under these circumstances the building has become considerably *colder* when the heating apparatus is used than it

was before the pipes were warmed or the fire
lighted. The only remedy is to place the openings
for the emission of hot air in such parts of the
building that it shall not meet with these opposing
currents of cold air. It is sometimes very difficult
to accomplish this, and it requires great judgment
efficiently to warm buildings under these circumstances.

(163.) One of the most efficient plans for warming large buildings, and at the same time the
most difficult to accomplish, except when directed
by great experience and judgment, is by combining
the two methods of placing part of the heating
surfaces in vaults or chambers, and part in flues
or drains. By this method perfect ventilation can
be secured, but the expense is considerably increased. Not only will the heating surfaces require
to be increased at least 50 per cent. beyond the
ordinary method of heating, but proper arrangements must be made for extracting the foul or
contaminated air from the ceiling or roof (see
Art. 360, &c.). But this plan of warming requires
likewise very special arrangements. The air in
this case must be admitted into a vault or chamber,
and there heated by means of coils, or pipes, or
other surfaces, on which the admitted air must
impinge and be kept sufficiently in contact to become heated. It is then to be conveyed into the
room or building which is required to be warmed
through drains or channels in which a certain
quantity of hot water pipes are placed, as the hot
air will not travel along cold *lateral* drains for any
considerable distance. To decide what proportion
of the whole heating surface shall be placed in the
vault, and what in the channels or drains, requires
great consideration and experimental knowledge.
If these proportions are erroneous, the heating may
become entirely neutralized in certain states of the

wind, and a reverse action may take place by which all the hot air may be discharged through the cold air drains when the wind blows from certain quarters. The form of the building and its exposed or sheltered situation have great influence in this case. In some buildings no such reverse action ever takes place. In perfectly plain straight buildings, this action of the wind is most likely to occur. In others of irregular form, and where there are numerous projections, or buttresses, or anything likely to break the force of the wind, the reverse action is the least likely to happen. In some cases less than a fourth of the entire heating surface laid in the trenches is sufficient to give motion to the air : in other cases half or two-thirds laid in the trenches will not be more than is necessary for the purpose of causing the heated air to flow readily through the drains. There is no plan so effectual as this for bringing warm fresh air into a building, and thus combining ventilation and warming, but the contingencies attending it are not inconsiderable.

In all those forms of the apparatus where the effect depends on bringing in a body of cold fresh air from the exterior of the building — and this applies also to all forms of hot air apparatus and cockle stoves—a very high wind will sometimes stop the action of the apparatus entirely. Suppose fig. 38D to represent the ground plan of a building to be warmed by any kind of heating apparatus placed in a vault or

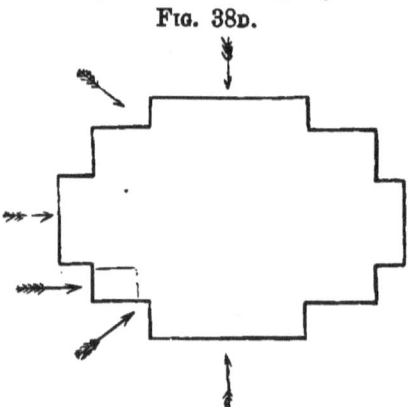

FIG. 38D.

chamber shown by the dotted lines. When the wind blows from any of the points shown by the arrows, the air will enter the vault or chamber without much difficulty if the channel is sufficiently large. But if the wind blows strongly from any other quarters than those shown by the arrows, the air, instead of being drawn into the chamber or vault, will experience a reverse action, and the air will be drawn or sucked out from the building. This arises from the greater pressure of the air on the *windward* side of the building, which causes a partial vacuum to be produced on the opposite side where the cold air drain is placed. Under these circumstances the whole action is reversed, and the heated air is sucked out of the building entirely.

When the force of the wind is much broken by the irregular form of the building, or by surrounding objects, this effect is scarcely likely to happen. Where it does occur, the best remedy is to make a second entrance for the air from the opposite side of the building and temporarily to close the one or the other, as the direction of the wind may render necessary.

The most effectual way of preventing the reverse action which is here described, is to bring the cold air into the hot air chamber by a very large flue or channel from above the roof. This flue must stand as high, or higher than the roof, and it should be surmounted by a cowl or turncap, balanced so as to turn *towards* the wind instead of from the wind, as usual for ventilating or smoke purposes. The large size, however, of this flue and the surmounting turncap is a very serious obstacle to its use. Thus, suppose a room containing 500 people. The quantity of air required to ventilate this room would be (Art. 360) 5,000 cubic feet per minute ; and this would require the flue or channel to have a capacity of 21 square feet. It would be almost

N

impossible to make so large a flue in any ordinary
building; and unless the flue was of sufficient
size, it would be of very little use. For a private
house, heated on this principle, of course a very
much smaller flue would be sufficient, but even a
flue of 2 feet square (or 4 square feet area) would
be a most inconvenient incumbrance if it passed
from the roof to the basement of the house to
supply air to the hot air chamber.

There is also another error to be guarded against
in adopting this mode of placing the heating surfaces in a separate vault or chamber. It is indispensable that the vault should not be too large.
If it is much larger than the heating surfaces, or
if the heating surfaces be placed too far apart from
each other, the entering air will pass by or escape
without sufficient contact, and will enter the room
or building as a cold draught. The more confined
the space of the heating chamber, the better will
be the effect, provided there is sufficient space to
allow the air a tolerably free passage: if it can
escape round the sides of the chamber without
coming in contact with the heating surfaces it will
certainly do so, as there is a natural repulsion
between these heated surfaces and the entering air.

(164.) The mechanical operation of fitting together the pipes is a subject known to most good
workmen who are acquainted with iron-work.
The usual and best kinds of pipes for the purpose
are those with socket-joints, those with flange-
joints having long ceased to be employed for hot-
water apparatus. And even for steam, where
flange-joints were formerly invariably used, the
socket-joints are now very frequently employed,
as they make a much neater and at the same time
an equally strong joint. In fact, when the joints
of socket-pipes are well made, the pipes themselves
will break before the joints will yield, or before the

faucet-end of one pipe can be drawn out of the
socket of the other.   The joints require to be
well caulked with spun yarn, and then filled up
with cement.   The strongest cement is that known
to engineers under the name of iron cement, consisting of iron borings, sal ammoniac, and sulphur;
but it requires to be used with great caution,
as it expands very much, and very frequently
bursts the pipes.   But other cements are frequently used, the principal of which are a mixture
of white and red lead, which makes a very good
cement; and a cement made of quicklime mixed
into a paste with boiled linseed oil* has occasionally been employed with considerable success.
Fire-clay sifted fine and mixed with boiled
oil also makes a good cement; and Portland
cement mixed with water has also been used
for jointing pipes where there is little or no
pressure.   But perhaps the hardest cement of all
is red and white lead mixed with a considerable
quantity of iron borings, which is very enduring
as well as an extremely hard cement.   For all
ordinary purposes, the best cement is red and
white lead; and a very small quantity of iron
cement applied as a means of finishing the joint
—the red and white lead being used at the bottom of the joint, and the iron cement at the top—
forms perhaps the best joints that can be made.
Many thousands of pipes have been burst at the
joints from improper modes of using the cement.
Whenever iron cement is used in too large quantities, and too strong in its chemical proportions,
the sockets are almost certain to burst after they
have been made a few months.   For a chemical

* This composition is very liable to spontaneous combustion,
and should therefore be used with caution, particularly by
mixing it apart from anything likely to ignite by the heating
of the materials.

action takes place in the iron cement, which slowly expands, and thus bursts the sockets of the pipes, if the quantity used is at all in excess ; and this slow chemical action has been found to go on for many months, and sometimes even for years, causing the most serious loss and inconvenience by the destruction of the pipes.

(165.) It may be desirable to make a few observations on the water that is to be used in this kind of apparatus. Sometimes the foulest and most filthy water is used in a hot-water apparatus, by which a thick coating of mud is deposited, and which must necessarily not only much reduce the effect of the apparatus, but also injure the boiler. But a far more general and, in fact, an extremely common error lies in using hard water, which contains a large quantity of earthy salts. Rain-water ought to be used when it can possibly be obtained, because all hard waters are impregnated with saline matter, which forms the sediment or incrustation so common in those vessels in which water is boiled. This incrustation always accumulates in the boiler of a hot-water apparatus in which hard water is used, and forms a coating, varying in substance from the thinnest lamina to three or four inches in thickness. When this deposit of saline matter occurs in a boiler, not only is less heat received by the water, in consequence of the conducting power being lessened by the interposed substance, but the boiler will be much injured by the increased heat of its external surface, and more fuel will be consumed.

The quantity of sediment formed in a hot-water apparatus, however, bears no comparison in ordinary cases with that of steam-boilers. In the latter, the quantity is so large as sometimes to require its removal at least once in every three or four days, and sometimes even oftener. But, as there is scarcely any evaporation from a hot-water boiler under ordinary circumstances, the quantity of sediment is so small, if the water is good, that the boiler does not require cleaning out even after many years' use. This remark, however, will not apply to such forms of apparatus as Mr. Rendle's or Mr. Corbett's (Art. 144 and 147); for in these the evaporation of the water is so great that the sediment in the boiler must necessarily be very considerable, and, without proper care, it may become a most serious evil. Owing to the extreme smallness of the boilers used for hot-water apparatus, it is very difficult to clean them from sediment, and with the majority of them it is impossible. When, therefore, such boilers are applied to an apparatus on the tank system, or to Mr. Corbett's open troughs, they will probably fail in a comparatively short time, unless the use of rain-water be strictly adhered to. The plan of open troughs cast on the pipes (Art. 147) is not of course liable to the same objection; for, however large the evaporation may be, the boiler cannot be affected by it, as the sediment will necessarily remain in the troughs, and can easily be removed.

(166.) This kind of sediment can only be removed from a boiler with great difficulty. It consists principally of carbonate of lime and sulphate of lime, together with the sulphates of soda and magnesia, and several other salts, varying considerably in different localities. A weak solution of muriatic acid (one part of acid by measure to 20 or 30 parts of water) will generally reduce this concreted sediment into a substance of less tenacity, which may then be removed with slight mechanical force. Sometimes this sediment is so hard as to bear polishing, when it exactly resembles marble, being beautifully speckled or veined. At other times it shows successive layers loosely aggregated. In some places where the water is obtained in mineral districts, large sediments of iron or even of copper may be found in the pipes and boilers of hot-water apparatus, causing the greatest possible annoyance. In this case the deposit is more usually found in the pipes than in the boilers, forming either a brown or a yellow deposit according as the iron may be in the state of a sulphate or a carbonate.* By using rain-water the inconvenience arising from these deposits will, however, be entirely avoided, and the apparatus will both last longer and be more efficient.

(167.) Some inconvenience has occasionally been experienced when a hot-water apparatus has been left for a long time without being used, and exposed to considerable degrees of cold, by the water becoming frozen in the pipes; for it is not only difficult in such cases to thaw the water, but sometimes also the pipes burst. To prevent the pipes from cracking, it will generally be sufficient to draw off a portion of the water, so that the

---

* There are forty-three ascertained substances, which are known as contaminating different waters. Of these, seven are gases, two are free acids, and thirty-four are neutral salts. The salts of lime most commonly found in water have been mentioned above in the text. The other impurities, with the exception of iron, are not of usual occurrence. Iron is found in water in two different states, the Carbonate and the Sulphate. In the state of carbonate, the iron gives a yellow ochre colour to the sediment. The carbonate of iron easily parts with its carbonic acid, and passes into an oxide of iron, which is insoluble in water. The sulphate of iron, which is very soluble in water, absorbs oxygen very easily and then becomes insoluble and is precipitated in the form of a brown oxide or ochre. These oxides are very destructive to the pipes and boilers of hot-water apparatus. The neutral salts thus formed, rapidly choke up the pipes and boilers, while the acids set free in their formation, seriously injure the metallic substance of the apparatus itself.

horizontal pipes shall not be quite full; for the cracking of the pipes arises from the sudden expansion which takes place in the water, at the moment of its passing into the solid state of ice. But when the apparatus is not likely to be used for a considerable time, it would be much better, if the weather be very cold, to empty the pipes entirely of water; for it is always troublesome to thaw the water when once frozen in the pipes.* In an apparatus used in a building of which the temperature is always above 32° this is obviously unnecessary, as the water cannot then be frozen. A plan, however, might be adopted which will effectually prevent the water freezing with any ordinary degree of cold, namely, by using salt water in the apparatus, instead of fresh water. This plan would certainly be somewhat injurious to the apparatus, on account of the action of the salt on the iron; but the injury would not be extensive, and would be very slow in its operation. Perhaps in this country such a plan is unnecessary; but should this kind of apparatus be adopted in colder climates, the suggestion might be useful. The larger the quantity of salt which a given portion of water contains, the greater is the degree of cold necessary to congeal it. Thus, the quantity of salt contained in sea-water is about three per cent.;† this re-

---

* It is perhaps not generally known, that water which has been boiled freezes sooner than that which has not been boiled. This circumstance was observed by Dr. Black, in 1775 (*Philosophical Transactions*, vol. lxv., p. 124). But it has since been remarked that Aristotle, Pliny, and others of the ancients, have noticed the same fact in their writings (*Memoirs of the Manchester Philosophical Society*, vol. i., p. 261).

† This quantity varies considerably in different localities. In the English Channel the quantity is as above stated; but on the coast of Spain it contains about 6 per cent., while the water of the Baltic only contains about 1½ per cent. Between

quires, according to Dr. Marcet, a temperature of about 28° to freeze it; but if the quantity of salt be increased to 4·3 per cent., the water will not freeze until the cold be reduced to $27\frac{1}{2}°$ of Fahrenheit, or $4\frac{1}{2}°$ below the ordinary freezing-point of fresh water. When the water contains 6·6 per cent. of salt, it will not freeze until the temperature be reduced to $25\frac{1}{2}°$ of Fahrenheit; and if it contains 11·1 per cent., the temperature must reach as low as $21\frac{1}{2}°$ before the water will congeal.*

The effect which would be produced on cast-iron pipes and boilers, by any of these quantities of salt, would not be of much importance; although, in process of time, it would certainly in some degree corrode the apparatus. †   When the apparatus has been once filled with salt water, the waste which occurs in the water, by evaporation, should only be supplied with fresh water; for, as the salt does not evaporate, the same quantity of salt will remain in the apparatus, and will combine with the fresh water when added.

(168.) As water at a medium temperature can hold in solution nearly 36 per cent. of common salt (chloride of sodium), and nearly 40 per cent. at the boiling temperature, there is no fear of any deposit forming in the boiler from this cause. ‡

the Tropics the quantity is very large; as much as 10 per cent. is stated to exist in some of the tropical seas and oceans. (Ure's "Chemical Dictionary," art. "Muriatic Acid.")

* Ure's "Chemical Dictionary," art. "Caloric;" and "*Blagden's Experiments*," *Philosophical Transactions*, vol. lxxviii., p. 279.

† A remarkable difference obtains in the rate at which oxidation acts on cast and on wrought-iron. Hard-cast iron will resist oxidation about three times as long as wrought-iron; and, according to the experiments of Mr. Daniell, the same difference exists in the length of time requisite to produce a given effect by acids. The effect on soft cast-iron will approach nearer to that of wrought-iron, varying with its hardness.

‡ Ure's "Chemical Dictionary," art. "Salt."

The reason of a deposit forming in boilers where hard water is used, is because the water leaves behind, on evaporation, the saline compounds which it held in solution; and as the water which is added to supply the place of that which has evaporated likewise contains the same extraneous matter, the quantity presently becomes larger than the water can hold in solution, and the residue is precipitated and hardened by the heat of the fire. All the salts of lime, which are usually contained in hard water, are likewise soluble in this fluid only in a very limited degree. For instance, sulphate of lime, one of the most common ingredients in hard water, is soluble in it only to the extent of ·2 or one-fifth of one per cent.; and carbonate of lime in a still smaller proportion; therefore the precipitation begins to take place so soon as the quantity exceeds this small amount.

It should also be observed that hot water sustains less of the salts of lime in solution than cold water, which latter deposits lime on being heated. If, therefore, the water in a hot-water apparatus is frequently or constantly changed, there is a rapid deposit of lime formed in the pipes or boilers or cisterns which contain it. This is particularly the case in all apparatus used for bath purposes, where from the constant change of the water, by the addition of fresh water to supply the place of that drawn off for use, the choking up of the boilers and pipes is extremely rapid unless constant attention is given to the removal of the sediment.*

A patent has recently been taken out by Mr. Stainton for the use of chloride of calcium mixed with water in certain quantities in order to prevent the water freezing, more particularly appli-

* See Chap. IX., art. 179* *et seq.*

cable to the high-pressure and small pipe system (Art. 119—133). It is favourably spoken of as a remedy for the pipes freezing, which is a casualty to which the small pipes are liable when used in colder climates than our own. It is likewise stated that the mineral oil used in Coffey's Patent (Art. 134) will not freeze, and retains its fluidity at all temperatures.

(169.) In Chapters XII. and XIII., some of the most important of the laws of heat, which are applicable to the general subject of this treatise, are described, and also the experimental data on which are founded several of the calculations given in the preceding chapters. Those who are desirous of investigating for themselves the accuracy of the various rules which have been given, will thus be enabled to judge of the subject in its more scientific bearings.

(170.) Before concluding these remarks, however, it may be proper to observe that buildings which are heated by hot water ought always to have some efficient mode of ventilation; for, even though the air may not be injured by the apparatus employed to heat it, the air of all inhabited rooms must necessarily be deteriorated by the respiration of the inmates. This remark would have been superfluous were it not that cases have occurred where the evils that have arisen from defective ventilation have been erroneously attributed to this plan of warming by hot water. The vapour which is given off from the lungs of the inmates of a room, under these circumstances, is condensed upon the windows, and has been supposed to arise from the water in the apparatus being converted into steam, and escaping through the joints of the pipes. If this were a solitary opinion, it might, like many others equally erroneous, be passed over in silence; but as this has been seriously objected against the invention by many who ought to know better, it may be worth while to state the cause more at length.

(171.) The quantity of vapour given off from the lungs, and also by exhalation from the skin, has been estimated at from twelve to thirteen grains per minute for each person. If, in consequence of imperfect ventilation of inhabited rooms, the air cannot escape after it has received this additional quantity of vapour exhaled from the body, it must, as soon as it has acquired a larger quantity of moisture than the temperature of the *external* air will support in the form of vapour (Art. 330), deposit a portion of it upon the glass; because, the glass being nearly of the same temperature as the external air, whatever quantity of the internal air comes in contact with it, its temperature is immediately lowered, and the excess of its vapour is condensed upon the surface of the glass. Thus, suppose the temperature of the air in a room to be 65°, and the dew-point 55°, then, if the temperature of the external air be only 35°, as much of the air in the room as comes in contact with the glass will deposit whatever vapour it contains above the quantity that a temperature of 35° will enable it to sustain. Under these circumstances, the amount deposited on the glass will be (Art. 330) about two grains for each cubic foot of air that is cooled by the glass; and the same effect, though in a less degree, will take place on all the other cold surfaces in the room. As each square foot of glass will cool one and a quarter cubic feet of air per minute, from the internal to the external temperature (Art. 101), we shall find that, under the circumstances we have supposed,— which is purposely taken as an extreme case,— the quantity of vapour deposited in this manner will amount to two and a half grains per minute on each square foot of glass.

(172.) We need be at no loss, then, to discover the cause of this accumulation of vapour on the windows and walls of rooms which are badly ventilated; and whenever the quantity of moisture thus condensed appears to be considerable, it may be taken as good evidence that the ventilation of the room is imperfect. That the same amount of condensation does not result from the use of hot air and cockle-stoves, is in consequence of a portion of the vapour being decomposed by the intense heat ; but when this method of avoiding the inconvenience is adopted, a worse evil is produced than that which is attempted to be removed, although, perhaps, it is not so obvious to the sight.

(173.) A similar error is very frequently committed in the construction of drying closets and rooms intended for drying various articles in manufactures and the arts. It is frequently supposed that a sufficient degree of heat is all that is required for this purpose, and the amount of ventilation is considered quite an unimportant matter. It need scarcely be observed that the reverse of this is the proper course, and that ventilation is, in these cases, of far more importance than the degree of heat maintained in the room.

(174.) The experiments of Dr. Dalton long since demonstrated that evaporation was independent of the air ; that the vapour arising from any liquid depends upon its temperature, and that the air retards the velocity of the discharge by its *vis inertiæ*.* If the air were wholly removed from the surface of the water, the vapour

* *Memoirs of the Manchester Philosophical Society*, vol. v., p. 575 ; and *Philosophical Magazine*, vol. xvi., p. 346.

proper to the particular temperature would be discharged instantaneously, instead of rising gradually, as when the water is exposed to the atmosphere. The *quantity* of evaporation is not affected by these causes. If water be placed under the receiver of an air-pump and the air exhausted, the full quantity of vapour which can be formed at that particular temperature will rise instantaneously; but if the air be allowed to fill the receiver, the same quantity of vapour will rise, and it will have slowly to filter its way and disperse itself through the air which fills the receiver. In these cases, we of course suppose that the vapour is not removed; but that, when once formed, it is allowed to remain in the receiver, and that the temperature is kept constantly the same. But when the vapour is allowed to escape, or to condense, the rate of its formation is very different in air and in vacuo; for Mr. Daniell's experiments proved that, in the same time, the quantity of vapour raised from a constant surface of water in vacuo and in air of atmospheric density was as 90 to 1 ; and that at all intermediate pressures the quantity of vapour was inversely proportional to the elasticity of the incumbent air.*

When, however, the air possesses motion, its effect on the quantity of vapour emitted from the surface of water over which it passes will be very different. The effect is, then, to drive off the vapour from the surface of the evaporating body as fast as it is formed; and this Dr. Dalton found to be proportional to the velocity of the air.† This latter case is precisely that with which we are concerned, while considering the effect of

---

* *Quarterly Journal of Science*, vol. xvii., p. 52.

† *Memoirs of the Manchester Philosophical Society*, vol. v., p. 576; and Watson's *Chemical Essays*, vol. ii., p. 57.

ventilation, and more particularly in the ventilation
of drying-rooms.

(175.) It is quite a mistaken notion to suppose
that a better effect will be produced by preventing too rapid a motion of the air through drying-
rooms, for the purpose of allowing the air to
imbibe more of the moisture by this means. The
infiltration of moisture into the air, or the absorption of moisture, as it is generally called, is extremely slow, as the experiments of Daniell and
of Leslie have proved; and therefore it is, by the
brisk motion of the air driving off the vapour as
fast as it is formed, that the principal effect is
produced in the ventilation of drying-rooms.*
The quantity of air which really becomes humid
by such means is extremely small; and Leslie
has estimated that, in a case of ordinary evaporation, where the air possessed but small velocity,
only the 184th part of the air that passed over
the wet surface became saturated.†

(176.) It is extremely difficult to estimate the
quantity of heating surface that is required for a
drying-room; for not only will the temperature
vary greatly with the nature of the substances to
be dried, but also with the degree of dampness
they possess, and likewise with the amount of
ventilation. The same quantity of heating surface
that would raise the temperature of a drying-
room to 80° or 90° with an imperfect ventilation,
might possibly not heat it above 60° or 70° if the
ventilation were perfect; and yet in the latter
case, probably, the drying effect would be greater

---

* The drier the air the more rapidly will it absorb the
moisture exhaled into it; but, under every condition of dryness, the absorption of moisture will be much less rapid by
the air than its discharge from the vaporizing body would be
were the pressure of the air entirely removed.

† Leslie *On Heat and Moisture*, p. 86.

than in the former. This, however, is not without certain limitations in its operations. Whenever the ventilation is sufficient to carry off the vapour as fast as it is formed, anything beyond this must be injurious, as it will lower the temperature of the room without promoting the dispersion of the vapour; and we have already seen that the quantity of vapour raised depends, in the first instance, upon the temperature of the vaporizing body. When, however, an atmosphere of vapour exists immediately around the body emitting the vapour, the emission is enormously reduced, and sometimes even wholly stopped; and this will occur, however high the temperature of the vaporizing body and of the air may be. In this case, therefore, even a very much lower temperature, accompanied by a brisk motion of the air, would be far more effectual than a high temperature with defective ventilation.

(177.) For common drying-rooms, where articles of various thicknesses are dried at the same time, and where the quantity of moisture to be evaporated is considerable, it has been found that 150 to 200 square feet of heating surface to each 1,000 cubic feet of space produces a temperature of about 120° when the room is empty and the ventilating apertures are closed. The temperature falls about 15° or 20° on opening the ventilating apertures; and when the room is filled with wet clothes the temperature falls to about 80° or 90°, and sometimes considerably lower, varying with the nature of the articles which are placed in the room. Mr. Buchanan states * that about 63 square feet of steam-pipe to 1,000 cubic feet of space produced a temperature of 100° in a drying-room; and in another instance 44 square feet of heating surface to 1,000 cubic feet of space

* Buchanan's *Treatise on Economy of Fuel*, &c., 211–218.

produced a temperature of 86° in a drying-room,
both being partially filled with damp goods. The
temperature of the steam is not stated in these
instances; but the pipes would naturally be hotter
than hot-water pipes, and therefore less surface
of them would be required; but the quantity of
pipe thus given is considerably less than was required in any instance which has come under the
author's own knowledge. In general, it would
not be safe to use much less than 150 square feet
of heating surface of hot-water pipes to 1,000 cubic
feet of space in any drying-room used for drying
linen, where quick drying is essential. For manufacturing purposes it will often happen that much
smaller quantities of heating surface will suffice;
for the goods often contain far less moisture, and
are not required to be dried so quick as in a
common closet for drying linen.

Tredgold has given rules for calculating the
required heating surface for drying-rooms for different purposes, but they are much too refined
for ordinary use.* A large amount of heating
surface is, however, absolutely indispensable
where quick drying is required; but attention
to the perfect ventilation of the room is of
more importance than its actual temperature.
It is also advisable to have some ready means
of altering the quantities of air admitted into
a drying-room; for not only is it desirable, if
possible, to alter the quantities of air admitted at
the different stages of the drying process, but in
different states of the weather the quantities of
air admitted ought to be likewise altered. When
the air is very humid, much less air ought to
be admitted than in dry bright weather. In
damp weather the drying process will go on much
faster, in proportion as a less quantity of air is

* Tredgold *On Heating Buildings by Steam*, p. 241 *et seq.*

admitted, and a higher working temperature maintained in the drying-room. In dry weather the opposite of this will often be desirable, and a much larger quantity of air may advantageously be admitted under these circumstances.

(178.) The physiological effects resulting from particular modes of warming and ventilating inhabited rooms form a most interesting subject of inquiry, and are not only interesting as matters of scientific research, but they closely concern every individual member of the community. It is a question which affects not merely the personal comfort of individuals, but, according to the opinion of the ablest pathologists, it influences the health, and affects the duration of life. In a subsequent chapter we shall endeavour to trace some of the physiological effects of various methods of artificial heat, and the most important consequences will be found to result from the use of some of these different inventions.

CHAPTER IX. ·

(179.*) WHEN large quantities of hot water are
required for baths, laundries, or other purposes,
considerable modifications of the apparatus become
necessary from that which is generally used for
warming buildings. From what has already been
stated Art. 166, it will be seen that in any apparatus
where the object is to supply large quantities of
hot water by continually drawing off the heated
water, and supplying its place by fresh cold water,
a large deposit of mineral substances constantly
occurs. When rain water alone is used this
deposit does not occur. But in every other case
it is absolutely necessary to provide some form of
boiler which can readily have the mineral deposit
removed. These deposits sometimes become of
enormous thickness unless frequently removed;
the destruction of the boiler is inevitable under
these circumstances. Some form of boiler different
to the usual forms used for warming buildings is
therefore indispensable.

There are many forms of boilers quite suitable

for this purpose, and so long as they admit of an easy removal of the deposit, the particular form is not very material. Any of the forms used for *steam* purposes answer very well, and the remarks on them need not be here repeated.

(180.*) It is usually the practice to heat the water by means of large cisterns connected with the boiler by a flow-pipe and return-pipe. A deposit very frequently occurs in these cisterns as well as in the boilers, and hence it is necessary to have ready access to the interior of the cisterns. The deposit in the cistern is not so serious as the deposit in the boiler, the principal inconvenience being the diminished capacity of the cistern, and the liability to stop up the various pipes connected with it. In public baths, wash-houses, hospitals, &c., constant inspection is necessary in order to watch, and if necessary remove these accumulations of sediment. In private houses, the apparatus may be used for long periods without inconvenience. Sometimes an apparatus may become choked and partially, or even wholly useless in a very few months or even weeks; and in other cases where the work is less severe, the deposit may be trifling even after years of operation.

(181.*) In erecting an apparatus intended to supply hot water in many different parts of the same building, very great difficulties frequently occur. It is not sufficient to heat a large tank and then draw off the heated water by various long or short lengths of pipe leading to the respective baths, taps, washing tubs, &c. The water in such cases generally becomes perfectly cold in the various pipes, and it frequently happens that many gallons of cold water must be drawn out of the pipes before any hot water can be obtained from the hot water cistern. The only effectual way to prevent this is so to arrange the system of pipes, that each one shall have a continuous circulation through it from the cistern to the boiler, or *vice versâ* from the boiler to the cistern. This of course much increases the expense of such an apparatus, but it is absolutely necessary in order to obtain success.

(182.*) The usual way of fixing this kind of apparatus is to place the hot water cistern on the *highest level* at which hot water is required; and then to draw off the hot water through a pipe or pipes passing through the various floors where hot water taps are fixed and finally descending to the boiler, thus to complete the circulation. But a far better plan is to place the hot water cistern near to the boiler *and below the draw-off taps*, and thus draw the hot water *upwards*; the upward action being produced by the pressure from the large cold water cistern placed higher than any of the draw-off taps, as in the figure annexed. The advantages of this plan are sometimes very important. For as it is generally necessary to draw off the hot water from the bottom of the cistern, when the cistern is fixed at an elevation above the draw-off taps, it is

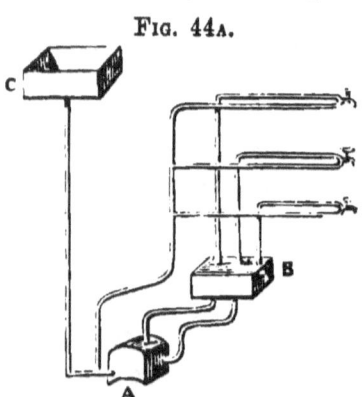

Fig. 44a.

most difficult to avoid drawing off the cold water instead of hot when thus fixed, the cold water naturally lying towards the bottom of the cistern. No such inconvenience arises when the hot water cistern is fixed as in fig. 44a, where b is the hot water cistern, and c the cold water supply.

(183.*) In many cases it is desirable to use steam as a means of heating water in large quantities for the purposes here described. This may be done either by blowing steam direct into the hot water cistern, or it may be made to pass through a coil fixed inside the cistern. In this case the action is produced by the colder water in the cistern condensing the steam as it passes through the coil, and the condensed steam then runs out at the other end of the coil. When the steam is blown direct into the hot water cistern there is some saving of expense, both in the cost of the apparatus and also in fuel, but this plan is attended with some serious inconvenience. Considerable noise is produced by steam suddenly blown into colder water in this way. This noise can be prevented to a great extent by closing the end of the pipe, and then allowing the steam to escape through a great number of very small holes drilled in the pipe; or it may be likewise prevented by making the steam pipe end in a small iron box filled with iron grape shot, about the size of a small marble, and then piercing the surface of the box with a great number of moderate sized holes. This latter plan was the subject of a patent many years ago. Both these plans are attended with some inconvenience. When the steam pressure in the pipe ceases, a vacuum is formed in the pipe, and the water in the cistern passes backwards into the boiler and may cause it to overflow. This result can be prevented by means of back-pressure valves, but an awkward complication is the result.

(184.*) To estimate accurately the proper length of a coil to heat a given quantity of water by means of steam is not very easy, as the effect varies every moment of time according to the temperature of the water in the cistern, and the pressure of the steam in the boiler. As a rough estimate for ordinary purposes, it may be stated that a coil made of iron pipe 50 feet long and 1 inch internal

diameter (which is equal to about 12 square feet
of surface for the inside diameter), and conveying
steam of 30 lbs. pressure per square inch, will boil
about 300 gallons of water per hour. If the steam
be only 15 lbs. pressure, it will take a coil of about
80 feet in length to boil the same quantity of water.

If it be intended to heat the cistern by means of
a coil of pipe with hot water instead of steam,
the coil will require to be about one-half longer
than a steam coil working at 15 lbs. pressure. A
very long coil has many objections, and therefore
when a coil of considerable length becomes necessary in order to give sufficient surface, it is better
to increase the diameter of the pipe than to extend
its length.

(185.*) To estimate the size of a boiler to heat
any given quantity of water, the rules already given
in Chapter IV., Art. 71 may be sufficient. But it
may be here stated, that as a general result 1 square
foot of boiler surface exposed to the *direct action of
the fire* will boil about 11 gallons of water from 52°
to 212° Fahrenheit per hour.

From these data it will not be difficult to calculate approximately the effect of almost any apparatus
that may be desired. Suppose a tank containing
1,000 gallons of water has to be boiled in (say)
three hours. The boiler, whether used for steam
or the circulation of hot water, must be of some
shape that can be easily cleaned out from the
deposit. In either case the boiler must be of such
size, that it shall expose a surface of 30 square feet
to the *direct action of the fire*. Such a boiler, supposing it were the waggon shape shown figure 16,
would be 5½ feet long, by 3 feet wide, and 2½ feet
deep. But if the water were only required to be
heated to about 140° instead of 200° or 212°, then
a similar boiler 4 feet long, by 2½ feet wide, and
2½ feet high would be sufficient.

(186.*) There is usually a considerable deposit formed in the cistern by all these several modes of heating water. It sometimes forms in great thickness on the outside surfaces of the coils used in this manner, whether they are heated by steam or by the circulation of hot water. It is therefore necessary in all these cases that the cisterns should occasionally be opened and cleaned out from any deposit, as otherwise the heating power of the apparatus is very seriously reduced. The deposit which occurs in the cisterns is not so destructive as that which takes place in the boiler. The deposit in the boilers unless frequently removed, is absolute destruction to them : in the cisterns the deposit only causes loss of power, though this may in some cases occur to a very serious extent.

(187.*) It may perhaps be useful to state the mode of estimating what will be the resulting temperature by mixing together any given quantities of water at different temperatures. Whether the quantities be large or small, the same rule will apply. Suppose we mix together two gallons of water at 200° Fahrenheit, and four gallons at 50° ; the resulting temperature will be found by multiplying each quantity by its respective temperature, adding these numbers together, and then dividing the product by the total quantity of water, thus—

$$2 \text{ gallons} \times 200° = 400$$
$$4 \text{ gallons} \times 50° = 200$$
$$\overline{}$$

Divide by 6 gallons) 600 = 100°

giving 100° Fahrenheit as the average temperature of the whole six gallons. Or suppose a plunge bath of 40,000 gallons at 60° Fahrenheit mixed with 2,000 gallons of hot water of 200° Fahrenheit. In this case 40,000 multiplied by 60°, and 2,000

multiplied by 200°, added together gives 2,800,000.
This number divided by 42,000, the total quantity
of water will give 66½° as the resulting temperature
of the entire quantity of water thus mixed together.
If steam and water are mixed together, the resulting temperature is calculated in the same way,
bearing in mind that the heat of steam, taking its
sensible and latent heat together as a constant
quantity (see Art. 257), is equal to 1200° Fahrenheit. Thus, if one gallon of water, converted into
steam, be mixed with ten gallons of water at 50°
Fahrenheit, the result will be—

$$
\begin{array}{ll}
\text{10 gallons water at } 50° = & 500 \\
\text{1 gallon of water in the} & \\
\quad \text{state of steam at } 1200° = & 1200 \\
\hline
& 1700
\end{array}
$$

Divided by 11 the whole quantity of water will
give 154½° as the average temperature. There are
many cases which occur in practice, where such
calculations as these are extremely useful.

(188.*) In some experiments on warming the
water in a large plunging bath containing 11,000
gallons of water, it was found that with an arched
boiler having (equal to) 30 square feet of direct
action surface exposed to the fire, it took 12 hours
to raise the temperature of the bath from 52° to 86°,
or to add 34° to the temperature with a moderate fire ;
or with a strong fire it raised the temperature 54° in
twelve hours. This heat was communicated to the
water by means of a coil of 75 feet of 4in. pipe, laid
at the bottom of the bath, which, in consequence of
the rapid conducting power of the water in the
bath, abstracted all the heat as fast as it could be
generated (see note to Art. 251). The water in
this case merely circulated through the coil of pipes,

and the same water returned again to the boiler.
In another case of a bath containing 40,000 gallons
of water, the estimated quantity of pipe laid as a
coil in the bath was 200 feet of 4in. pipe ; the
arched boiler was 10 feet long × 3½ feet wide
outside, and the time of heating was 40,000 gallons
of water raised 20 degrees in seven hours.

CHAPTER X.

ON HEATING BY STEAM.

Variableness of Temperature—Alterations in the Level of Pipes—Removal of condensed Water—Proper Steam-pressure to be employed—Mode of Calculating the Heating-surface required—Size of Boilers—Quantity of Water condensed—Boiler Surface required—Forms and Power of Boilers—Quantity of Coal—Size of Furnace-bars—Mode of supplying Water to the Boiler—Safety Valves, Gauges, and other Appendages required—Air-vents and Outlets for the condensed Water.

(179.) Since the introduction of the invention of heating buildings by hot water, the use of steam for this purpose has been comparatively neglected; still there are many situations where its employment may be advantageous, particularly in buildings used for manufacturing and commercial purposes. No work containing rules for its application has been published for many years. The only books on the subject which the author has met with are Buchanan's *Treatise on Heating by Steam*, last published in 1815, and Tredgold's *Principles of Warming and Ventilating Public Buildings, &c.*, last published in 1836. Both these works have been out of print for many years, and it therefore appears desirable in the present work to give such rules as may be necessary for erecting steam-heating apparatus for such ordinary purposes as it is likely now to be applied to.

(180.) The principal difficulty with an apparatus for heating by steam is the constant attendance it requires, not only to ensure its safety, but to secure a continuance of its heating power. Unlike the hot-water apparatus, which retains its heating power for many hours, the steam apparatus ceases to afford heat as soon as the fire fails to possess sufficient force to raise steam in the boiler. The risk of explosion in inexperienced hands is also a serious objection to its use. But when the apparatus is of sufficient magnitude to warrant the continued superintendence of an experienced person, it may be considered perfectly safe.

(181.) One great convenience in a steam apparatus is the facility with which the pipes can be conveyed in almost any direction. Provided the condensed water be carried off, and the pipes thereby be kept clear, it matters not whether the pipes are carried above or below the boiler, and that to almost any extent. But in carrying out this principle, it must be borne in mind that wherever there are considerable alterations of level, the condensed water must have a proper outlet from the lowest part; for otherwise the pipes will very quickly become choked with water and cease to act; and hence any sudden and considerable alterations in level become extremely difficult to deal with in practice, although *in principle* there is nothing whatever to hinder the most rapid, multifarious, and even eccentric alterations in level that can possibly be contrived. But as every sudden dip or alteration of level in the pipe will require a separate outlet for the condensed water, it becomes exceedingly difficult, and very undesirable in practice, to make these sudden alterations in level. Like the hot-water apparatus, it is therefore found most advantageous to lay the pipes with a very small but gradual inclination, so

as to allow the condensed water to flow out at one
place, or at least from as few places as possible.
The condensed water must not only be carried off
in order to keep the pipes clear for the passage of
the steam, but likewise to prevent the pipes from
breaking. For if the condensed water remains in
the pipes after· they become cold, the next time
the pipes are heated they are extremely likely to
crack, if they are of cast iron, in consequence of
the great difference of temperature which there
would then be between the upper and the lower
sides of the pipes.

(182.) In arranging a steam apparatus, the first
thing is to calculate the proper quantity of heating
surface. In order to do this, it is necessary to
determine at *what pressure* the boiler is to work.
Formerly it was considered that $2\frac{1}{2}$ lbs. pressure
per square inch of the boiler was quite as much as
could safely be used for this purpose, and the old
writers on the subject recommend this pressure to
be employed. But of late years the use of high-
pressure steam has so accustomed engineers to
regard boilers as perfectly safe with pressures of
80 lbs. or 100 lbs. and upwards to the square inch,
that no one would now object to employ steam of
5 lbs., 10 lbs., or even 20 lbs. pressure per square
inch, for heating buildings of any description,
whenever such pressures are deemed desirable.
The higher the pressure in the boiler, the higher
will be the temperature of the pipes or other
radiating surfaces ; and therefore the higher the
pressure the smaller the size or less the quantity
of the radiating surfaces which will be required.

(183.) In heating manufactories and other commercial buildings, it is often desirable to use the
surplus steam from a steam-engine boiler, or the
waste steam after it has passed through the
cylinder of a high-pressure engine. In other

cases, where a separate steam-boiler is used for heating the apparatus, it is seldom desirable to use steam of much more than about 5 lbs. to 10 lbs. per square inch. Whatever pressure, however, is fixed upon, the quantity of radiating surface necessary may be found in the following manner :—

In Chapter VI. (*ante*) rules are given for calculating the radiating surfaces necessary to heat most kinds of buildings with hot-water pipes. If, then, by these rules the required amount of radiating surfaces be calculated, the *reduced quantity* that is required when steam-heat is to be applied may be ascertained as follows: In Table I., Appendix, the temperature of steam at all pressures is given. Suppose now that to heat a building to 60° with hot water, it has been found (by rules given in Chapter VI.) that 300 feet of four-inch pipe is required. If it is proposed to use steam of 5 lbs. pressure,—that is, 5 lbs. in addition to the 15 lbs. which is necessary to prevent a vacuum in the boiler,—if we look in Table I., Appendix, for the nearest pressure to 20 lbs., we shall find the temperature of the steam is 225°. Now the temperature of hot-water pipes being reckoned at only 180°, or 120° above the proposed temperature of the room, and the temperature of the steam-pipes (at this pressure of 20 lbs.) being 225°, or 165° above the proposed temperature of the room, the quantity of steam-pipe will be found by a common rule-of-three sum ; namely, as 165° is to 120° so is 300 feet of hot-water pipe to the quantity of steam-pipe required to produce the same effect, which in this case will be 218 feet. Again : suppose the pressure of the steam is 30 lbs.,—that is, 30 lbs. above the 15 lbs. necessary to prevent a vacuum. If we look in the Table for the temperature of steam at 45 lbs. pressure, we shall find it is 275°, or 215° above the proposed temperature of the building. In this case the calculation will be, as 215° is to 120° so is 300 feet the quantity of hot-water pipe required to the quantity of steam-pipe necessary to give the same heat, which in this case is 167 feet.

(184.) These calculations are not absolutely correct; but they are sufficiently accurate for every purpose of practical utility. For it will be shown in another chapter (Chapter XII.) that the radiation of heat is proportionately greater at high temperatures; and therefore while the above mode of calculation is quite sufficient for all practical purposes, such as those here contemplated, the rule must not be construed to include scientific exactness.

It may here be observed that the rule already given (Art. 41), for the elongation and expansion of hot-water pipes, must be observed wherever steam-pipes are applied to any heating purposes.

(185.) In apportioning the boiler for heating any given quantity of pipe, both Tredgold and Buchanan assume that the *steam-space* in the boiler ought to be equal to the whole contents of the pipes, so as to fill the pipes easily with steam when the valves are opened; and, further, that the steam-space ought to be about half the entire capacity of the boiler. These proportions, however, do not appear suited to modern practice, now that steam of much higher pressure is used. And although exposing a large surface of the boiler to the fire is always economical, a boiler of *large capacity* is not always convenient, nor, indeed, is it necessary for successfully applying steam to the heating of buildings.   .

(186.) Tredgold deduces from his experiments that with steam of $2\frac{1}{2}$ lbs. pressure above the atmosphere (or $17\frac{1}{2}$ lbs. actual pressure), 182 superficial feet of iron pipe will condense as much steam as will produce one cubic foot of water per hour, when the temperature of the room is 60° Fahrenheit. In this case he assumes the average temperature of the entire surface of the pipe to be 200° Fahrenheit. From this datum the proper size of the boiler can be calculated.

It appears (by Art. 71) that four square feet of boiler surface exposed to the *direct* action of the fire will evaporate one cubic foot of water per hour. The boiler for a steam apparatus ought therefore to evaporate as much water per hour as the pipes will condense in the same time ; hence a boiler possessing four square feet of surface exposed to the *direct* action of the fire will just supply sufficient steam to heat

182 sq. ft. of pipe when the pressure is 2½ lbs. above the atmosphere.
161   „   „   „   „   10 „   „   „
149   „   „   „   „   20 „   „   „
135   „   „   „   „   30 „   „   „

and supposing that in each case the temperature of the room is 60° Fahrenheit. It follows also that each of these several quantities of pipe will give off the same amount of heat in equal times.

(187.) It by no means follows that these proportions of boiler surface are desirable to be adopted. Although they certainly will heat the pipe when skilfully applied and carefully attended, it is desirable to allow about one-half more boiler surface to meet the common contingencies of most apparatus, and to prevent the necessity for heavy firing. Watt allowed eight square feet of boiler surface to evaporate one cubic foot of water per hour, and in the Cornish engines even a larger surface than this is allowed.* It will therefore be desirable to assume that six square

See note to Art. 71.

feet of boiler surface exposed to the *direct* action
of the fire will be necessary to heat the several
quantities of pipe here mentioned.

(188.) Having ascertained the surface of the
boiler to be exposed to the fire, it becomes necessary to determine its shape and capacity. The
forms of steam-boilers are almost as numerous
as those which have been proposed for the hot-
water apparatus. But for the ordinary purposes of
heating buildings by steam, it is absolutely indispensable that the boilers should be simple in form,
tolerably easy to clean out, of rather large capacity
in proportion to their fire surface, and so durable
as to require very little repair. For many purposes, these qualifications are frequently sacrificed,
and necessarily so in order to obtain lightness,
rapid draught, and high evaporating power.

(189.) For low-pressure boilers where the steam
does not reach beyond 8 lbs. or 10 lbs. per square

| Fig. 39. | Fig. 40. | Fig. 41. |

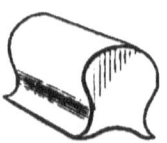  

inch, there is no better form than the old waggon boiler, fig. 39, which can be used of any size.
The Cornish boiler, fig. 40, is only suitable for
larger apparatus, and is scarcely desirable for
boilers of less size than about ten feet long and
four feet diameter. The cylindrical boiler with
hemispherical ends, fig. 41, is very suitable for
high pressures, as it is one of the strongest forms
for a boiler, and is therefore always regarded as
one of the safest boilers that can be made.

Any of these boilers have sufficient capacity, in

proportion to their surfaces, to render them very desirable for the purpose of applying steam-heat to buildings. In calculating the proper size for a boiler to produce a given effect, the same rule must be applied that has already been given (Art. 71) for calculating the power of boilers for the hot-water apparatus, namely, that the flue-surface of the boiler is only to be taken at one-third its actual measurement, when we want to estimate the *extent of surface exposed to the direct action of the fire.* Thus in fig. 39, suppose the boiler to be 6 feet long, 2½ feet wide, and the flue-surface 2 feet deep all round the boiler; in this case the bottom of the boiler will have 6 feet × 2½ feet = 15 feet of surface. The flue-surface measured entirely round the boiler by 2 feet deep, will be equal to 34 square feet; and one-third of this latter measurement will be equal to 11⅓ feet, which, added to the former quantity of 15 feet, will make the boiler-surface equal to 26⅓ square feet *exposed to the direct action* of the fire. This boiler (by Arts. 186 and 187) would be sufficient to heat 798 square feet of steam-pipe at a pressure of 2½ lbs. per square inch, when the temperature of the room is kept at 60° Fahrenheit.

(190.) The quantity of coal that will be necessary to produce a given effect with a steam apparatus may be ascertained as follows:—The quantity of water condensed by the pipes must first be ascertained as already pointed out. Suppose in the case already stated, that 182 square feet of pipe, in a room of 60° temperature, condenses one cubic foot of water per hour, when the pressure is 2½ lbs. per square inch: then, by the data already given (Arts. 94 and 95), it will be found that about 9 lbs. of coal is required to evaporate one cubic foot of water; and this must be the quantity required to replace the loss of

P

heat by condensation of steam in the pipes; therefore the coal burnt would be 9 lbs. per hour, if there were 182 feet of steam-pipe in a room of 60°, and the steam at the pressure of $2\frac{1}{2}$ lbs. per square inch. Whatever quantity of water the pipes will condense per hour,—and which quantity can be estimated by the preceding rules,—the consumption of coal will be 9 lbs. for every cubic foot of water thus condensed.

It has been already explained (Arts. 71 and 95) that a very great difference exists in the heating effects of coal burned in different furnaces. The quantity above stated is a full allowance, and much more than some steam-boilers will require when very carefully worked.

(191.) The size of the furnace-bars should be one square foot to burn about 10 lbs. to 11 lbs. of coal per hour; and the surface of the boiler exposed to the direct action of the fire will generally be about five times the area of the furnace-bars. The furnace of a steam apparatus will very seldom allow the use of a door to the ash-pit, as it requires a stronger draught than a hot-water apparatus, and the ash-pit door is very liable to reduce the draught. Unlike the hot-water apparatus, the heating effect of a steam apparatus ceases directly the fire fails to produce its full effect on the boiler; and therefore a brisk fire and constant attention are absolutely necessary for a steam apparatus, in order to keep up the proper action in the pipes.

(192.) The various apparatus necessary to secure a proper supply of water to the boiler, and likewise to prevent danger by any undue pressure in the boiler, as well as the contrivances required to carry off the air from the apparatus, and to remove the condensed water from the pipes, are all much more complicated and expensive than anything required for these purposes in a hot-water apparatus.

The boiler is generally supplied with water by means of a stone-float apparatus, of which there are several varieties (Figs. 42 and 43). A stone placed in the boiler, by a wire passing through a stuffing-box balanced by a lever with an attached weight, is made to turn a cock which communicates by a pipe from the boiler with an elevated cistern of water. When the stone falls in the boiler by the partial evaporation of the water, the cock is opened by the lever, and as much water comes into the boiler as will again float the stone, and thus shut

Fig. 42.   Fig. 43.

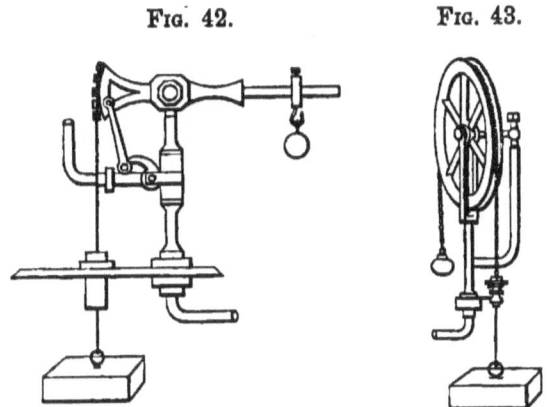

the cock. The cistern must always be high enough to have the column of water of greater weight than the pressure of the steam; otherwise on opening the cock either water or steam would escape up the pipe, instead of the water coming down from the cistern. The pressure from the water in the cistern should be not less than from one-fifth to one-fourth greater than the pressure of the steam in the boiler; and as every two feet *vertical height* of the water in the cistern and pipe is equal to 1 lb. pressure of the steam in the boiler, the supply cistern ought to be not less than the following height above the boiler :—

6 feet high to supply a boiler working 2½ lbs. steam,
12 „   „   „   „   5   „
24 „   „   „   „   10   „
75 „   „   „   „   30   „

and generally a little more than this may be desirable in order to secure certainty of action.

These elevations are very difficult to obtain for the higher pressures, and hence one of the difficulties in working very high pressures except in conjunction with a steam-engine; in which case a force-pump is used to inject the water into the boiler without the aid of the elevated cistern. A modern invention of a self-acting injector has been introduced, but it is doubtful how far a complicated, though ingenious, apparatus can be safely trusted, unless it is under the constant and daily inspection of experienced and careful super- · intendents.

(193.) A safety-valve is an indispensable appendage to a steam-boiler. Whatever form of safety-valve be used, the size of its clear opening ought to be not less than one circular inch for every six square feet of the boiler exposed to the direct action of the fire; or, in other words, one circular inch for each cubic foot of water per hour which the boiler will evaporate. A valve of this size is sufficient, whatever the pressure of the steam may be. When the boilers are large, two or more valves may be used, so as to give in the aggregate this amount of opening for the escape of steam.

(194.) A steam-gauge is a very desirable adjunct to a steam-boiler. For low-pressure boilers a mercurial steam-gauge is frequently used. But a very convenient gauge for either high or low pressures is the ingenious instrument known as

"Bourdon's Pressure Gauge," which shows on a small clock-face the correct pressure of the steam.

(195.) A water-gauge is also another very useful addition to the fittings of a steam-boiler. This gauge consists of a glass tube, about 1 foot long, fixed between two cocks communicating directly with the boiler, generally in the front, near the furnace-door, and so placed that the water-level in the boiler can be seen in the glass· gauge, which is so fixed that the water-level is nearly about the middle of the glass tube. The water always stands at the same height in the tube as in the boiler. Any variation, therefore, in the height of the water in the boiler can be easily seen by looking at the height in the glass tube. The use of the two cocks of the glass gauge is merely to shut off the communication with the boiler, in case the glass tube should break: on all other occasions the cocks are to be left open.

(196.) Every steam-boiler ought also to be supplied with two gauge-cocks, placed at the front of the boiler, at such height that one is always above the water-level, and the other about two or three inches below the water-level. By these cocks (as well as by the instrument last described), the proper height of the water in the boiler can be ascertained. If the upper cock, when opened, gives off water instead of steam, the boiler is too full of water: while if the lower cock, when it is opened, should give off steam instead of water, then the boiler is deficient of water, and danger may be apprehended.

(197.) A large cock is also required to be fixed in the boiler, called a "blow-off cock." It is for the purpose of emptying the boiler, and cleaning out the sediment, and requires to be frequently used. The boiler should be emptied once every two or three days, less or more, according to the hardness of the water, and the amount of work which the boiler performs. The pipe for the " blow-off cock " may be either fixed at the lowest point of the boiler, and be used simply to run the water off; or the pipe may enter the top of the boiler, and descend within a couple of inches of the bottom of the boiler; in which latter case the water is forced out through the pipe and cock by the pressure of the steam, taking care that the fire is so much slackened when this operation is performed that the boiler shall not be injured by the heat of the fire while the boiler is thus being emptied.

(198.) All these various fittings for steam-boilers can so easily be obtained ready made, that it is not considered necessary here to describe them more particularly, as it is only in their application and general use that they require explanation. There are, however, still some further appliances necessary for working a steam-heating apparatus which must not be passed over in this description. A cock is generally placed between the boiler and the pipes which heat the building. On opening this cock the steam drives before it all the air which was contained in the pipes; and unless a ready outlet for this air is provided, a portion of the pipes will remain full of air, and continue nearly cold. If the pipes are laid horizontally, there is but little difficulty in getting rid of this air, through a "blow-pipe" at the farther extremity, which may be opened when the pipes are about to be heated; and when they are quite hot the blow-pipe may be nearly closed, in order to prevent the waste of steam which would necessarily occur after the air was expelled. But when the heating surfaces occupy several different levels, and particularly when they assume the form of coils, steam-plates, or ornamental columns, it is not always very easy to discharge the air without a good deal of care and contrivance. The air is considerably heavier than steam, and therefore it naturally falls to the lower levels; and it is from these lower levels that the exit for the air must therefore be provided. These eduction pipes ought never to be wholly closed, even when the air has been expelled; but the management of these air-pipes is often a serious difficulty. In very extensive apparatus it has occasionally been found so difficult to drive out the air from the pipes and other heating vessels, that the employment of an air-pump to exhaust the air has sometimes been found necessary each time that the apparatus is warmed, in order to prevent the great delay in heating the apparatus which otherwise would occur.

(199.) It is also necessary to provide for the escape of the condensed water. When steam of very low pressure is used, and the apparatus is upon a small scale, an inverted siphon-pipe is frequently used for this purpose. This inverted siphon-pipe must be of sufficient depth to resist the pressure of the steam, and frequently this becomes extremely inconvenient. Thus if the steam is 2 lbs. pressure, the siphon will require to be about 5 feet deep; if the steam be 5 lbs. pressure, the siphon will require to be about 12 feet deep. These siphon-pipes for the condensed water are only suitable for steam of very low pressure. When steam of higher pressure is employed, there are several contrivances for stopping the escape of steam, and at the same time allowing the waste water to run off. One of these plans is shown in fig. 44. This is a close box or vessel about 10

Fig. 44.

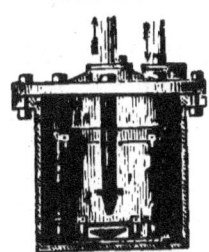

or 12 - inch diameter and the same in height, in which is placed an extremely light copper pan or cup about 2 inches smaller in diameter and in height than the outer box. A pipe passes through the lid of the outer box, and dips down to within about 1 inch of the bottom. The condensed water runs into the box, between the outer case and the inner copper vessel. As the water gradually runs in, it soon floats the light copper vessel, which is empty, and thus presses it up against the open end of the pipe, which passes through the lid of the outer box. This goes on until the space between the outer box and the inner copper vessel becomes quite full of water, when having no other escape, the water runs over into the empty copper vessel which sinks by its increased weight. When it sinks, it opens the outlet in the pipe passing through the lid of the outer box; and then the pressure of the steam acting on the surface of the water in the copper vessel, forces all the water in the vessel upwards through the pipe. When this occurs, the copper box again floats, being relieved of the increased weight, and in rising stops the exit through the central pipe. This apparatus thus becomes self-acting, discharging the water, but stopping the escape of the steam.

Another plan is by means of a very thin and small copper cylinder, about an inch and a half or two inches diameter, filled with æther, and which floats on an outer case of copper, into which the condensed water runs. When steam enters the outer case, the heat causes the small cylinder to elongate in a slight degree, sufficient to close a small opening against which it is thus pressed. The condensed water being colder than the steam, when the condensed water fills the outer case, the small cylinder again contracts and opens the outlet for the escape of the water.

If the condensed water from the apparatus is to
be used for any purpose, either as distilled water,
or to supply the boiler again with fresh water, it
can be collected as it flows out from the valve-
box; or it may be collected from the siphon-pipe
previously described.

(200.) The advantages of a steam apparatus
over a hot-water apparatus appear to consist
principally in the greater ease of application of
the heating surfaces, where great inequalities and
frequent alterations in the levels occur, and particularly where the boiler is necessarily placed
considerably higher than the places which are to
be heated; and likewise in the somewhat smaller
extent of heating surface which is required. The
disadvantages of the steam apparatus are its
greater danger, and the absolute necessity which
exists for constant attendance and watchfulness;
its want of permanence of temperature, except
when the greatest possible care is bestowed; and
its somewhat increased cost of fuel above that of
a hot-water apparatus. The first cost incurred for
the erection of the two kinds of apparatus will
differ but little when the work is done in an
equally substantial manner; but the wear and
tear, and repairs, of a hot-water apparatus will
be less than that of a steam apparatus, as in the
former there is absolutely nothing that can wear
out except the boiler; while in a steam apparatus
there are various things which constantly require
attention and repair, in addition to the greater
amount of wear in the steam-boiler itself, caused
by the large quantities of sediment which require
to be constantly removed, otherwise the boiler is
very speedily destroyed.

CHAPTER XI.

ON HEATING BY HOT AIR.

Sylvester's Cockle-Stove—Improved Cockle-Stove—Estimate of Heating-power—Hot-air Stoves with Projecting Plates, and their Heating-power—Bernhardt's, Hazard's, and other Hot-air Apparatus—Difficulty of Horizontal Flues—Multi-tubular Hot-air Apparatus—Hot-blast Apparatus.

(201.) THE application of fire-heat to metallic surfaces on which atmospheric air is made to impinge directly, without the intervention of either water or steam, is usually called a HOT-AIR APPARATUS. This distinguishes it from any other apparatus in which air, although frequently quite as highly heated, receives its heat from pipes or vessels heated by steam or hot water. This method of heating atmospheric air possesses considerable utility. Previous to the introduction of the hot-water apparatus, this method of heating was much more used than it is ever likely again to be. The hot-water apparatus is so much more wholesome than the hot-air apparatus, that the latter can only be recommended where either partial vitiation of the air is of no great moment, or else where the object is merely to produce a drying and warming effect of considerable intensity for certain manufacturing or economic purposes. In such cases as these, the hot-air apparatus frequently presents considerable advantages, as a drying effect of greater intensity can be thus obtained than by any of the apparatus which have been already described.

(202.) The hot-air apparatus in the form in which it is now best known, was employed at the close of the last century by Mr. Strutt, of Belper, near Derby, for warming his cotton mills. It was subsequently improved, and extensively introduced into general use by Mr. Sylvester, who, in 1819, published his description of the apparatus, with experiments on its heating powers. The construction recommended by Mr. Sylvester, however, appears to be defective in one particular: for in his apparatus the furnace or fire-grate was placed so far below the surface of the cockle-case, that its effect was considerably diminished. In his apparatus the fire-grate was placed about two feet *below the bottom* of the cockle-case, or about from 7 feet to 8 feet below the surface which was to be heated by the fire. In some recent cases in which this form of apparatus has been used by the author, the furnace and fire-bars are placed entirely within the cockle-case, so as to make the fire-bars lie about one foot *above* the bottom of the cockle-case, and about five feet higher than those used by Mr. Sylvester. The figs. 45 and 46 show the apparatus in the last-mentioned form, but in all other respects it is the same as that proposed by Mr. Sylvester.

Fig. 45 is a section of the furnace, constructed of fire-brick inside the cockle-case. The cockle-case is of wrought iron one-fourth of an inch thick, and made arch-shape and closed both back and front, except at the furnace-door, which is fixed to the front. Between the sides of the furnace and the inside of the cockle-case is a vacant space about 4 in. to 5 in. wide, and the flame, smoke, and heated gases all pass downwards between these spaces and enter the longitudinal

flues, which extend along the bottom at both sides
of the cockle-case. These smoke-flues should
be about 9 in. by 6 in., and their *upper* sides
are partially covered by two plates of iron, which
so cover them as to leave a narrow open slit from
three-fourths of an inch to one inch wide all
the length of the cockle-case on both sides. These
slits are the only places where the smoke can
escape into the flues. These side flues communicate with another flue at the back, similar in size,
but *entirely covered* on the top. This last flue

FIG. 45.   FIG. 46.

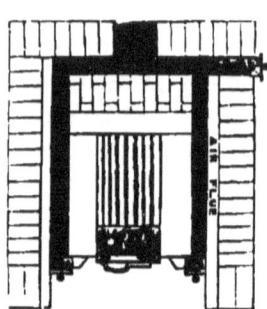

leads into the chimney. There must be a soot-
door at one or both ends of this back flue, to
enable it to be cleaned out. To clean out the
longitudinal or side flues, there is a soot-door in
the front, at each side of the furnace-door: and
a rod with a handle at one end, and a small rake
at the other end, is made to slide along the
narrow slit already described, which keeps it
constantly clear of soot, by merely drawing out the
handle occasionally when the fire is replenished
with fuel.

(203.) The fire is entirely contained within the
cockle, without any other outlet whatever for the

heated gases, flame, and smoke, except the two narrow slits left in the horizontal flues, already described. The cockle-case thus becomes heated. Between the *outside* of the cockle-case and the brickwork which surrounds it, a vacant space is left of from 4 in. to 6 in., extending entirely round the iron case ; and at the lower part of this vacant space or air-chamber, a triple row of small pipes about 2 in. diameter, or 2 in. square, passes through the brickwork, and extends to within three-fourths of an inch of the iron surface of the cockle-case. These pipes require to be numerous. The cockle-case here shown and described is supposed to be about 5 feet wide, 5 feet long, and 5 feet high. Such a cockle would require about 150 to 200 such pipes. The use of them is to bring fresh air into immediate contact with the surface of the cockle ; and when once it is brought in through these pipes, it cannot escape or return back again, as these pipes, when thus placed, form as it were a trap, which prevents the return of the air, and projects it against the highly-heated sides of the cockle. The air thus brought in passes over the entire surface of the cockle into the upper part of the air-chamber. Through the top of the brickwork, in the upper part of this hot-air chamber, any required number of pipes of large diameter lead the heated air into the different rooms to be warmed, passing through the necessary flues or channels provided for the purpose. These large hot-air pipes are so fixed in the brickwork as to extend quite through the hot-air chamber, until they reach within one inch of the surface of the iron-cockle case (fig. 45). By these means the air is prevented from too easy an escape from contact with the heated cockle which otherwise would occur. The numerous small tubes at the lower part of the cockle which project the cold air against the heated iron are also essentially necessary. They were not used when the cockle-stove was first invented; and Mr. Sylvester states that, in some stoves to which they were subsequently added, the heating power was just doubled when these induction pipes were used.

(204.) It is obvious that this apparatus, when the furnace or fire-box is built inside the cockle-case, can only be used on a rather large scale. It can scarcely be made less than $4\frac{1}{2}$ to 5 feet square; and perhaps the best form that can be used is a square base with a top in the form of a groined arch, so that the air may be easily brought in from all the four sides; and a cockle-case of five feet square and five feet high would have a heating power equal to nearly 650 feet of four-inch pipe heated by hot water. A smaller apparatus of this kind could be made if the furnace were constructed of cast iron, with merely a lining of fire-brick; and in this way a cockle of about $3\frac{1}{2}$ feet square might be used, but the furnace itself would probably not be very durable, in consequence of the thinness of the fire-brick.

(205.) To estimate the power of this cockle stove the experiments of Mr. Sylvester afford a very good guide. From these experiments, it appears that with his improved hot-air apparatus or cockle-stove, 17 square feet of heating surface warmed 344,600 cubic feet of air 56° in twelve hours, with the consumption of 60 lbs. of coal. This is equal to 1574 cubic feet of air raised 1° per minute, when the consumption of coal was 5 lbs. per hour, the heating-surface being 17 square feet; or it is also equal to 321,626 cubic feet of air raised 1° by the combustion of 1 lb. of coal. If we compare these results with the calculations (Art. 99) for estimating the power of the hot-water apparatus, we shall find that each square foot of the cockle is equal in heating power to 7 square feet of hot-water pipe. Therefore, in calculating the proper size of the cockle, we can either estimate it by the quantity of air required to be heated per minute, by Mr. Sylvester's rule, or we can ascertain by the rules given (Arts. 99 to 111) the surface of hot-water pipe required, and one-seventh of this quantity will be the proper superficies of the iron case of the stove to produce the same effect.

(206.) The actual temperature of the cockle-stove itself can also be estimated without any great difficulty. If the radiating power of the stove is seven times that of hot-water pipes heated to 200° Fahrenheit, we shall find by the experiments on cooling by Petit and Dulong, in Chapter XII. (Art. 227, *et seq.*), and by reducing the figures there given to Fahrenheit's scale, that 4·3 times the temperature of hot-water pipes will give seven times the effect. Therefore, hot-water pipes being (we will assume) at 200° (or 140° above the temperature of the air), the stove will be of the actual temperature of about 560° Fahrenheit to produce seven times the effect of hot-water pipes. As this kind of stove burns with rather a slow draught, it is probable that this statement of the temperature is not far from the truth.

(207.) It must be borne in mind that in this form of apparatus the back and front of the cockle, as well as the sides, can be made available for heating the air, if properly constructed and placed at a distance from any wall, and so arranged as to bring a supply of fresh air to act upon it on all sides. It should also be observed that this estimate of the heating power of the cockle, compared with hot-water pipes, must be taken with this limitation:—That the usual mode of fixing hot-water pipes allows the air already in the room or building to be heated repeatedly by the radiating surfaces; while with the cockle-stove it is always (except in some very peculiar cases not here contemplated) the *external* air which is brought into contact with the stove, to be heated to the required temperature. This will make a very considerable difference in the effect produced in heating any given building (Art. 162), whatever may be the nature of the heating surface.

(208.) Another form of the cockle-stove was invented by the son and successor of Mr. Sylvester,

Fig. 47.

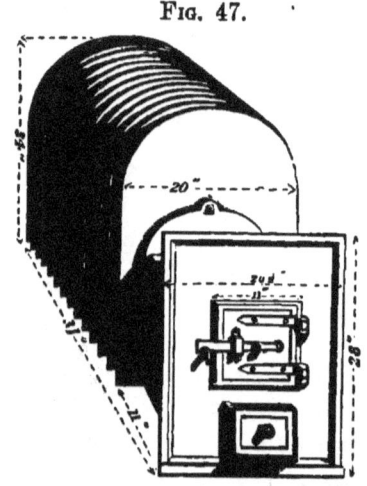

about the year 1830. It was subsequently adopted by Mr. Goldsworthy Gurney, and called by his name. It can be made of almost any size, and usefully employed in many cases. The invention consists of a cast-iron cockle, on the outer side of which are a great number of projecting plates extending to as much as eight or ten, or even twelve inches, beyond the surface of the cockle, which is exposed to the direct action of the fire; and by this means the radiating surfaces are greatly increased in extent, and their temperature proportionally lower. Fig. 47 shows this kind of cockle-stove, which requires to be cased in brickwork, with a very free admission of air to the external projecting plates. The air is admitted below the level of the furnace-door, and, after receiving the required warmth from the heated plates, is conveyed from the upper part of the cockle into channels or hot-air flues, and from thence conveyed into the various places to be heated.

(209.) To estimate the heating power of these stoves, it is only necessary to ascertain the surface of the inner arch, which is exposed to the direct action of the fire. The projecting plates add nothing whatever to the heating power of the stove. They merely conduct the heat of the inner arch, and thereby rapidly lower its actual temperature, and thus to a great extent prevent the deleterious effects which result from the air coming into contact with intensely heated surfaces. The actual heating effect, however, would be the same whether these projecting plates were present or not, provided the air were made to impinge with sufficient force and rapidity against the intensely heated sides of the stove immediately in contact with the fire.

(210.) The temperature of the cockle-stove with its slow combustion and isolated fire-box having been estimated at 560° Fahrenheit (Art. 206), we may well suppose that this stove, with the fire in actual contact with the sides of the inner arch of the stove, the temperature of the arch itself would be not less than 700° Fahrenheit, if the heat were not rapidly conducted away by the projecting plates. This temperature will make the heating effect of this stove ten times that of hot-water pipes; and therefore, if we measure the actual surface of the inner arch of this stove (leav-

Q

ing out the back, which is not used for heating the air), and multiply the number of square feet thus found by *ten*, we shall have the number of square feet of hot-water pipe that will be equal to the surface of this stove.

It is difficult, however, to estimate the real power of this stove, because by making a very fierce fire the effect may be increased so much as to invalidate any ordinary calculation of the power. This is not the case with a hot-water apparatus, because there the limit to the temperature of the heat-distributing surface is absolutely fixed, and can never exceed the heat of boiling water. But with all hot-air stoves, there is no absolute limit of this kind; by skilful or unskilful treatment of the fire, the heating power may be so greatly altered as to render any calculation uncertain.

(211.) Another form of hot-air apparatus was some years since rather extensively used. In the year 1834 a patent was taken out by M. Bernhardt for a peculiar form of hot-air apparatus. The inventor failed to carry out the invention beyond a very limited extent; and in the year 1842 the same plan was again patented, with very little variation, by Mr. Hazard, of Clifton. The general form of this apparatus is shown, fig. 48.

In this apparatus the smoke and products of combustion pass from the furnace directly into a set of cast-iron tubes, generally five or six in number, which tubes form a long coil as shown in the figure. By the peculiar form given to this coil, the various pipes, after passing in a straight line for 5 or 6 or 8 feet, enter into a cast-iron box (shown in the figure), and from thence through another set of pipes, again pass into another box like the first, and so on, passing backward and forward until five or six or more alternations have been made, and the smoke and products of combustion have thus passed through a coil of pipes of the aggregate length of 300 feet, or upwards, and usually about two or three inches diameter.

FIG. 48.

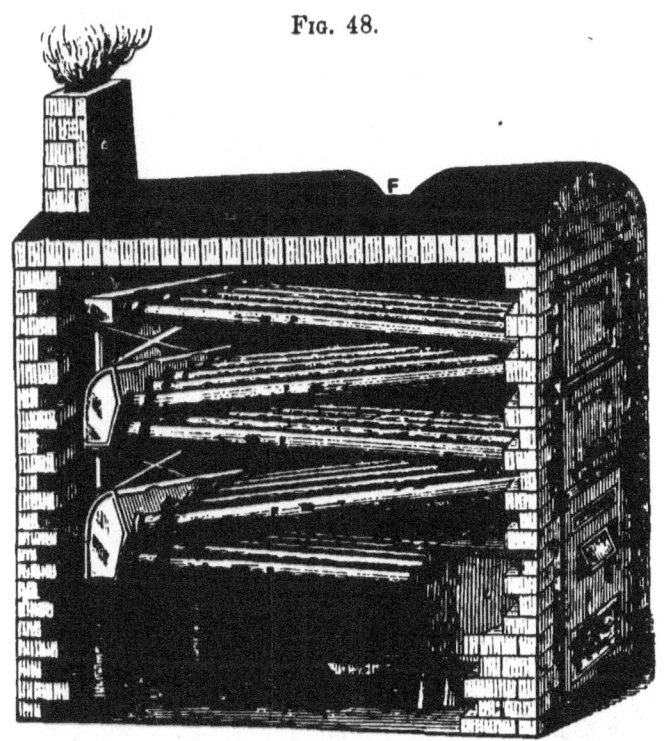

The whole of these pipes are then enclosed in a brick casing, into which cold air is admitted at the bottom, which, after passing upward, and successively coming into contact with the various pipes, finally escapes at the top of the case, highly heated, and is from thence conveyed into any building or room where it is required.

(212.) This plan of heating air is no doubt very economical; and for many manufacturing and

Q 2

drying purposes it would possess very great advantages, if it were not for the extreme difficulty of cleaning the pipes from the soot. The cast-iron boxes which connect the pipes can be made to open at the ends, and possibly (though with much more difficulty) in the front also; but it scarcely appears possible effectually to clean out the pipes themselves, and the constant dirt and extreme annoyance of the efforts to keep the pipes clear necessarily disqualifies this apparatus for general use. Like many other patent inventions, this apparatus could boast of numberless certificates of high approval from parties of unquestioned veracity; but, notwithstanding these testimonials, the use of this form of apparatus appears now to have entirely ceased.

(213.) These several forms of hot-air apparatus are all subject to one general objection to their application. Whenever the furnace and heating-surfaces are placed at a distance from the place to be heated, and are not directly under it, there is great difficulty in conveying the heated air *laterally* through horizontal flues, unless the flues have first given to them a very great *vertical* height. This vertical height of flue gives velocity to the current of heated air; and without this it is most difficult to make the current travel *horizontally* along the flues which are requisite to convey it to the different rooms. Mr. Sylvester, in his description of Mr. Strutt's hot-air cockle, states, that it is necessary to have the cockle and furnace from 20 to 30 feet below the rooms which are to be heated, in order to obtain sufficient velocity of the air before it passes in a *horizontal* direction. It is almost impossible to obtain this depth for the furnace below the rooms which are to be heated in any ordinary buildings; and unless, therefore, the furnace can be fixed immediately under the place to be heated, the hot-air apparatus becomes extremely difficult of application. The only way to overcome the difficulty is to warm the horizontal flues by some extraneous means. When this can be done, the hot air travels readily through the horizontal flues, and when once the current has been established, it continues readily to flow. This is now sometimes done by means of hot-water pipes, which are laid in the horizontal flues; and these pipes can be heated by means of a coil of wrought-iron pipes passing through the furnace of the cockle-stove. The proper size of this subsidiary apparatus for warming the flues may be easily ascertained. In very large flues of from three to four feet square (or from nine to sixteen square feet area) two hot-water pipes of four inches diameter, laid the whole length of the flue, are sufficient; and in smaller flues two pipes of two inches diameter will be large enough for the purpose. The size of the coil to heat this pipe can be ascertained by the rules given (Art. 71) for apportioning the size of boilers; by which it will be seen that one superficial foot of coil placed in the furnace will be sufficient to heat from 40 to 50 feet of 4 inch pipe in the flue, or twice that length of 2 inch pipe.

(214.) Another apparatus for heating air to a moderate extent, and well suited for warming and ventilating purposes, is shown in figs. 49 and 50. It consists of a number of iron pipes of 10 to 15 feet long, and about 2 or 2½ inches diameter, fixed at each end into a tube-plate (fig. 49), and as close together as they can be easily placed. This series of tubes, with its two end-plates, is then built into a furnace in such manner that the flame and heated gases shall act as extensively as possible upon the *outside* of the tubes. The air to be heated enters the tubes above the furnace-door at one end, and passing through the tubes, is delivered at the opposite end into a channel or

Fig. 49.

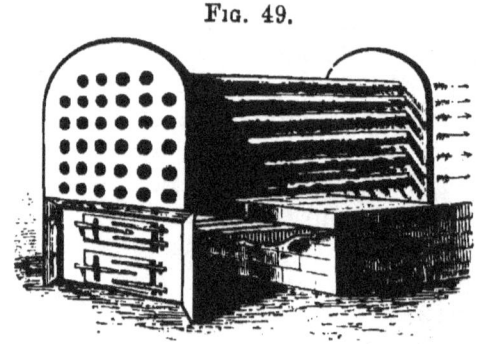

Fig. 50.

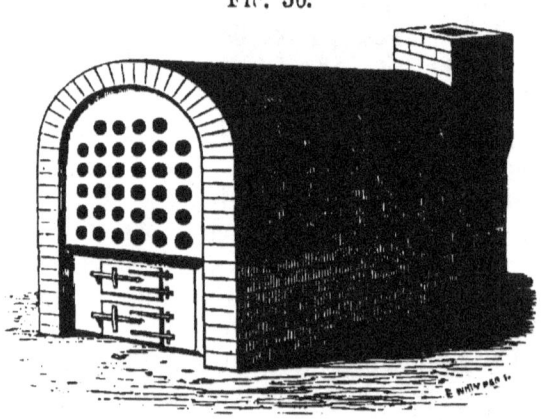

flue, by which it is conducted into the room or building which is required to be heated.*

(215.) The power of this form of apparatus can be estimated by reckoning every square foot of pipe thus exposed to the fire as equal to about seven square feet of hot-water pipes. In point of

---

* This form of apparatus, by no means new, though not much used, has recently (in the year 1873) been made the subject of a Patent by Messrs. Truswell of Sheffield.

heating power, therefore, it will be about equal to the Belper hot-air cockle; but in economy of fuel it will be decidedly inferior. The apparatus will also require to be frequently cleared from soot, which will speedily condense upon the surface of the pipes, and must be removed through proper openings made in the sides and ends of the brick casing in which the tubes are enclosed.

(216.) All these several forms of hot-air apparatus require a very free admission of air into the air-channels of the apparatus. Unless this be attended to in fixing the stoves, the heating power will be greatly curtailed; and therefore a small close furnace-room, or stoke-hole, will obviously be improper for any of these previously described forms of apparatus.

(217.) There is yet another form of hot-air apparatus which it may be desirable to mention, although it is solely applicable to manufacturing and drying purposes, and would be altogether unfit for warming buildings used for habitation. This apparatus has long been known and used in the manufacture of iron, under the name of the "hot-blast apparatus." It is very much like the apparatus described as "Bernhardt and Hazard's Patents," only that instead of the smoke and heated gases passing through the pipes as in that arrangement, in this case it is the *air* which passes through the pipes, and the pipes themselves are heated in a furnace very much after the fashion of a gas retort. The form of the apparatus has undergone many changes, and it has been used in an infinite variety of shapes. It has sometimes been used as a continuously coiled pipe, of rather large diameter; sometimes a number of small pipes branch out from a much larger pipe, and form an arch; and sometimes a form almost similar to Bernhardt's patent has been used.

In all cases the air is forced through the pipes by mechanical means, and the pipes themselves are heated externally in a sort of reverberatory furnace. The pipes thus become red hot, and the air in passing through them is heated to any required temperature, varying from 300° to 600° Fahrenheit.* For manufacturing purposes, this form of apparatus may sometimes be extremely useful, where motive power by a fan or otherwise can be obtained; and enormous quantities of air can be heated by this means. The apparatus is subject to a considerable amount of wear and tear, in consequence of the great heat of the furnace; but by proper arrangements of the parts this may be considerably modified.

(218) The proportions of this form of apparatus, to suit it for any of the ordinary purposes to which it may be applied, are not easily deduced from the experiments published respecting its application to iron manufacture. The velocity of the air passing through the pipes when the apparatus is used in the blast-furnaces, is from 350 to 400 miles per hour; the volume of heated air from 2,000 to 3,000 cubic feet per minute; and the heated surfaces of the pipes from 1,200 to 1,800 square feet. These quantities are so enormous, and the alteration of any one of them so completely invalidates any calculation founded upon their joint effect, that no useful data can be drawn from them to apply the apparatus to any other purpose. Nevertheless, it is clear that where a motive-power for forcing air through the heated pipes can be employed, a very useful apparatus in this form might be obtained, which would be applicable for drying purposes, as the

* See Dufrenoy's " Report on the Use of Hot Air in the Iron Works of England ; " and also a Translation, published by Murray. London, 1836.

force and temperature of the blast may be regulated to almost any extent.

(219.) The best mode of producing a blast of air for the purpose here described is by the fan described (Art. 363, *et seq*.). Where steam-power is used, this application will present no difficulty; but unless some mechanical means of forcing the air through the tubes with considerable velocity can be employed, this form of apparatus would be utterly unsuitable. When used in this way, there is no apparatus that can compare with it in power, wherever large volumes of intensely heated air are required for desiccating purposes. For drying timber in very large quantities; for the rapid evaporation of moisture in many processes; for curing-stoves for animal substances; and for many other purposes of manufacturing economy—this mode of heating and drying presents advantages of no inconsiderable extent, though, up to the present time, it has not been used in this manner.

The deterioration of the air by passing over highly-heated metallic surfaces must not be lost sight of when considering any of the plans for hot-air apparatus. This part of the subject is fully treated on in Chapter III., Part II., of this work, to which the reader is referred.

CHAPTER XII.

ON THE LAWS AND PHENOMENA OF HEAT.

Radiation and Conduction—General Law of Cooling Bodies in Air and other Gases—Different Law for Conduction and for Radiation—Effect by the Incidence of the Rays—Effects of Surface on Radiation—Effects of Colour—Effects of Roughness—Absorptive Power of Bodies for Heat—Conducting Power of Metals—Conducting Power of Wood—Conducting Power of Liquids—Cooling Influence of Water and Air—Reflective Powers of Bodies—Specific Heat of Bodies—Latent Heat—Spontaneous Evaporation—Heat and Cold by Condensation and Rarefaction of the Air—Motion of Liquids influenced by Heat—Effect of Heat on Strength of Materials.

(220.) However various are the methods by which artificial heat is distributed in the warming of buildings, they are all reducible to certain rules, which constitute the primary laws of heat. These laws are very numerous, and some of them extremely complicated; but they possess a very high degree of philosophic interest. They are far too extensive, however, to allow in this work even a bare outline of all the various phenomena to be given, which this branch of science exhibits. But in the present chapter such of the laws of heat as relate to the subject more immediately before us shall be stated, in order to afford a more ready and convenient reference for those who wish to study the scientific principles of warming buildings by artificial heat.

(221.) There are four distinct properties of heat, which all bodies possess in a greater or less degree, which we shall first consider. These are, *radiation*, *absorption*, *conduction*, and *reflection*. There are also others, which will be subsequently mentioned, such as the specific heat of bodies, their change of state, and other subjects connected with their chemical constitution.

(222.) Heated bodies give off their caloric by two distinct modes—radiation and conduction. These are governed by different laws: but the rate of cooling by both modes increases considerably in proportion as the heated body is of a greater temperature above the surrounding medium. This variation was long supposed to be exactly proportional to the simple ratio of the excess of heat; that is to say, supposing any quantity of heat given off in a certain time at a specified difference of temperature, at double that difference twice the quantity of heat would be given off in the same time. This law was originally proposed by Newton, in the *Principia*, and, although rejected as erroneous by some philosophers, it was followed by Richmann, Kraft, Dalton, Leslie, and many others; and was usually considered to be nearly accurate, until the masterly and elaborate experiments of MM. Petit and Dulong proved that, although approximately correct for low temperatures, it becomes extremely inaccurate at the higher degrees of heat.

(223.) The cooling of a heated body, under ordinary circumstances, is evidently by the combined effects of radiation and conduction. The conductive power of the air is principally owing to the extreme mobility of its particles, for otherwise it is one of the worst conductors we are acquainted with; so that when confined in such a manner as to prevent its freedom of motion, it is a most useful non-conductor.

(224.) The proportions which radiation and conduction bear to each other have, in general, been very erroneously estimated. Count Rumford considered the united effect, compared with radiation alone, was as five to three; and Franklin supposed it to be as five to two. Dr. Murray also considered a certain relation existed between radiating and conducting powers.

No such general law, however, can be deduced; for the relative proportions vary with the temperature, and with the peculiar substance or surface of the heated body. For while *the cooling effect of the air by conduction is the same on all substances, and in all states of the surface of those substances*, radiation varies materially, according to the nature of the surface.

(225.) The elaborate experiments of Petit and Dulong have placed on record a vast amount of most valuable information on this subject;* and by their researches some of the most important of the laws of heat have been deduced. The following abstract will give such of their deductions as are most applicable to the subject under inquiry.

(226.) The influence of the air by its power of conduction varies with its elasticity or barometric pressure. The greater the elastic force, the greater also is its cooling power, according to the following law: *When the elasticity of the air varies in a geometrical progression whose ratio is 2, its cooling power changes likewise in a geometrical progression whose ratio is 1·366.*

The same law holds with all gases, as well as with atmospheric air; but the ratio of the progression varies for each gas.

---

* *Ann. de Chimie*, vol. vii., p. 113, *et seq.*; and *Annals of Philosophy*, vol. xiii., p. 112, *et seq.*

(227.) To show the relative velocities of cooling at different temperatures, the following Table, constructed from the experiments of Petit and Dulong, is given. The first column shows the excess of temperature* of the heated body above the surrounding air; † the second column shows

TABLE VI.

Excess of Temperature of the Thermometer above that of the Air; Centigrade Scale.	Total Velocity of Cooling of the naked Bulb.	Total Velocity of Cooling of Bulb covered with Silver leaf.	Amount of Cooling due to Conduction of the Air alone.
260°	24·42	10·96	8·10
240	21·12	9·82	7·41
220	17·92	8·59	6·61
200	15·30	7·57	5·92
180	13·04	6·57	5·19
160	10·70	5·59	4·50
140	8·75	4·61	3·73
120	6·82	3·80	3·11
100	5·57	3·06	2·53
80	4·15	2·32	1·93
60	2·86	1·60	1·33
40	1·74	·96	·80
20	·77	·42	·34
10	·37	·19	·14

* The temperatures *in all these experiments* of Petit and Dulong are expressed in degrees of the Centigrade thermometer. As the zero of this thermometer is the freezing-point of water, and from that to the boiling point of the same fluid is 100°, in order to find the number of degrees of *Fahrenheit's* scale, which answers to any given temperature of the *Centigrade*, multiply the number of degrees of *Centigrade* by nine and divide the product by five; add 32 to the quotient thus obtained, and this sum will be the number of degrees of *Fahrenheit* required. As, however, in the above Table, the temperatures given are only the *excess*, and not the absolute temperatures, the 32° to be added by this rule must be omitted.

† In these experiments, the temperature of the air was at 0° Centigrade, therefore these temperatures are both the excess and also the absolute temperatures.

the rate of cooling of a thermometer with a plain bulb; and the third column gives the rate of cooling when the bulb was covered with silver-leaf. The fourth column shows the amount due to the cooling of the air *alone*; and by deducting this from the second and third columns respectively, we shall find what is the amount of *radiation* under the two *different states of surface,* noticed at the top of the second and third columns.

(228.) Some very remarkable effects may be perceived by an·inspection of the above Table. It appears that the ratio of heat lost by contact of the air alone is constant at all temperatures; that is, whatever is the ratio between 40° and 80°, for instance, is also the ratio between 80° and 160°, or between 100° and 200°. This law is expressed by the formula—

$$v = n \cdot t^{\,1\cdot233}$$

where t represents the excess of temperature, and n a number which varies with the size of the heated body. In the case represented in the foregoing Table, $n = 0\cdot00857$.

(229.) Another remarkable law is, that the *cooling effect of the air is the same for the like excess of heat on all bodies without regard to the particular state or nature of their surface.* This was ascertained by Petit and Dulong, in a series of experiments not necessary here to detail, but which abundantly prove the accuracy of the deduction.*

(230.) By comparing the second and third columns in the preceding Table, it will be immediately perceived that the loss of heat by *radiation* (deducting the cooling by conduction of the air, given in the fourth column) varies greatly with the nature of the radiating surface; though, whatever be the nature of the surface, *the loss of heat follows the same law in all cases, though in a different ratio.*

* *Annals of Philosophy,* vol. xiii.

It should be observed that in this Table the second, third, and fourth columns show the number of degrees of heat which were lost per minute by the body which was the subject of experiment; and, therefore, these numbers represent the *velocity* of cooling.

When the numbers in the last column are deducted from those in the second and third columns, the difference will show the loss of heat by *radiation* for the plain and silvered bulb respectively; the fourth column being the loss by conduction of the air, which is the same for all surfaces. It will immediately be perceived, therefore, that the loss of heat by conduction and by radiation bear no constant ratio to each other. But, while *conduction proceeds by a regular geometrical progression*, radiation follows another law, viz., *when a body cools in vacuo, surrounded by a medium whose temperature is constant, the velocity of cooling, for excess of temperature in arithmetical progression, increases as the terms of a geometrical progression, diminished by a constant quantity.* This law is represented by the formula—

$$V = m. \, a^{\theta} \, (a^{t} - 1)$$

where a is the constant quantity for all bodies $= 1 \cdot 0077$; t the excess of temperature of the radiating body: θ the temperature of the surrounding medium ; and m a co-efficient, which varies with the size and nature of the radiating body, to be determined for each particular case. It will likewise appear that, when we compare the *total cooling* of two different surfaces, the law is more rapid at low temperatures, and less rapid at high temperatures, for the body which *radiates the least*, in comparison with that which radiates with greater power.

(231.) But the cooling of a body by conduction

of the air differs from the effect of radiation in a
remarkable manner, in this particular—that while
*the ratio of loss by conduction continues the same
for the same excess of temperature, whatever be the
absolute temperatures of the air and heated body;
radiation increases in velocity, for like excess of temperature, when the absolute temperatures of the air and
heated body increase.* The following Table shows
the law of *cooling by radiation* for the same body at
different temperatures :—

TABLE VII.

Excess of Temperature of the Thermometer.	Velocity of Cooling when the surrounding Medium is at the undermentioned Temperatures.			
	0°	20°	40°	60°
220°	8·81	10·41	11·98	—
200	7·40	8·58	10·01	11·64
180	6·10	7·04	8·20	9·55
160	4·89	5·67	6·61	7·68
140	3·88	4·57	5·32	6·14
120	3·02	3·56	4·15	4.84
100	2·30	2·74	3·16	3·68

It will be observed in this Table that, when
the absolute temperatures of the surrounding
medium and radiating body are increased 20° of
Centigrade, *the difference between their temperatures
continuing the same,* the velocity of cooling is
multiplied by 1·165, which is the mean of all the
ratios in the above Table, experimentally determined.

(232.) The total cooling of a body by radiation
and conduction, then, we shall find to be represented, under all circumstances, by this formula—

$$m.\ a^\theta\ (a^t - 1) + n.\ t^b$$

The quantities a and b are constant for all bodies,
and under all circumstances; the first being =
1·0077 and the latter = 1·233. The co-efficient

m will depend on the size and nature of the heated surface, as well as upon the nature of the surrounding medium. The co-efficient n is independent of the absolute temperature, as well as of the nature of the surface of the body; but will vary with the elasticity and nature of the gas in which the body is plunged: t is the excess of temperature of the heated body, and θ the temperature of the surrounding medium.

(233.) The fact, already adverted to, that the ratio of cooling of those bodies that radiate least is more rapid at low temperatures, and less rapid at high temperatures, than those bodies that radiate most, is perhaps one of the most remarkable of the laws of cooling. It was first deduced experimentally by Petit and Dulong, and it may be mathematically proved from their formulæ.* It appears, however, that when the total cooling of two bodies is compared, the law is more rapid at low temperatures for the body which radiates least, and less rapid for the same body at high temperatures; though separately, for conduction and for radiation, the law of cooling is, for the former, irrespective of the nature of the body, and for the latter, that all bodies preserve, at every difference of temperature, a constant ratio in their radiating power.

(234.) To revert to the first Table in this chapter. We find the total cooling at 60° and 120° (of Centigrade) to be about as 3 to 7; at 60° and 180°, as 3 to 13; and at 60° and 240°, as 3 to 21: whereas, according to the old theory of Newton, they should have been respectively as 3 to 6; as 3 to 9; and as 3 to 12. But we find that the deviation increases greatly with the increase of temperature, and that when the *excess* of temperature of the heated body above the

* *Annals of Philosophy*, vol. xiii., p. 335.

R

surrounding air is as high as 240° of Centigrade (432° of Fahrenheit), the real velocity of cooling is nearly double what it would appear to be by the old and imperfect theory, varying, however, with the nature of the surface.

(235.) But radiant heat is subject to other laws besides those we have yet considered. Rays of heat diverge in straight lines from every part of a heated surface, and likewise from extremely minute depths below the surface of hot bodies, being subject to the laws of refraction, the same as light. The intensity of these rays *decreases as the square of the distance,* and the emission of the rays is greatest in a line perpendicular to the surface. The same law obtains here, also, as with light—that the effect of the ray is *as the sine of the angle* which it forms with the surface from which it emanates.* This *law of the sines,* first discovered experimentally by Leslie, suggests a practical caution connected with the subject before us, namely, that the shape of the pipes used to warm a building is not wholly unimportant; for, if flat pipes be used, and they be laid horizontally, the major part of the *radiated* heat from the upper surface will be received on the ceiling, and therefore will produce but little beneficial effect. The loss sustained in this way will be greater in proportion to the higher temperature of the pipes; for it will be seen by the Table at the beginning of this chapter, that the relative proportion which radiation bears to conduction increases with the temperature: at the ordinary temperature of hot-

* This law is thus stated by Fourier: "The rays of heat which issue under different angles from the same point of the surface of any body, have an intensity which decreases proportionally to the sine of the angle formed by their direction with the plane tangential to the surface at the point of emission."—*London and Edinburgh Philosophical Magazine,* vol. ii., p. 104; also, *Report, British Scientific Association,* vol. iv. (1835), p. 22.

water pipes, about one-fourth the total cooling is due to radiation.

(236.) The radiation of heat, we have already seen, is greatly either increased or diminished according to the nature of the surface of the radiating body. Professor Leslie has given the following as the relative powers of radiation by different substances :—*

TABLE VIII.

Lamp-black . . . 100	Tarnished Lead . . 45
Water (by estimate) . 100	Thin Film of Jelly (one
Writing Paper . . 98	quarter of former) . 38
Resin 96	Tin scratched with Sand-
Sealing-wax . . 95	paper . . . 22
Crown Glass . . . 90	Mercury . . . 20
China Ink . . . 88	Clean Lead . . . 19
Ice 85	Iron, polished . . 15
Red Lead . . . 80	Tin Plate . . . 12
Isinglass . . . 80	Gold, Silver, and Copper 12
Plumbago . . . 75	Thin Laminæ of Gold,
Thick Film of Oil . 59	Silver, or Copper Leaf,
Film of Jelly . . 54	on Glass . . . 12
Thinner Film of Oil . 51	

(237.) It is very generally supposed that *colour* has a considerable influence on radiant heat, and also upon the absorption of heat, the two effects being similar and equal. Sir Humphrey Davy, by exposing surfaces of various colours to the heat of the sun, proved experimentally† that the absorbing power of different colours was in this order : black, blue, green, red, yellow, and white; black being the best, and white the worst absorbent. In this order, then, we should expect to find the radiating powers of different colours, and that by painting a body with a dark colour we should increase its power of radiation. This,

* Leslie, On Heat, pp. 81-110.
† Beddoe's *Contributions*, p. 44; also, Dr. Stark's " Re-searches on Heat," *Philosophical Transactions*, 1834; and *Repertory of Arts*, 1834, pp. 257, 312.

however, is not the case ; and there are the strongest
reasons for supposing that the absorption and
radiation of *simple heat*—that is, heat without light,
or heat from bodies below luminosity—are wholly
irrespective of colour, and depend upon the nature
of the surface.

(238.) By comparing the results given in the
above Table, it will appear that the radiation of
heat bears no relation to colour, when the radiating
body is below the temperature of boiling water.
By the Table it appears that lamp-black and white
paper are nearly equal in power; while Indian ink
is much less, and black-lead still lower in the scale ;
though, as far as colour only is concerned, these
last are nearly the same as lamp-black.   Professor
Powell considers, as also did Leslie, that *softness*
may probably tend to increase the radiation of *simple*
heat ;* and the former found that a thermometer-
bulb coated with a paste of chalk was affected (*by
this kind of heat*—that is, heat below luminosity)
even more than a similar one coated with Indian
ink ; but the same result does not occur with
luminous hot bodies.†   Professor Bache has likewise made an extensive series of experiments on
this subject, which confirm this result.‡   The
experiments of Leslie proved that radiation proceeds
not only from the surface of bodies, but also from
small depths below the surface ; and therefore the
thickness of coating of any good radiating substance
materially affects the results, as may be observed by

---

* It is necessary to distinguish particularly between *simple
heat* from bodies of a limited temperature and that which is
given off from *luminous* hot bodies. From these latter, the
experiments of Nobili and Melloni prove the existence of two
distinct kinds of heating rays given off at the same time from
the same body.

† Professor Powell's " Report on Heat," *British Scientific
Association*, vol. i., p. 279.

‡ *Ibid.*, vol. ix. (1840), p. 18.

the above Table.* The thickness which produces the greatest effect, however, probably varies with different substances; and it is therefore necessary to separate this effect from anything merely resulting from the colour of the heated body. Professor Powell, after an elaborate examination of all the phenomena attending the heat received from the sun, is of opinion that there is no *simple radiant heat* received by us from the sun's rays; and that the simple radiant heat, which no doubt is initially radiated from the sun, is absorbed by the atmosphere of that luminary, some small portion, perhaps, which escapes being stopped in the higher regions of our own atmosphere.† The experiment of Sir H. Davy, on the absorption and radiation of *solar* heat by different colours, is therefore not applicable to the case of *simple heat,* or such heat as is given out by bodies below luminosity. And in conformity with this view is the experiment of Scheele, in which he found that if two thermometers filled with alcohol, one red, and the other colourless, were exposed to the sun's rays, the coloured one would rise in temperature much more rapidly than the other; but if they were both plunged into the same vessel of hot water, they rose equally in equal times.

(239.) We are fully justified, then, from these and other analogous experiments, in drawing the conclusion that *the radiation of* SIMPLE HEAT *is not influenced by the colour of the heated body.* Any difference which appears to obtain in this respect is, therefore, solely referable to the *nature* of the colouring substance.

(240.) The effect of roughness of the surface was also investigated by Sir J. Leslie; and he found that either tarnished surfaces, or such as

* Leslie, On Heat, pp. 106–110.
† *Report of the British Scientific Association,* vol. i., p. 290.

are roughened by emery, by the file, or by drawing streaks or lines with a graving tool, always had their power of radiation considerably increased.* The accuracy of this deduction had not been questioned until some recent experiments of M. Melloni, by which it has been ascertained that this increased effect from roughened surfaces is not a general law, but is only a particular result, for which another explanation must be sought. M. Melloni experimented with four plates of silver, two of which, when cast, were left in their natural state, without hammering, and the other two were planished to a high degree under the hammer. All the plates were then finely polished with pumice-stone and charcoal ; and after this one of each of the pairs of plates was roughened by rubbing with coarse emery paper in one direction. The quantity of heat radiated from these plates was as follows :—

Hammered and polished plate . . .	10°
„ and roughened „ . .	18
Cast and polished plate . . .	13·7
„ and roughened „ . . .	11·3

In comparing these effects, it appears that the hard hammered plate increased in radiating power four-fifths by roughening its surface; while the soft cast plate lost nearly one-fifth of its power by the same process. M. Melloni, therefore, draws the conclusion that the roughness of the surface merely acts by altering the superficial density, and that this will vary according as the body is of a greater or less density previous to the alteration of its surface by roughening.†

* Leslie, On Heat, p. 81, &c.
† Melloni, "On Emissive Power of Bodies," &c., *Comptes Rendus de l'Académie des Sciences*, and *Edinburgh Philosophical Journal*, 1838.

(241.) It was deduced from experiments by Leslie that the *absorptive* power of bodies for heat was very nearly proportional to the *radiative* power ;[*] and Dr. Ritchie has subsequently proved that these effects are precisely equal to each other.[†]

(242.) The velocity with which heat enters into and quits any body is supposed to be equal, though this velocity is different for each different body. On this property of bodies with regard to heat, many of the experiments have been founded which constitute the laws of heat. It has also been established, by the experiments of MM. Melloni and Nobili, that *the radiating powers of surfaces, for simple heat, are in the inverse order of their conducting powers.* It follows, therefore, that neither the radiating powers nor the conducting powers of bodies will discover their actual rate of cooling comparatively with any other body.

(243.) We might be led to conclude, from all that precedes, that those metals which are the worst conductors would be the most proper for vessels or pipes for radiating heat; because we find that the heat lost by contact of the air is the same for all bodies, while those which *radiate most,* or are the worst conductors, give out more heat in the same time than those bodies which *radiate least,* or are good conductors. Such would be the case if the vessels were *infinitely* thin; but, as this is not possible, the slow conducting power of the metal opposes an insuperable obstacle to the rapid cooling of any liquid contained within it, by preventing the exterior surface from reaching so high a temperature as would that of a more perfectly conducting metal, under similar circumstances ; thus preventing the loss of heat, both by contact of the air and by radiation, the

---

[*] Leslie, On Heat, &c., pp. 19–98.
[†] *Journal of the Royal Institution,* vol. v., p. 305.

effect of both being proportional to the excess of heat of the *exterior* surface of the heated body. If a leaden vessel were *infinitely* thin, the liquid contained in it would cool sooner than in a similar vessel of copper, brass, or iron: but the greater the thickness of the metal, the more apparent becomes the deviation from this rule; and, as the vessels for containing water must always have some considerable thickness, those metals which are the worst conductors will oppose the greatest resistance to the cooling of the contained liquid, although apparently in opposition to the result of the preceding experiments.

It is difficult on these grounds to account for the effect which lead-paint has in preventing the free radiation of caloric from bodies coated with it; because, in this case, the lead must be extremely thin, and ought, therefore, to increase the amount of radiation. The effect probably arises from the total change of state which the lead undergoes by its chemical combination with the carbonic acid in the process of making it into white lead. Practically, it is found to have an injurious tendency on the free radiation of heat from most bodies, varying, however, with their radiating powers. On a good radiator, its effect is the most injurious; on a bad one, less so: but its use should be avoided as much as possible in all cases where the free radiation of heat is the object in view.

(244.) Various experiments have been made by Richmann, Ingenhausz, and Dr. Ure, to ascertain the conducting power of metals. Dr. Ure's results, which differ but little from the others, place the metals in the following order as regards their conducting power, namely—silver he found by far the best conductor; next copper; and then brass, tin, and wrought iron, nearly equal; then cast iron and zinc; and lead he found by far the worst of all.*

* Ure's Dictionary of Chemistry, Art. "Caloric."

(245.) The only accurate experiments, however, which have been made on this subject are those by M. Despretz,* which give the following results :—

TABLE IX.

Gold	.	.	. 100·0	Tin	.	.	.	30·39
Platina	.	.	. 98·10	Lead	.	.	.	17·96
Silver	.	.	. 97·30	Marble	.	.	.	2·36
Copper	.	.	. 89·82	Porcelain	.	.	1·22
Iron	.	.	. 37·43	Fire-brick	.	.	1·14
Zinc	.	.	. 36·30				

This Table gives a very useful practical enunciation of the value of different substances as conductors of heat. But to ascertain the absolute conducting power of the various substances is extremely difficult; the preceding Table obviously only shows their relative conducting powers. Experiments, however, have been made on the absolute conducting powers of some substances, which are of considerable practical value, although they leave much yet to be desired.

(246.) M. Biot ascertained the conducting power of a bar of iron, by plunging one end of it into a bowl of mercury heated to $102\frac{1}{2}°$ Centigrade ($216°$ Fahrenheit), and ranging along the bar eight thermometers at various distances from each other. The observations were made after the temperature became permanent; the air during the experiment was $16\frac{1}{4}°$ Centigrade ($61°$ Fahrenheit), and the results were as follow :—

* Despretz, *Traité de Physique*, p. 201; and *Quarterly Journal of Science*, vol. xxv., p. 220. Professor Daniell (*Chemical Philosophy*, p. 107) places platina much lower in the scale than M. Despretz, namely, 38·1. See also experiments by Calvert, *Proceedings of Royal Society*, vol. ix., p. 170 (for March, 1858), and Quetelet's *Natural Philosophy* (by Wallace), p. 144.

TABLE X.

No. of Thermometer.	Distance from Mercury, Decimeters.	Excess of Temperature above the Air, Centigrade.	Distance from Mercury, Inches.	Excess of Temperature above the Air, Fahrenheit.
0	0	86·25°	0	155°
1	2·115	29·375	8·326	53
2	3·115	17·5	12·263	31½
3	4·009	11·25	15·783	20
4	4·970	7·1875	19·566	13
5	5·902	4·6875	23·236	8
6	7·777	2·1875	30·618	4
7	9·671	1·25	38·074	2
8	11·556	Insensible	45·495	Insensible

In this Table, the first column gives the numbers of the thermometers in their regular order; the second column gives the distance of each thermometer from the source of heat, viz., the bowl of mercury; and the third column gives the excess of temperature of the thermometers above that of the atmosphere, measured by the Centigrade scale. The fourth and fifth columns are the same as the second and third, only the measures are given according to the English scales, instead of the French.

(247.) A similar experiment,* in which the source of heat was melted lead, is given in the Table XI. The temperature of the air was 18·125° Centigrade (64½° Fahrenheit), and that of the iron bar was not taken until it had been exposed to the heat of the lead for several hours, in order to ensure permanence of temperature.

---

* For these experiments, see Biot's *Traité de Physique*, tome iv., p. 670, *et seq*. Also, *Reports, British Scientific Association*, vol. x. (1841), p. 15, *et seq*.

TABLE XI.

No. of Thermometer.	Distance from Extremity, Decimeters.	Excess of Temperature above the Air, Centigrade.	Distance from Extremity, Inches.	Excess of Temperature above the Air, Fahrenheit.
1	2·230	76·875°	8·78	138·37°
2	3·230	47·187	12·71	84·93
3	4·120	29·375	16·22	52·87
4	5·081	17·812	20·00	32·06
5	6·028	10·625	23·73	19·12
6	7·899	3·750	31·09	6·75
7 -	9·783	1·562	38·51	2·73

(248.) A third experiment, also by M. Biot, was made with a bar of copper, plunged at one extremity into melted lead. It had fourteen thermometers ranged along it, of which, however, only eleven were available. The unit of distance was 101 millimeters,* and the temperature of the air was 15·75° Centigrade (60½° Fahrenheit).

TABLE XII.

No. of Thermometer.	Distance from Extremity, Decimeters.	Excess of Temperature above the Air, Centigrade.	Distance from Extremity, Inches.	Excess of Temperature above the Air, Fahrenheit.
4	5·25	80·50°	20·87	144·9°
5	6·25	65·75	24·85	118·3
6	7·25	53·75	28·82	96·7
7	8·25	43·75	32·80	78·7
8	9·25	35·50	36·77	63·9
9	11·25	24·00	44·73	43·2
10	13·25	15·70	52·68	28·2
11	15·25	11·00	60·63	19·8
12	17·25	7·50	68·58	13·5
13	19·25	5·25	76·53	9·4
14	21·25	3·75	84·49	6·7

* The millimeter is ·03937 of an inch English measure, and therefore the 101 millimeters are equal to 3·97637 inches.

(249.) Some experiments by M. Despretz,* similar to the preceding, and extending also to other substances, are contained in the following Table. In these experiments the distance between each of the consecutive thermometers was 10 centimeters (3·937 inches English), and the temperatures are given by the Centigrade scale.

TABLE XIII.

Excess of Temperature above the Air, Centigrade Scale.	No. of the Thermometers.						Temperature of the Air, Centigrade.
	1	2	3	4	5	6	
Copper .	66·36	46·28	32·62	24·32	18·63	16·18	17·08
Iron .	62·9	36·69	20·52	12·32	8·19	6·61	17·34
Pewter .	63·41	35·17	21·52	15·52	—	—	17·34
Zinc .	64·17	38·02	25·43	17·93	—	—	5·62
Lead .	65·13	29·42	14·93	9·99	—	—	17·12
Marble .	63·91	6·08	1·95	1·47	—	—	17·15

In all these experiments the substances were exposed to the cooling influence of the air. In Mr. Kellard's *Report on the Laws of Conduction of Heat*, made to the British Scientific Association (vol. x.), the mathematical formulæ of MM. Fourier, Libri, Poisson, and others, are given at considerable length, for estimating the conducting powers for heat; but they are not suitable for insertion here.

(250.) Experiments on the same plan as the preceding were made by MM. Delarive and Condolle on the conducting power of wood: the results are given in the following Table.† They show the difference in conducting power, accord-

---

* *Traité de Physique*, par M. Despretz.

† *Ann. de Chimie*, vol. xl., p. 91; and *Quarterly Journal of Science*, vol. xxvii., p. 188.

ing to the direction of the fibre. The bars of wood were about five inches long, one and a half inch broad, and one inch thick. The first thermometer was three centimeters (one and one-eighth inch English) from the end; and the others were two centimeters (three-quarters of an inch English) apart from each other. The temperatures given are the excess above the atmosphere, at the respective distances from the heated end.

TABLE XIV.

Names of the Woods.		No. of Thermometers, Centigrade Scale.				
		1	2	3	4	5
Walnut	Longitudinal direction of the fibre.	80·13	43·0	19·63	9·19	5·13
Oak		81·7	41·2	17·5	7·2	3·7
Fir		84·0	39·25	20·6	8·5	3·7
Poplar		79·8	34·2	14·2	6·2	2·8
Walnut	Transverse direction of the fibre.	99·5	37·43	13·9	6·0	3·25
Oak		79·3	22·75	7·5	3·6	2·4
Fir		70·9	13·8	4·5	2·5	1·9
Poplar		78·5	13·75	3·44	1·56	1·0

(251.) Fluids, both in the liquid and aeriform state, are bad conductors of heat, unless they have perfect freedom of motion, in which case they become far better conductors than solids on account of the extreme mobility of their particles.* Water, under these circumstances, is a far better conductor of heat than any of the metals (except mercury); and air, particularly when charged with moisture, is also an excellent conductor, under like

---

* In aeriform fluids, radiant heat is transmitted with extreme rapidity without any visible motion of the particles; but this is a very different case to that stated in the text.

circumstances. Count Rumford made some experiments to ascertain the cooling power of different fluids, based upon the principle that their conducting power and absorbing power are equal.* The mode of operating was to enclose a thermometer in a glass balloon filled with the particular substance of which the conducting power was to be ascertained; and to place it first in freezing water, and then plunge it into boiling water. The respective times required to raise the thermometer from 0° to 70° Réaumur were supposed to indicate the conducting powers of the different substances, and would therefore be their cooling powers upon any heated body placed in them. The following Table shows the results of these experiments, the times being given in minutes and seconds.

TABLE XV.

Substances.	Times of Cooling.	Ratio of Conducting Power.
Mercury	0′·36″	1000
Moist Air	1 ·51	330
Water	1 ·57	313
Common Air . . .	7 ·36	80·41
Rarefied Air, density ¼ .	7 ·37	80·23
Do. do. $\frac{1}{24}$	7 ·51	78
Torricellian Vacuum .	10 ·53	55

These experiments, however, are by no means conclusive; and there is every reason to believe that, by varying the method of the experiment, a vast difference would be found in the results. The glass balloon which surrounded the thermometer was only one inch and a half diameter, the thermometer itself being half an inch diameter.

* Rumford's *Essays*, vol. ii., p. 425.

This space was far too small to allow a free and rapid motion among the particles of the various media which were the subject of experiment; and therefore the conducting powers of the liquids, which have less facility of motion among their particles than the aeriform fluids, appear far less than they really are. Dr. Osborne ascertained the refrigerating power of water compared with air of the same temperature to be as 14 to 1,[*] while Rumford's experiments show it to be less than four to one: but there is reason to believe that, when due provision is made for allowing perfect mobility among the particles of water, the cooling power of water will be very much greater than the amount even which is stated by Dr. Osborne ;[†] and Leslie states that water at the boiling-point conducts heat five times more rapidly than the same fluid when near the freezing-point. [‡]

(252.) The power of *reflection* in all bodies is

---

* *Reports of the British Scientific Association,* vol. iv. (1835), p. 96.

† Some experiments of the author led him to conclude that the conducting power of boiling water, compared with air, was fully double the amount stated by Dr. Osborne ; and these agreed so well with the experiments of Mr. Parkes, already quoted (Art. 73, note), that the author was induced to adopt the proportion of 28 to 1 as the cooling power of water and air, in an extensive and peculiar apparatus which required the cooling powers of these media to be properly adjusted. The result has proved that the cooling power of water is fully equal to this estimate ; but it is probable that the relative conducting power varies with the temperature of the heated body. For unless the temperature be sufficiently high to give free motion to the particles of water, the cooling power will be reduced ; while the same difference can scarcely occur in the air, in consequence of the extreme mobility of the particles. The diminished adhesion of liquids by increased temperature has been experimentally determined by Dr. Ure. (See Art. 264.)

‡ Leslie, Heat and Moisture, p. 17.

inversely as their radiating power, as was experimentally determined by Leslie.* The following Table shows the reflective power of different substances.

TABLE XVI.

Brass	.	.	.	100	Lead		60
Silver	.	.	.	90	Tinfoil softened by Mer-		
Tinfoil	.	.	.	85	cury		10
Block Tin	.	.	.	80	Glass		10
Steel	.	.	.	70	Glass coated with Wax		5

(253.) The *specific heat* of different substances is a subject of considerable importance, as connected with the heating of buildings; for on the theory of specific heat are based many of the calculations for ascertaining the proportions of the various apparatus employed.

(254.) Every substance contains a certain distinctive quantity of heat, called the specific heat of that particular body, which can be ascertained by mixing together (with proper precautions) known quantities of different substances. Thus, if a certain weight of quicksilver, one pound for instance, of the temperature of 40°, be mixed with the same weight of water, of the temperature of 156° the resulting temperature will be 152·3°; so that the water will lose 3·7° of temperature, and the quicksilver will be raised 112·3°. Or, if a pound of water at 100° be mixed with a pound of oil at 50° the resulting temperature will be 83·5°, and not 75°, which would be the mean temperature of the two. Thus it appears that the same quantity of heat that will raise a pound of water 1° will raise the temperature of a pound of oil 2°, or a pound of mercury 23°; and so of other substances, which all possess

* Leslie, On Heat, pp. 20 and 98.

a capacity for caloric, each peculiar to itself. The following Table shows the specific heat of some of the principal substances, which have been ascertained, chiefly from the experiments of Berard and Delaroche, and Petit and Dulong. They are all referred to water, as the standard, and are supposed to be the quantity of heat contained in equal weights of the several substances.*

Table XVII.

Water	. . . 1·0000	Oxygen	. . . 0·2361
Aqueous Vapour	. 0·8470	Carbonic Acid	. 0·2210
Alcohol	. . . 0·7000	Carbonic Oxide	. 0·2884
Ether	. . . 0·6600	Charcoal	. . . 0·2631
Oil	. . . 0·5200	Sulphur	. . . 0·1850
Air	. . . 0·2669	Iron (wrought)	. 0·1100
Hydrogen	. . 3·2936	Mercury	. . . 0·0330
Azote	. . . 0·2754	Platinum	. . . 0·0314
Oxide of Azote	. 0·2369	Gold	. . . 0·0298

(255.) It has, however, been much questioned whether the several substances possess the same capacity for heat at all temperatures; and MM. Petit and Dulong, as also Dr. Dalton, appear to have established the fact that the capacity for heat of every substance increases with the temperature; or, in other words, that the quantity of heat given out by any body, in cooling a given number of degrees, is greater at high temperatures than at low temperatures. The following Table exhibits the specific heat of several bodies between the temperatures of 0° and 100°, and also between 0° and 300° of Centigrade.

* A further list will be found in Table V., Appendix.

TABLE XVIII.

Substances.	Mean Capacity between 0° and 100° Centigrade.	Mean Capacity between 0° and 300° Centigrade.
Iron	·1098	·1218
Mercury . . .	·0330	·0350
Zinc	·0927	·1015
Antimony . . .	·0507	·0549
Silver . . .	·0557	·0611
Copper . . .	·0949	·1013
Platinum . . .	·0335	·0355
Glass . . .	·1770	·1900

(256.) But independent of the sensible heat of bodies as ascertained by direct measurement, when any change of state occurs, as from the solid to the liquid, and from the liquid to the aeriform, or *vice versâ*, certain quantities of heat enter into or quit the respective substances, which, not being directly measurable by the thermometer, have been termed *latent heat*. Ice at 32° requires 140° of heat to enter into combination with it before it assumes the liquid state; and it then becomes water, though still only at the temperature of 32°. And water at 212° requires 1000° of heat to enter into combination with it before it assumes the state of steam, though in the latter state it still only shows a thermometric temperature of 212°. The following Table gives the quantity of heat rendered latent when certain solids assume the liquid state and certain liquids assume the state of vapours: the degrees are those of Fahrenheit's thermometer.

Table XIX.

TABLE OF LATENT HEAT.*

Of Liquids: by Dr. Irvine.	Of Vapours: by Dr. Ure.
Water . . . 140°	Vapour of Water at 212°, 1000°
Sulphur . . . 143·7	„ Alcohol . 457°
Spermaceti . . 145	„ Ether . 312·9
Lead . . . 162	„ Oil of Turp. 183·8
Bees'-wax . . 175	„ Nitric Acid 550
Zinc . . . 493	„ Ammonia 865·9
Tin . . . 500	„ Vinegar . 903
Bismuth . . 550	„ Petroleum 183·8

(257.) Many important facts arise from the theory of latent heat. One of the most important, in regard to the arts, is the peculiar properties of steam. But without entering at any length into this important and extensive subject, it may be observed that experiments have clearly proved that at all temperatures and pressures steam contains exactly the same absolute quantity of heat. For while under increased pressure, steam can be made to exhibit almost any thermometric temperature, the latent heat of high-pressure steam always decreases exactly in the same ratio as its sensible heat increases, so that its latent and sensible heat together always amount to 1180° above the freezing point of water. Thus a certain weight of steam at 212°, when condensed into water at 32°, gives out

Sensible Heat . . . 180°
Latent Heat . . . 1000
———
1180
———

* Dr. Ure's Dictionary of Chemistry, Art. "Caloric."

s 2

And the same weight at 400°, when condensed into
water at 32°, gives out

Sensible Heat . . .	368°
Latent Heat . . .	812
	1180

The same holds good with steam at all other temperatures, and extends from 212° to the highest
range of steam pressures.

(258.) The phenomena of spontaneous evaporation presents some important facts connected with
the subjects of our present inquiry. To Dr. Dalton
we owe much of the knowledge which we possess
of the laws that govern the vaporization of liquids;
and his numerous experiments on the subject rank
among the most valuable contributions to science.
Evaporation is distinguished from ebullition by
this circumstance:—that during the latter phenomenon, the liquid which is converted into vapour
maintains an invariable temperature, provided the
pressure upon it does not change; while, with the
former, the temperature is constantly subject to
change, the quantity of liquid evaporated being
proportional to the temperature and to the surface
exposed. Evaporation is wholly independent of
the pressure of the air, and is the same in a vacuum,
in the natural atmosphere, or in a condensed medium: but this is only true as regards the absolute
quantity which can exist in a given space, for the
air is found greatly to impede evaporation, by its
inertia opposing a mechanical obstruction, and
thereby retarding the time required for the operation, when such obstruction does exist. In a
vacuum, the evaporation is almost instantaneous;
but, in a space containing air, the time required for
the evaporation of the maximum quantity of vapour

is longer in proportion to the density of the air, though ultimately the quantity is the same.*

(259.) The quantity of liquid discharged into free space by spontaneous evaporation is, under these circumstances, much influenced by the motion of the air, which thus carries off the successive strata of vapour that rise from the liquid, a very strong wind causing about twice as much vapour to be discharged as a still atmosphere. Dr. Dalton's experiments enable us to estimate the evaporating force under all circumstances. He ascertained that from a circular vessel of six inches in diameter, kept at the temperature of 212°, the quantity of water evaporated was 120 grains per minute in a still atmosphere, 154 grains per minute with gentle motion of the air, and 189 grains per minute with a brisk motion of the air. The temperature of the air does not influence the evaporation; and he found that, at any other temperature of the water, the evaporation was exactly proportional to the elastic force of the vapour at that temperature.

The following Table was constructed by Dr. Dalton from his experiments:—

---

* Professor Daniell made some experiments on this subject, by which he was led to conclude that "the amount of evaporation is *cæteris paribus* in exact inverse proportion to the elasticity of the incumbent air." (*Quarterly Journal of Science*, vol. xvii., p. 52.) This remark, however, applies only to the rate of evaporation, and not to the quantity which can exist in a given space. Even from ice, the evaporation is very considerable; and some experiments of M. Schuebler show, that during the very coldest weather, the spontaneous evaporation of ice in the open air is sometimes nearly 1–40th of an inch in depth from each square foot of surface every twenty-four hours.—*Quarterly Journal of Science*, vol. xxvii., p. 187.

TABLE XX.

Showing the force of Vapour and the full Evaporating Force for every Temperature, from 20° to 212° Fahrenheit, expressed in grains of water, that would be raised *per minute*, from a vessel six inches in diameter, supposing there were no vapour already in the atmosphere.

Temperature. Fahrenheit.	Force of Vapour.	Evaporating Force in Grains.			Temperature. Fahrenheit.	Force of Vapour.	Evaporating Force in Grains.		
		Still.	Gentle.	Brisk.			Still.	Gentle.	Brisk.
20°	·129	·52	·67	·82	68°	·676	2·70	3·47	4·24
22	·139	·56	·71	·88	70	·721	2·88	3·70	4·53
24	·150	·60	·77	·94	72	·770	3·08	3·96	4·84
26	·162	·65	·82	1·02	74	·823	3·29	4·23	5·17
28	·174	·70	·90	1·10	76	·880	3·52	4·52	5·53
30	·186	·74	·95	1·17	78	·940	3·76	4·83	5·91
32	·200	·80	1·03	1.26	80	1·000	4·00	5·14	6·29
34	·214	·86	1·11	1·35	82	1·07	4·28	5·50	6·73
36	·229	·92	1·18	1·45	85	1·17	4·68	6·07	7·46
38	·245	·98	1·26	1·54	90	1·36	5·44	6·98	8·56
40	·263	1·05	1·35	1·65	95	1·58	6·32	8·11	9·95
42	·283	1·13	1·45	1·78	100	1·86	7·44	9·54	11·71
44	·305	1·22	1·57	1·92	110	2·53	10·12	12·98	15·93
46	·327	1·31	1·68	2·06	120	3·33	13·32	17·09	20·97
48	·351	1·40	1·80	2·20	130	4·34	17·36	22·27	27·34
50	·375	1·50	1·92	2·36	140	5·74	22·96	29·46	36·16
52	·401	1·60	2·06	2·51	150	7·42	29·68	38·08	46·74
54	·429	1·71	2·20	2·69	160	9·46	37·84	48·56	59·59
56	·458	1·83	2·35	2·88	170	12·13	48·52	62·26	76·41
58	·490	1·96	2·52	3·08	180	15·15	60·60	77·77	95·44
60	·524	2·10	2·70	3·30	190	19·00	76·00	97·53	119·7
62	·560	2·24	2·88	3·52	200	23·64	94·56	121·35	148·9
64	·597	2·39	3·07	3·76	210	28·84	115·36	148·04	181·6
66	·635	2·54	3·27	3·99	212	30·	120·	154·	189·

In this Table,* the first column gives the temperature of the water; the second, the elastic force of the vapour at that temperature; and the third, fourth, and fifth columns show the number of grains' weight of water evaporated per minute, from a vessel six inches diameter (or 28·274 square

* See Dr. Dalton's experiments, *Memoirs of the Manchester Philosophical Society*, vol. v., p. 579, *et seq.* The Table given by Dr. Dalton only extends to the temperature of 85°: the temperatures above that are calculated from his data.

inches, equal to ·196 of a square foot), when the air is still, or in gentle or in brisk motion respectively.

(260.) In order to know exactly the quantity of vapour that will be evaporated per minute, we must know the quantity which already exists in the atmosphere. For this purpose we must find the dew-point of the air,—that is, the temperature at which the vapour in the air just begins to condense.* By referring to the Table we shall then find the quantity of vapour in the air at that time; and this, deducted from the quantity shown by the Table to be given off at the ascertained temperature of the evaporating liquid, will give the exact quantity of water that will be evaporated per minute. Thus, suppose the dew-point of the air to be 48°, and the temperature of the air and of the evaporating liquid to be 70°, with a still atmosphere : the vapour in the air, as shown by the Table, at the temperature of 48°, is 1·4 grains; which, subtracted from that at 70°, viz. 2·88 grains, gives 1·48 grains per minute for the quantity of vapour given off from a surface of six inches diameter.

It will be observed, that if the temperature of the air be lower than the temperature of the evaporating fluid, the vapour will again condense; but the rapidity of this condensation will depend upon the relative temperatures, and the quantity of moisture contained in the air.

* Dr. Dalton used for this purpose a very thin glass vessel, into which he poured cold water, and noted the temperature. If the vapour was instantly condensed, he poured out this water and applied some a little warmer, and so on, until he obtained the proper temperature at which he could just perceive a slight dew deposited on the glass. This temperature then was the dew-point. Daniell's Hygrometer is a much more elegant instrument, but more difficult to manage. See Daniell's " Chemistry," art. " Hygrometer."

(261.) The heat and cold caused by the condensation and rarefaction of the air produce important effects, both in natural phenomena, and in the mechanical application of this agent in the arts. When the air is partially exhausted in a receiver enclosing a thermometer, the temperature sinks considerably; but if the air be again suddenly admitted to the original pressure, the temperature rises considerably higher than in the first instance. The reverse of this occurs when the air is first condensed, and then allowed to resume the original pressure; in which case the temperature sinks considerably lower on the original pressure being restored, than the thermometer indicated before the condensation.*

This result, however, appears to be produced by an unavoidable imperfection in the experiment. For the vessel that contains the air communicates its heat in the one case, and abstracts it in the other, which causes the apparent difference; and this difference is therefore dependent upon the rapidity of the operation, thereby allowing more or less time for the interchange of heat between the air and the vessel containing it.

In Dr. Dalton's experiments,† the thermometer which sank 2° on rarefying the air rose 4° on again allowing it suddenly to resume its original pressure. When the air was first condensed and then suddenly allowed to resume its original pressure, the thermometer rose 2° on the condensation, and then sank 3½° on diminishing the pressure suddenly to its original amount. These inequalities, however, although an accompaniment of the experiment, are not due to the mechanical expansion or condensation of the air; and the results will vary

---

* *Manchester Memoirs*, vol. v., p. 515; and *Nicholson's Journal*, vol. iii., p. 160.
† *Ibid.*

with every alteration in the condition of the containing vessel, and of the quantity of air.

(262.) Professor Leslie investigated this subject, both experimentally and theoretically, and he found that the quantity of heat evolved under these circumstances was very great; but as the specific heat of air increases as its density diminishes, only a small portion of the heat thus produced becomes appreciable. When air of the following densities is restored to its original pressure, represented by 1, the heat evolved is as follows :—

TABLE XXI.

Density of the Air.	Heat Evolved (Fahrenheit).	Density of the Air.	Heat Evolved (Fahrenheit).
1·0	0°	0·4	94·5°
0·8	20·6	0·2	216·
0·6	48·	0·0003	13500·

This evolution of heat he considered to be owing to the diminution of the specific heat of the air by condensation. When air was condensed so as to double its density, the mathematical formula which he constructed from his experiments showed that the sensible heat which would be evolved would be equal to 67·5° Fahrenheit.*

(263.) Mr. Ivory also investigated the mathematical law of the heat produced under these circumstances; † and he found " that the heat extricated from air, when it undergoes a given condensation, is equal to three-eighths of the diminution of temperature required to produce the

* Dr. Thomson " On Heat and Electricity," p. 126, *et seq.*
† *Philosophical Magazine* (Second Series), vol. i., p. 89; and *Quarterly Journal of Science*, vol. xxiii., p. 228.

same condensation, the pressure being constant." Air, under a constant pressure, diminishes 1-480th of its volume for every degree of depression of Fahrenheit's scale; and therefore 1° of heat will be extricated from air when it undergoes a condensation equal to $\frac{1}{480} \times \frac{3}{8} = \frac{1}{180}$. If a mass of air were suddenly reduced to half its bulk, the heat evolved would be $\frac{1}{2} \div \frac{1}{180} = 90°$.[*] M. Despretz made some experiments to ascertain whether water evolved heat when subjected to great pressure; and he found that a compressive force equal to 20 atmospheres caused the disengagement of only one sixty-sixth part of a degree of heat.[†] This result might reasonably be expected from the small degree of compression to which water can be subjected.

(264.) Dr. Ure instituted some experiments to ascertain the influence which heat had in increasing the fluency of various liquids; and he found in all cases a great increase in the velocity of their efflux through a given aperture, by raising their temperature. In some liquids of a viscid nature, the increase appears to be very great. Such, for instance, is the case with some kinds of oil, which, by raising the temperature from 65° to 254°, had the velocity of efflux increased nearly six times. With those which were less viscid, the increase was less. Water, by having its temperature raised from 60° to 164°, had its velocity increased about one-sixth.[‡] These results are wholly independent of any effect from alterations of pressure, but appear to arise from the particles of the fluids possessing less adhesiveness among themselves by the increased temperature.

* *Ibid.*
† *Quarterly Journal of Science,* vol. xxiv., p. 20.
‡ *Reports of the British Scientific Association,* vol. viii., p. 23. (See also Art. 251.)

(265.) The effects of heat in producing variations in the strength of materials is a subject
which has received but little attention; and with
the exception of some experiments made by the
Franklin Institute of Pennsylvania, on the strength
of iron and copper, little appears to be known.
By these experiments * the maximum strength of
wrought iron, estimated by its resistance to a
longitudinal strain, was obtained by heating the
iron to 572° Fahrenheit; at which temperature
the strength is 15·17 per cent. greater than at the
ordinary mean temperature of the air. The rule
deduced for the strength of iron at high temperatures is, that " the thirteenth power of the temperature above 80° Fahrenheit is proportionate to
the fifth power of the diminution from the maximum tenacity." And it appears that, at the temperature of about 1050°, iron loses about one-half
of its maximum strength; at 1240° it loses about
two-thirds; and at 1317° seven-tenths of its maximum tenacity is overcome.

The following Table shows the diminution of
tenacity at various temperatures. The maximum
tenacity of the fourth column is a calculated
amount, obtained by adding 15·17 per cent. to
the strength of the metal when tried cold. The
seventh column shows that the mean irregularity
of structure is 10 per cent. of the strength when
tried cold.

* " Report on the Explosion of Steam Boilers, by a Committee of the Institute;" *Journal of the Franklin Institute*,
vols. xix. and xx.

TABLE XXII.

No. of the Experiment.	Temperature observed (Fahrenheit).	Tenacity observed. In lbs.	Maximum Tenacity at the point of Fracture. In lbs.	Manner of obtaining the Maximum.	Diminution by Heat, in parts of the Maximum Tenacity.	Irregularity of the Metal in parts of the original Strength.
1	520°	58451	·63267	experiment	·0761	·0992
2	570	60398	66146	—	·0869	·1125
3	596	57682	63386	calculation	·0899	·2401
4	600	56938	63086	—	·0964	·2401
5	630	60010	67033	experiment	·1047	·1440
6	662	58182	65785	—	·1155	·0644
7	722	54442	64483	calculation	·1436	·0507
8	732	53378	62736	experiment	·1491	·1310
9	734	57903	68407	calculation	·1535	·0644
10	766	54819	65176	experiment	·1589	·1563
11	770	54781	65445	calculation	·1627	·0234
12	824	55892	70080	—	·2010	·0413
13	932	45531	68202	—	·3324	·0413
14	947	42401	66193	experiment	·3593	·0446
15	1030	37587	68071	calculation	·4478	·0460
16	1111	27603	61531	—	·5514	·0330
17	1155	21967	54992	—	·6000	·0330
18	1159	25620	64234	—	·6011	·1102
19	1187	21913	60102	—	·6352	·0330
20	1237	21298	63065	—	·6622	·1147
21	1245	20703	63065	—	·6715	·0347
22	1317	18913	63065	—	·7001	·1147

(266.) In the experiments made with copper, at various temperatures, the Committee found that unlike iron, the maximum strength of copper was at a very low temperature; and that it increased in strength at every reduction of temperature down to 32°, which was the lowest at which they could try it. The mean strength of copper at ordinary temperatures was found to be 32,146 lbs. per square inch; and the following Table exhibits the diminution of strength at high temperatures. To obtain the actual temperature of each .experiment, 32° must be added to the temperatures given in the second column.

TABLE XXIII.

Number of the Experiment.	Temperature above 32° Fahrenheit.	Diminution from the Strength at 32°.
1	90°	·0175
2	180	·0540
3	270	·0926
4	360	·1513
5	450	·2046
6	460	·2133
7	513	·2446
8	529	·2558
9	660	·3425
10	769	·4398
11	812	·4944
12	880	·5581
13	984	·6691
14	1000	·6741

(267.) The strength of cast iron at high temperatures has not been ascertained with the same accuracy. It has, however, been estimated that its maximum strength is at the temperature of about 300° Fahrenheit, which is considerably below that of wrought iron.

(268.) The effects of heat, which have here
been mentioned, are a few of the laws and phenomena of this important branch of science. They
appear to be those which are most closely connected
with the subject of the present inquiry : but it must
not be supposed that they are given as an epitome
of the general laws of heat, as the subject would
be far too extensive for the present work. Many
most important and interesting phenomena have
therefore been entirely omitted ; but those which
are here mentioned, and others that are alluded to
and interspersed in various parts of this treatise,
appear to be all that are requisite for the illustration
of the subject now before us, and will afford useful
suggestions to those who wish to investigate the
principles of heating buildings theoretically and
scientifically, as well as to learn the mere practical
details.

CHAPTER XIII.

EXPERIMENTS ON COOLING.

(269.) FROM what has been stated in the preceding chapter, it is evident that the velocity with which a heated body cools depends upon various circumstances; and experiments are necessary, in order to obtain data for the calculations which the known laws of heat enable us afterwards to apply to the subject now under investigation.

No experiments on cooling are extant, that appear to be suitable to the present purpose, except some that were made by Tredgold,* and these are erroneous in the application he has made of them. For he has neglected all considerations of the thickness of the body on which he experimented, and has therefore estimated that the rate of cooling of a very thin sheet-iron vessel, containing a heated fluid, is the same only as a cast-iron pipe, though the latter is fully six or eight times the thickness of the former. The same error also occurs in his experiments on the cooling of glass; and, consequently, his conclusions on the dispersion of heat, as applied to the warming of buildings, are erroneous to a considerable extent. Another source of error lies in his having estimated the quantity of water which

---

* Tredgold "On Heating Buildings by Steam," &c.

the vessel contained at rather too large an amount, in allowing for the specific heat of the vessel; which latter could not possibly have had quite the same temperature as the water, owing to loss by imperfect conduction, and to the cooling influence to which it was exposed. The effect of each of these errors is to make the rate of dispersion appear more rapid than the true velocity; and the result is, that in some of the calculations founded on these experiments the errors amount to upwards of 16 per cent.

(270.) To ascertain the velocity of cooling for a surface of cast iron, a pipe thirty inches long, two inches and a half diameter internally, and three inches diameter externally, was used in the following experiments. The ends of the pipe were closed by corks which entered into the pipe one inch and a half at each end; and the bulb of a thermometer was inserted into the water about three inches from one end, the temperature of the water being the same in every part of the pipe. The exposed surface of the pipe (including the surface exposed by the thickness of the metal at the ends) was 287·177 square inches. The quantity of water contained in it was 132·534 cubic inches; and the equivalent to be added to this for the specific heat of the pipe is 39·341 cubic inches; making the estimated quantity of water 171·875 cubic inches.* The rates of cooling were tried with different states of the surface: first, when it was in the usual state of cast-iron pipes covered with the brown surface of protoxide of iron; next, it was varnished black; and finally,

* In estimating the equivalent of water which was required to represent the specific heat of the pipe, the difference between the external and internal temperatures necessarily required an allowance to be made on account of the thickness of the pipe being considerable. This diminution of specific heat was estimated to be equal to a superficies of the pipe one-

the varnish was scraped off, and the pipe was painted white with two coats of lead paint. The following Table shows the observed time of cooling, corrected and reduced to the same excess of temperature above the circumambient air.

TABLE XXIV.

TABLE OF THE COOLING OF IRON.

Temperature of Room, 67°. Maximum Temperature of Thermometer, 152°.

Thermometer Cooled.		Rusty Surface.		Black Varnished Surface.		White Surface.	
from	to	Observed Time.	Calcu-lated Time.	Observed Time.	Calcu-lated Time.	Observed Time.	Calcu-lated Time.
152°	150°	2′ 30″	2′ 21″	2′ 16″	2′ 16″	2′ 19″	2′ 24″
152	148	5 0	4 44	4 38	4 36	4 53	4 51
152	146	7 45	7 12	7 28	7 3	7 28	7 22
152	144	10 15	9 44	9 45	9 27	10 13	9 57
152	142	12 45	12 15	12 2	11 54	12 57	12 36
152	140	15 0	15 0	14 32	14 32	15 22	15 22

Minutes.

The ratios of Cooling 1° are therefore, { Black Varnished Surface . . . 1·21
Iron Surface . . . . 1·25
White Painted Surface . . . 1·28*

These ratios are in the proportion of 100, 103·3, and 105·7; but as the relative heating effect is the inverse of the time of cooling, we shall find that 100 feet of varnished pipe, 103¼ feet of plain

sixteenth of an inch thick. The total equivalent of water which would represent the specific heat of the pipe, supposing it to be exactly of the same temperature as the water contained in it, would be 52·455 cubic inches; from this must be deducted 13·114 cubic inches as above stated, leaving the equivalent specific heat of the pipe equal to 39·341 cubic inches of water, as stated in the text.

* These ratios of cooling, it will be observed, are for pipes of three inches in diameter; but the cooling of any other size can be calculated from the data here given.

T

iron pipe, or 105¾ feet of iron pipe painted white, will each produce an equal effect.

In these experiments, it might have been expected to find greater differences between the effects of the various states of the surface, than appears really to obtain. The greatest difference only amounts to about 5¾ per cent., but it would probably be greater in proportion, with an increased thickness of the coating of paint.

(271.) To ascertain the effect of glass windows in cooling the air of a room, the following experiments were made, with a vessel as nearly as possible of the same thickness as ordinary window-glass. The temperature of the room, in these experiments, was 65°; the thickness of the glass was ·0825 of an inch; the surface of the vessel measured 34·296 square inches, and it contained 9·794 cubic inches of water, including the equivalent for the specific heat of the glass. The time in which this vessel cooled, when filled with hot water, is shown as follows :—

TABLE XXV.

TABLE OF THE COOLING OF GLASS.

Thermometer Cooled		Observed Time of Cooling.	Calculated Time of Cooling.	Average Rate of the Observed Time of Cooling.
from	to			
150°	140°	6' 40''	6' 54''	1·176° per minute, at an excess of 65° above the Temperature of the air.
150	130	14 15	14 43	
150	120	23 30	23 40	
150	110	34 0	34 0	

From the average rate of cooling which is here given, the effect of glass in cooling the air of a room may easily be calculated. As the specific heat of equal volumes of air and water * is as one to 2990 the above average will show that each square foot of glass will cool 1·279 cubic feet of air 1° per minute, when the temperature of the glass is 1° above that of the external air.

But by this we shall only find the effect of glass in a still atmosphere; and therefore, to ascertain the cooling effect of external windows, when exposed to the action of winds, further experiments are necessary.†

(272.) In some researches of Leslie's, on the cooling power of wind, he used a bright metallic ball filled with hot water, and noted the time of cooling when it was exposed to wind at different velocities. The result he obtained was, that the cooling effect on the ball was very nearly in a direct ratio with the velocity. But it will be obvious, by referring to the experiments of Petit and Dulong, in the preceding chapter, that the relative cooling of heated bodies, when exposed to air moving at different velocities, must depend upon the nature of the surfaces. For, while the quantity of heat which is abducted by the air is proportional to the number of particles of air which pass over the heated body in a given time, the heat that is lost by radiation is not only independent of this effect, but the relative proportion

---

* See Art. 97.

† In some experiments by Wyman, on the cooling influence of glass windows, he obtained the following results, in a room of 1930 cubic feet, and having 531 square feet of walls, and 33·21 square feet of windows. The room was allowed to cool about 10°, first with the windows uncovered, and then with the windows covered with double blankets. The average of the experiments gave—

Room cooled 0·150° per minute, with windows uncovered;
    ”   ” 0·106° per minute, with windows covered;
or 1° in 6·6 minutes, with windows uncovered, and 1° in 9·4 minutes with the windows covered.—Wyman "On Ventilation." Boston, 1846.

of heat lost by radiation differs for each particular substance. As the bright metal ball that Leslie employed in his experiments would lose only an extremely small proportion of its heat by radiation, it might naturally be concluded that the rate of cooling would be nearly in a direct ratio with the velocity of the air. But with a surface of glass, the result must be very different, because the radiation is then very considerable; and, therefore, the total cooling will be much slower than the simple ratio of the velocity. For while a surface of glass of the temperature of 120°, and at an excess of 52° above the surrounding medium, loses about two-thirds of its heat by radiation, a bright metallic surface of the same temperature will only lose one-eleventh part of its heat by the same cause.

(273.) In the following experiments it appears that the cooling effect of wind, at different velocities, on a thin surface of glass, is very nearly as the square root of the velocity. In these experiments, the velocity of the air was measured by the revolution of the vanes of a fan; the temperature of the air was 68°; the time required to cool the thermometer 20° was noted for every different velocity, and the maximum temperature of the thermometer, in each experiment, was 120°. In still air it required 5′ 45″ to cool the thermometer this extent of 20 degrees; and the annexed Table shows the time of cooling by air in motion.

Table XXVI.

TABLE OF THE COOLING OF GLASS BY WIND.

Velocity of the Wind, in Miles, per Hour.	Times of Cooling the Thermometer 20°, from 120° to 100° of Fahrenheit.		
	Observed Time of Cooling.	Time reduced to Decimals of a Minute.	Calculated Time; being the Inverse of the Square Root of the Velocities; in Decimals of a Minute.
3·26	2′ 35″	2·58	2·58
5·18	2 10	2·16	2·04
6·54	1 55	1·91	1·82
8·86	1 40	1·66	1·56
10·90	1 30	1·50	1·41
13·36	1 15	1·25	1·27
17·97	1 5	1·08	1·10
20·45	1 0	1·0	1·03
24·54	0 55	·91	·94
27·27	0 48	·81	·88

(274.) In consequence of the large quantity of glass in buildings used for horticultural purposes, the cooling effect of wind is of considerable importance. We see, however, that with an increased velocity the cooling effect is considerably less, in proportion, on glass than on metal. And it will be very much less on window-glass than even what is here stated: for, as glass is an extremely bad conductor of heat, the increased thickness which window-glass possesses over that which composes the bulb of a thermometer, will make a material difference in the quantity of heat that is lost by the abduction of the air, as there will be, in this case, a greater difference between the temperature of the external and the internal surface. The cooling effect of wind is therefore not near so considerable on glass as is generally supposed; and it will probably be nearly one-half less on window-glass than what is shown by the

preceding experiments. Exact experiments on
this subject, however, are extremely difficult,
owing to the unequal action of the air on different
parts of the surface. Thus, if a cylindrical or
spherical vessel be employed, the action of the
wind will be such that the particles of air striking
the surface in front will cause a vacuum at the
back part of the vessel, from which latter part the
heat will be given off by radiation alone. It thus
becomes impossible to ascertain accurately what
would be the effect, if the whole of the surface
could be acted upon by the wind; and the same
remark will apply to most other forms of surface
which could be employed experimentally.

(275.) The following Table shows the results of
Tredgold's experiments on cooling, already referred
to. They are here given, principally, to show the
cooling power which thin sheet-iron will have in
any building in which it is used ; and they may
therefore be useful now that corrugated iron is so
frequently employed in many descriptions of buildings. In these experiments the specific heat of the
vessels was added to the quantity of water they contained. The vessels were always cooled from 180°
down to 150° of Fahrenheit ; the time required for
this cooling, and the other particulars of the experiments, being noted in the Table.

TABLE XXVII.

Nature of Surface.	Surface Exposed. Square Inches.	Equivalent Quantity of Water. Cubic Inches.	Temperature of the Room (Fahrenheit).	Time of Cooling. Minutes.	Average Excess of Temperature (Fahrenheit).	Loss of Heat per Minute. (Fahrenheit).
Tinned plate .	79	62·28	55½°	46	124½°	·759
Sheet Iron .	76·7	61·7	57	29	123	1·18
Glass . . .	71	61·2	56½	31½	123½	1·075

In these experiments, as already stated, the cooling of the sheet-iron will not afford a fair criterion for the effect of cast-iron pipes; but it will be perceived that thin sheet-iron has, in a still atmosphere, the same cooling power as glass; and therefore its effect in cooling the air of any building in which it is used will be very great, and in high winds it will produce a far greater cooling power than the same extent of glass (Art. 272), as its cooling power will be nearly in the direct ratio of the velocity of the wind.

These remarks on the cooling power of different substances when employed in buildings must of course be understood merely to apply to those cases in which the interior temperature of the building is higher than that of the external temperature. Under other circumstances these experiments would not be applicable.

281

PART SECOND.

ON THE VARIOUS METHODS OF WARMING AND

VENTILATING BUILDINGS.

THE COMBUSTION OF FUEL, ETC.

CHAPTER I.

Early Methods of Warming Buildings—The Romans: their
Stoves—Baths—Flues—Mode of Preparing their Firewood
—the Persian Method—Chinese Method of Flues—Method
of the Ancient Britons—Invention of Chimneys—Burning
of Coals in England—Early Writers on the subject—
Improvements in the Form and Construction of Stoves and
Fire places.*

(276.) THE various methods of warming buildings have consisted, in all countries and in all ages,
until a very recent period, of the rudest appliances
and the most inartificial inventions. At a very
early period, it is true, the Romans were acquainted
with the method of heating rooms and buildings
by flues; and these were elaborate in their construction, and complicated in their arrangements.

* In this and the following chapters, the articles on
"Stoves" and "Ventilation," written by the author, and
published in the "Encyclopædia Metropolitana," have been
incorporated with such further observations as the want of
space prevented him from making in the work in question.

But they were so expensive in their construction,
and so wasteful in their expenditure of fuel, that
this method of warming buildings could only be
adopted by very few of even that rich and luxu-
rious nation.   The comparatively late invention of
chimneys fully accounts for the immense size and
peculiar construction of the flues used by the
ancients; for unless a large space were provided
for the combustion of the fuel and the entrance of
the air, the heat could not have been conducted
through the flues, owing to the absence of the
necessary draught produced by the use of a high
chimney.   The *hypocaustum* of the Romans was
this plan of flues.   It appears to have consisted of
a long furnace; and a number of narrow arches
(*testudines alvei*) received the fire of the hypocaustum, and conducted it along and underneath the
floor of the room to be warmed.*   The whole
of the hypocaustum was immediately below the
room which was to be heated.   Sometimes a great
number of short columns or pillars supported the
floor instead of these arches.   They were set in four
rows very close together, and the flame of the furnace
passed between them, as appears by some very perfect specimens which have been discovered.†   Pliny
the younger, in his letter to Gallus, giving the
description of his villa Laurentinum, mentions
that his bedchamber was warmed by a small
hypocaustum:‡   and this plan was generally
adopted in heating the baths.§   For this latter
purpose, however, an improved method was
adopted when the Thermæ of Rome were built,

---

* Castell's " Illustrations of the Villas of the Ancients,"
pp. 8, 9.
† *Philosophical Transactions*, vol. xxv., p. 2225, and vol. xli.,
p. 855.
‡ Castell's " Illustrations," p. 13.
§ *Ibid.*, p. 9.

and which has been described by Seneca.* The
water of the bath was heated by passing it through
the fire in a brass pipe of a serpentine form, thence
called *Draco*. The most approved mode was to
employ the *Miliarium*,† which appears to have been
a leaden vessel of large circumference, the middle
part being open for the spiral pipe, and for the
draught of the fire to pass through. This vessel
of water that surrounded the flame was also placed
upon part of the same fire, and for that reason the
bottom was obliged to be made of brass, as' were
also the pipes.

(277.) But the method of warming by the
hypocaustum was far too expensive for general use.
The Romans used portable furnaces, containing
embers and burning coals, to warm the different
apartments of their houses, which were placed in
the middle of the room.‡ These were sometimes
made to contain water, which was heated by the
fuel of the furnace, and probably they were also
used for cooking. One of these boilers and furnaces,
found at Herculaneum, was in the shape of a castle
with four towers.§ The usual kind of stoves,
however, were nearly on the plan of our braziers.
They were mostly elegant bronze tripods, supported by satyrs and sphinxes, with a round
dish above for the fire, and a small vase below to
hold perfumes, which were thrown into the brazier
to correct the smell of the coals. A square stove
of bronze, of the size of a moderate table, found
at Herculaneum, rested on lions' paws, and was
ornamented upon the border with foliage. The
bottom was a strong iron grating, walled up with
bricks above and below, so that the fire could not

* "Seneca, Nat. Quæst.," lib. 3, cap. 24.
† "Palladius," lib. i., tit. 40.
‡ Adams's "Roman Antiquities," p. 454.
§ Fosbroke's "Archæology," 4to, vol. i., p. 233.

touch the sides of the stove, nor fall through the bottom. It was similar to those still used in large rooms in Italy.* But the smoke from these stoves was so considerable, that the furniture of the winter rooms was different from the summer rooms; and Vitruvius expressly states that these winter apartments had plain cornices, and were without carved work or mouldings, in order to allow the soot to be easily and frequently cleaned away.† At great entertainments it was usual to have watchmen stationed to be ready to extinguish any fire which might happen; the smoke issuing from the kitchen windows being so great on these occasions, that it was common to speak of this great smoke as synonymous with a great entertainment.‡ The utmost care, however, was taken to prevent the smoke as much as possible, and to procure wood which gave the smallest quantity of smoke in combustion. A great deal of the firewood used by the Romans was procured from Africa.§ The bark of the wood was peeled off; the wood was then suffered to lie a long time in water, and afterwards dried and anointed with the lees of oil, which was considered the most effectual way to prevent it from smoking.‖

(278.) In the time of Seneca (A.D. 64), another method of heating buildings was adopted; which consisted of pipes built in the walls, that conveyed the heat from a furnace, constructed in the earth under the edifice. These pipes or flues were conducted to the different rooms; and the upper end was often ornamented with the representation of

---

* Fosbroke's " Archæology," vol. i., p. 237.
† " Vitruvius," lib. vii., cap. 5. Beckmann's " History of Inventions," vol. ii., p. 74.
‡ " Beckmann," vol. ii., p. 71.
§ *Ibid*, vol. ii., p. 300.
‖ Adams's " Roman Antiquities," p. 454. " Beckmann," vol. ii., p. 79.

a lion's or a dolphin's head, or any other figure,
and it could be opened or shut at pleasure. These
pipes, however, were liable to become full of soot;
and as they were very likely to catch fire by being
over-heated, laws were made forbidding them to
be brought too near to the wall of a neighbouring
house.*

(279.) The Persians used a stove consisting of
an iron vessel sunk in the earth in the centre of
the apartment. After a fire had been kindled, and
had well warmed the place, a wooden top, like a
small low table, was placed over the hole in the
floor which contained the stove, and this top was
then spread with a large coverlet quilted with
cotton, which hung down on all sides to the floor.
Those people who were not very cold, only put
their feet under the table or covering; but those
who required more heat put their hands under it
also, or crept under it altogether.† The Jews
likewise used such stoves in their houses, and the
priests had them also in the Temple:‡ in fact,
throughout the East this mode of warming apartments appears to have been commonly adopted.

(280.) In China, a very elaborate system of flues
has been long in use, by which the floors of the
rooms are heated by a furnace constructed below,
with a moderate expenditure of fuel. A very
equable temperature appears to be maintained by
these means, notwithstanding the winter temperature of some parts of China is so low that the
thermometer nearly reaches the zero of Fahrenheit's
scale. Father Gramont described this mode of
heating in 1771,§ but the date of its introduction
does not appear to be known.

(281.) Although the Romans must have intro-

* "Beckmann," vol. ii., p. 90.
† Ibid., vol. ii., p. 83.   ‡ Ibid, vol. ii., p. 85.
§ Philosophical Transactions, 1771, p. 61.

duced their methods of warming buildings into
England at a very early period, as appears by
various remains which have been excavated in
recent times,* the inhabitants of Britain long
contented themselves with contrivances of the
rudest and simplest character. Among the ancient
Britons, in each dwelling there was only one place
for a fire, which was merely a hole in the centre
of the floor. In the time of the Anglo-Saxons,
the ordinary plan was to place the ignited fuel on
the hearth in the middle of the floor, and an
opening in the roof, immediately above the hearth,
permitted the escape of the smoke. In the better
class of buildings, an ornamented turret was
erected in the centre of the roof for carrying off
the smoke, while in ordinary houses, the opening
in the roof was merely defended from the weather
by louvre boards, in the manner now practised in
many of our commonest buildings used for manufactories.

(282.) The invention of chimneys necessarily
made a great alteration in the mode of heating
buildings. The date of their introduction has been
much debated: but there appears to be no positive
evidence of their existence before the middle of
the fourteenth century, the earliest record being,
that an earthquake at Venice, in 1347, threw down
a great many chimneys.† Twenty years after this
they appear to have been unknown at Rome; for
in that year Francesco da Carraro, lord of Padua,
came to Rome, and, finding no chimneys at the inn
where he lodged, he caused two chimneys to be
built by workmen whom he had brought with him;
and over these chimneys, the first ever seen at
Rome, he caused his arms to be affixed.‡ This

* *Philosophical Transactions*, vols. xxv. and xli.
† Beckmann's " History of Inventions," vol. ii., p. 98.
‡ *Ibid.*, p. 99.

slow communication of such an important invention, so closely connected with health and comfort, contrasts most strangely with the rapid promulgation of every discovery and improvement of the present time.

The introduction of chimneys into England appears to have been in the reign of Richard II.; and one of the first is supposed to have been at Bolton Castle, built in this reign.* It was long before they came into general use; but in the reign of Elizabeth, most rooms in respectable houses were furnished with them, and apologies were made to visitors if they could not be accommodated with rooms with chimneys.†

(283.) It is uncertain at what time stove-grates were first used, though probably they were not invented till coals became the ordinary fuel. For, though coals were known to the Britons before the arrival of the Romans, their use was barely tolerated in England till the seventeenth century, as it was supposed that the air was rendered unwholesome by their use.‡

After the improved method of burning fuel under open chimneys was introduced, they were used not only as the receptacle for the fire, but they also became the ordinary place of resort for conversation and conviviality for all the inmates of the house. The chimney corner was the post of honour; and the custom of the whole family sitting under the chimney-breast is not even yet entirely exploded in some of our rural districts.

(284.) The earliest writers who endeavoured to improve the construction of stoves were Keslar, of Frankfort, in 1614; Savot, in 1625; Glauber, in

* Fosbroke's "Archæology," vol. i., p. 113.
† *Ibid.*, vol. i., p. 112.
‡ See chapter vi., Art. 382. Also Fosbroke's "Archæology," vol. i., p. 72, &c.

1669; and Delesme, in 1686. In 1713 (or perhaps even in 1709), M. Gauger published a most excellent treatise on the construction of fireplaces, which, in 1715, was translated and published in this country by Dr. Desaguliers.* This treatise, which is now scarce, contains a most lucid explanation of the methods of economizing fuel, based on the soundest principles of philosophy. It was the first attempt which had been made to apply the known laws of heat to the construction of fireplaces; and though, in consequence of wood being the fuel universally used in France at that period, and this fuel being always burned upon the hearth, the author made no mention of *stoves*, but merely of *fireplaces*, the translator, Dr. Desaguliers, added a chapter on the *stoves* to be used in these improved fireplaces; and the work in that new form was a complete epitome of all those principles which Franklin, and, after him, Count Rumford, so successfully brought under the public notice, and which, if strictly carried out, would form, even at the present day, the best guide to the proper construction of stoves and fireplaces. An epitome of this little work would be, in fact, a recapitulation of all the most approved methods of constructing fireplaces, stoves, and chimneys; but most of these principles are now too well known to require explanation, and others

---

* Dr. Desaguliers, in his "Experimental Philosophy," vol. ii., p. 557, published in 1744, mentions this book, and states that the author concealed his name, but that he knew him to be Monsieur Gauger, of Paris; and he mentions a curious circumstance of a Frenchman coming over to this country, who, although so ignorant that he could scarcely read three pages of the book, professed himself its author, and applied to the king to grant him a patent, free of expense, on account of the great value of the stoves described in the book, and that he was too poor to pay for the patent. The real author of this book still remains a subject of conjecture, and it has been sometimes attributed to the Cardinal de Polignac.

which are less so, will be touched upon in another form in the course of this treatise.

(285.) The word *stove* is used in this treatise to signify either a close or an open firegrate to burn fuel in ; and, in general, this is what the word is now supposed to mean. In horticulture, the building itself which is heated, and not the place which holds the fire, is called a stove, and this expression is employed by many old writers. Anciently, however, the term *hothouse*, which we now use to signify a building for horticultural purposes, was descriptive of a sudorific bath, the use of hothouses for the purposes of horticulture being an invention of comparatively recent date. At the beginning of the seventeenth century hothouses were used for the cultivation of orange trees, and were considered a mark of royal magnificence.*

(286.) The various elegant forms given to the stove grates of the present day are quite a modern invention. Formerly they were called "cradles of iron for burning sea-coal,"† from which we should suppose them to be very different in construction to ours ; and even those described in Dr. Desaguliers' work, as late as the beginning of the eighteenth century, are nothing more than a few bars bent into a semicircle and fastened into the back. How far the utility of stove grates has been affected by the modern alterations of form, we shall endeavour to show in the following chapter ; and subsequently we shall inquire into the physiological effects produced by some of the modern methods of distributing artificial heat.

* Fosbroke's "Archæology," vol. i., p. 275.
† *Ibid.*, vol. i., p. 268.

U

CHAPTER II.

Forms of Fireplaces and Chimneys—Should be made to reflect heat — Contraction of Chimney-breast — Hollow Hearths and Backs for Fireplaces—Rumford's Principles of Construction—Errors in the Construction of Stoves—Register Stove in a case—Jeffrey's Stove—Franklin's Pennsylvania Stove—Cutler's Torch Stove—Sylvester's Radiating Stove—Russian and Swedish Stoves—Cundy's Stoves—German Stoves — Hot-air Stoves—Cockle Stoves — Dr. Nott's Stove—Dr. Arnott's Stove—Franklin's Vase Stove —Gas Stoves—Joyce's Stoves—Beaumont's Stove—Palmaise Stove.

(287.) PREVIOUS to the publication of M. Gauger's treatise (already alluded to, Art. 284) chimney fireplaces were generally made in the form of a large square recess, and the breast of the chimney was of the same size as the recess itself. The error of this construction was pointed out, in the work by M. Gauger, by a reference to the known laws of heat. Radiant heat is subject to the same law as light,—that the angle of reflection is equal to the angle of incidence. Hence it follows that a ray of heat, falling perpendicularly on the flat sides of the chimney recess, will be reflected back upon itself: if the ray forms an angle vertically with the side, it must be reflected up the chimney; and if the ray forms an angle horizontally with the flat side of the chimney, this angle must necessarily be so small that it cannot be reflected forward beyond the jambs of the mantel-piece. On these considerations, M. Gauger recommended that the back of the recess should be contracted so that the sides should incline outwards at a considerable angle, by which means the radiant heat would be reflected into the room, and much more effect produced. The fig. 51 will explain this: it is supposed to be the ground plan of a fireplace with straight sides; A B C are the jambs, sides, and back of the fireplace, and D the body which radiates heat. If now a ray *w* falls on the side, it will be reflected at the same angle, and fall

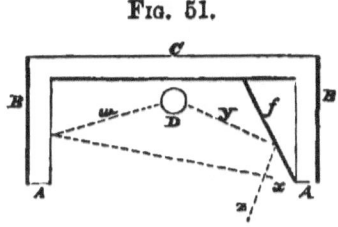

Fig. 51.

just within the jamb at *x*; but if the side be inclined, as shown by the line *f*, then a ray *y* falling upon it, will be reflected into the room in the direction of *z*, and of course a vast difference in the effect will be experienced.

But this author likewise recommended other improvements. He advised a considerable contraction of the breast of the chimney, by which less heat would escape through the funnel or shaft of the chimney, while at the same time he proved that the smoke would escape with equal facility as before. Another improvement which he suggested was in making the hearth and back of the fireplace hollow, by means of metal plates, so that by having these hollow spaces to communicate with the external atmosphere, the air in passing through them would be warmed before it entered the room, and would prevent the cold currents of air which otherwise would enter through the crevices of the doors and windows. This construction of fireplaces was intended for the burning of wood, and as the fire. was therefore merely laid on the hearth, the effect of these hollow spaces, and particularly the hollow hearth, would of course be very considerable, by warming a large quantity of air.

(288.) Subsequent inventors have been much indebted to this work by M. Gauger. Franklin acknowledged the great assistance he had derived from it; and the methods of economizing fuel, afterwards so successfully introduced by Count Rumford, in his improved stoves, are all similar in principle to the plans recommended by the French author.

(289.) In the year 1796, Count Rumford published his *Essays on the Management of Fire and the Economy of Fuel*; and he there describes those improvements which have ever since that time been followed in the construction of stove grates. The error shown by M. Gauger to exist in the construction of fireplaces, by making the sides parallel to each other, was, at the time that Rumford wrote his treatise, still continued. In pointing out the error of this mode of construction, he showed that, in order to obtain the greatest effect from the fuel, the sides of the fireplace ought to be placed at an angle of 135° with the back of the grate, or (which is the same thing) at an angle of 45° with a line drawn across the front of the fireplace. This angle must necessarily reflect the greatest number of rays into the room; the difference of effect between this mode of construction and that of the parallel sides being very great. The reduction in the size of the throat of the chimney was likewise another improvement which he effected, though this also had been recommended by M. Gauger nearly a century previous. The angular covings for the sides of the fireplaces, Rumford considered should not be formed of iron, but of some non-conducting substance, such as fireclay, in order that more heat might be reflected from them into the room.

A circular form for these sides or covings, he considered produced eddies or currents, which would be likely to cause the chimney to smoke; and he likewise objected to the old form of registers or metal covers to the breast of the chimney for the same reason, and because by their sloping upwards towards the back of the fireplace, they caused the warm air from the room to be drawn up the chimney, and thus impeded the passage of the smoke. These registers are now made so as to be lower at the back than at the front of the stove; but, in general, they are placed far too high up. And the very same reasons which decide the angle of greatest effect for the cheeks or sides of the fireplace to be 45°, also apply to this case; and a large quantity of heat now lost by the ordinary register stoves would be saved if the register top were placed at this angle, and sufficiently low down to allow it to reflect the heat from the fire into the room. The dimensions of the firegrate itself Count Rumford recommended to be much less than formerly; and the best proportions for the chimney recess, he stated, were that the width of the back should be equal to the depth from front to back, and the width of the front, or the opening between the jambs, should be three times the width of the back.

(290.) Although the best form for register stoves has now for several years past been adopted, the desire for novelty has caused the true principles of construction to be frequently departed from; and we accordingly find, in the most modern stoves, considerable deviations from these principles. Fig. 52 is a section of a register stove constructed on the best possible plan for diffusing

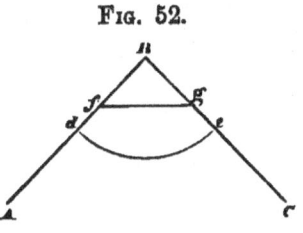

FIG. 52.

heat into the room. The two sides are a right angle of 90° A B C; and the bars *d e* describe a quadrant of a circle whose radius is just half the length of the side A B. If now we wish to follow Rumford's rule of making the back one-third the width of the front, we obtain this by taking one-third of the length A B, which will give B *f*; and then, if we draw the line *f g*, we shall obtain exactly the required dimensions. By this arrangement it will be perceived that the sides of the stove form an angle of 135° with the back; and all the rays of heat which fall upon these sloping sides will therefore be reflected into the room, directly in front of the stove, in right lines. The falling cover, or register top, should also form an angle of 135° with the back, by which a large portion of heat will be radiated downwards into the room. These proportions however, cannot well be adopted in stoves of very large size, as they will be found to throw the stove too far back; but for all moderate-sized stoves no form can be adopted which will produce so good an effect.

(291.) Various methods have been contrived to render available some further portion of the heat which is given off by the fuel during combustion in these stoves, in addition to that which is obtained by radiation. These contrivances are nearly all of them mere modifications of that pointed out by M. Gauger, by means of the double back and hearth of his improved fireplaces. Notwithstanding this invention is at least 150 years old, it has, during the last thirty years, been repeatedly brought forward, and more than once patented, by different persons, as a new invention. The principle has been the same in all the different cases, with but very little difference in the mode of applying it. A current of air is brought from the external atmosphere, and is made to pass through a small flat box at the back of the stove; the back of the stove itself forming one side of the box; and in order to prevent the air from escaping into the room before it is sufficiently warmed, the box is divided horizontally by several partitions so as to check the passage of the air, which is carried successively through them all before it escapes into the room. This mode of warming the air is exceedingly useful, and is capable of various modifications. It is not only economical in fuel, but frequently is very efficacious in remedying smoky chimneys, and in preventing those cold draughts from doors and windows, which are so exceedingly unpleasant and unhealthy.

(292.) A most efficient mode of applying this principle is to enclose the whole of the register stove in an ornamental case. In this way the stove stands forward in the room a few inches, and the case forms an air-chamber entirely round the back and sides of the stove, by which a very large heating surface is obtained, which moderately warms the air, while the whole effect of the radiant heat is obtained from the open fire, the same as in an ordinary register stove. When the air-chamber communicates with the external air, this forms a very excellent stove, and the atmosphere is purer than when the air of the room only is made to pass over the heated surface.

(293.) A stove invented by Mr. Jeffrey (the inventor of the well-known instrument called the Respirator) accomplishes the same object in a somewhat different manner. This stove stands extremely prominent in the room, without being at all sunk in the wall as fireplaces usually are. The entire back of the stove above the fire consists of a series of flat tubes, one inch wide, and about nine inches deep from back to front, which are placed edgeways, one inch apart from each other, so as to present alternately a close and an open space of one inch in width. These tubes are about eighteen inches long, and reach quite to the top of the stove, and pass just through the mantel-piece. The only way by which the smoke can escape into the chimney is by passing through the openings left between these flat tubes, which are so connected at the top as to prevent the smoke passing into the room. The lower end of these tubes opens into a small air-chamber which communicates with the external atmosphere, and the air therefore passes from this chamber through the flat tubes, and escapes at rather an elevated temperature into the room from the upper part of the tubes; having been warmed in its passage through the tubes by the heat which has been abstracted from the smoke passing between them. Nothing new in principle is obtained by this arrangement; the only difference between this plan and that of the many previous contrivances for the same purpose is, that the heat which warms the air is derived from the smoke, in the others it is obtained by bringing the air into contact with the surface that is heated directly by the fuel contained in the stove itself. The method employed by Mr. Jeffrey appears very likely to cause the chimney to smoke, unless the draught is particularly good ; for the smoke must have so large a portion of its heat abstracted by passing between the flat tubes, that its power of ascending must necessarily be very much reduced. The stove also presents some practical difficulties in removing the soot from the chimney, for which purpose a part of the jamb has to be removed.

(294.) Dr. Franklin, in 1744, invented a stove for burning wood, which he called the Pennsylvania stove; in which he introduced the principle of heating the air much in the manner first recommended by M. Gauger, by means of a double or hollow back. This stove* was in the form of an oblong box with the front removed. At about three or four inches from the back of this box, a flat close chamber was fixed, three inches deep, the whole width of the stove, and reaching to within about four inches of the top. The smoke escaped over the top of this flat chamber, and passed downwards between it and the real back of the stove, and thence passed into the chimney. This hollow chamber communicated underneath the stove with a tube opening into the external atmosphere, and a considerable quantity of air thus passed through the flat chamber, and escaped into the room through small holes left at the sides, after traversing the length of the chamber three or four times by means of divisions placed across it for that purpose. The heated surface of the stove itself also warmed the air of the room, and a large quantity of radiant heat was also given off from the burning fuel. This stove is very economical. It was a good deal used in America, and some of them have been used in England, in those parts of the country where wood is abundant.

(295.) An ingenious stove was proposed in 1815, by Mr. Cutler, and called the torch stove.† It possessed an open fire, exactly like an ordinary register stove, but it was made to consume its own smoke on a very ingenious principle. Below the bars of the grate there was a deep box which sank down into the hearth, and which contained the fuel. The fire was lighted at the top, and as the fuel burned away, the box was wound up by a chain passing over a rack and pinion, placed out of sight, in the interior of the stove. By this means the fuel burned only at the top; and the quantity of heat could be regulated at pleasure,

* Franklin's "Works," vol. ii., p. 225, et seq.
† Repertory of Arts, vol. xxviii. (second series), p. 203.

according to the height to which the fuel was supplied from the box below. For, as the sides of the box were solid, and therefore no air could pass through the fuel it contained, the combustion took place only in that part of the fuel which was raised up to the level of the firebars : and as this part of the fuel always burnt clear, the raw coal below was gradually and slowly heated, and the gaseous parts were consumed while passing slowly through the ignited fuel at the top. This stove was rather cumbrous in appearance, and there was an inconvenience attending its use, arising from the trouble of relighting the fire and filling the box with fuel, when the latter did not contain a sufficient quantity to last the entire day ; and this alone is almost sufficient to prevent its general use. The principle, however, of lighting the fire at the top, and supplying fresh fuel from below, is undoubtedly good, as it affords the most perfect control over the intensity of the fire, and consumes nearly the whole of the smoke. To obviate the inconvenience of supplying the coals in this way, and still retaining the advantage of burning the fuel from the top, downwards, Mr. Dowson patented a plan* for supplying the fresh fuel below the ordinary bars of the grate, whilst the fire was burning, by a peculiar apparatus, which delivered the fresh coals at the bottom of the grate ; but neither this, nor the plan proposed by Mr. Cutler, ever came at all into extensive use, as both are troublesome and expensive. Dr. Arnott has very recently (1854) again revived Mr. Cutler's plan of the torch stove, and has proposed it as a remedy for the inconveniences, real and imaginary, which are attributed to our ordinary open grates or fireplaces. There is not the least probability that this plan will ever come into gene-

* *Repertory of Arts*, vol. xxix. (second series), p. 76.

ral use; for its greater cost, its complicated management, its practical inconvenience in replenishing the fuel, and its somewhat difficult adaptation to existing fireplaces, all militate against it, even were its advantages greater than they really are. In fact the plan, somewhat modified, has been several times brought forward. Several years ago it was proposed in America, by simply making each side of the stove consist of a hollow box, of which the hob formed the lid. The bottom of the box was level with the bottom grate of the stove; and an opening from the side of the box, close to the bottom, allowed the coals which were put in these side boxes to be pushed *under* the burning fuel in the grate, and thus produced the same effect as the winding box of Cutler's stove. This plan, like each of the others, is attended with difficulties in practice; for, although descriptively it appears very easy, in practice it is found extremely difficult to force any sufficient quantities of fuel under a mass of burning coals.

(296.) Another modification of the register stove is the invention of Mr. Sylvester. In this stove the hearth consists of a great number of hollow bars fitted into an appropriate frame. The hollow bars not only form the hearth, but at their extreme end the fuel is placed, and the bars thus form the grating on which the fuel is burned. The air to support the combustion passes through the hollow bars, and also through the front of the fire. The ends of these bars on which the fire is placed become hot, and the remaining part of the bars also becoming heated by the conducting power of the metal, a radiation of heat into the room is produced by the bars which form the hearth. The fire being placed as it were on the hearth, the cheeks or sides of the stove are of much greater length than usual, and a larger quantity of heat is radiated into the room. The smoke escapes through openings in the back, which are placed something like louvre boards. The ashes from the stove fall into a box below the hearth, which requires to be occasionally emptied by removing the loose bars forming the hearth.

The principle of this stove is extremely good, but the arrangement is much too expensive; and there are several inconveniences which must prevent its general use, not the least of which is the extensive alteration in the brickwork and woodwork which surround the fireplaces, in order to adapt them to rooms not purposely built for them. But there is ample room for improvement in the ordinary open fireplaces, and many ingenious inventions already exist for this purpose. The great obstacle, however, to their general use, even supposing their merits greater than they really are, is the extensive alterations generally required when they are made to replace stoves of the old construction. For while several of the modern stoves economise fuel to a considerable extent, their expense puts them beyond the reach of those to whom this economy of fuel would be important; and to those who can afford the expense of their application, the economy of fuel is a matter of indifference. The old principle of open stoves is undoubtedly wasteful of fuel, as a large portion of heat escapes up the chimney; but it is doubtful whether any of these modern inventions yet accomplish great economy of fuel, accompanied with moderate cost and easy application.*

(297.) The stoves which come now to be mentioned are of a different character, and form that class (the only one known in many parts of

---

* In the Report of the Commissioners presented to Parliament, 25th August, 1857, "On Warming and Ventilating Dwellings," there is a very good epitome of all the patents, 91 in number, taken out for stoves between the years 1781 and 1857.

Europe) which heat the air by contact with their surfaces, and not by radiation directly from the burning fuel itself. In most of these stoves the fire is wholly concealed from view; while in a few of modern invention, a part of the fire is dimly seen through talc placed in the front.

(298.) In the north of Europe, close stoves are alone used for heating buildings. In Russia and Sweden, the stove is generally made of brick, tiles, or stone, and it occupies a large space at one end of the room. It is usually either square or oblong, and is divided by partitions into different compartments, so as to increase the surface over which the smoke and heated gases pass before they finally escape into the chimney. The materials which compose the stove being slow conductors of heat, they retain the heat for a long time when once warmed; and these stoves seldom require replenishing with fuel more than once a day. They usually burn wood for fuel, and are supplied by a firedoor exterior to the room intended to be warmed. M. Guyton endeavoured to introduce these stoves into France at the close of the last century, and paid much attention to the best form of construction and the comparative cost of fuel, the results of which were published in the *Annales de Chimie*.[*]

Mr. Cundy took out a patent in this country, in 1844, for a stove made of fireclay, which it was supposed would combine the advantages of an open fire with the mild heat from earthen or pottery surfaces. The large size which such a stove requires to be, in order to afford sufficient heating surface for a building of any considerable extent, must necessarily prevent its general application; and it has not hitherto proved generally successful where large areas are required to be heated.

[*] See also *Repertory of Arts*, vol. xvi., p. 254, *et seq.*

(299.) In Germany, iron stoves are used, which heat the air by contact with their surfaces. A very superior stove is also used, made of glazed earthenware, which is very similar to the Swedish stoves just described, and which gives a mild and agreeable heat. This stove necessarily occupies much room; but it is decidedly the best construction for a close stove, as the quality of the air is less injured by it than by the heated metal from ordinary close stoves. With all these close stoves it is usual to employ a vase of water to supply moisture to the air, to prevent the unpleasant effects which would otherwise be experienced, and which will be noticed in the following chapter.

(300.) In this country, hot-air stoves constructed of iron have been usually employed for warming large buildings, until the introduction of the plan of warming by steam and by hot water.

The hot-air stove is too well known to need description. It may be observed, however, that the great defect of these stoves is that they heat the air too highly, and thereby render it unwholesome; and whatever plan of construction tends to increase the heated surface exposed to the air without increasing the size of the firebox itself, will of course lower the temperature of the surface which heats the air, and thereby render the stove less objectionable. This has been accomplished in a very ingenious manner by Mr. Sylvester, by means of covering the cast-iron case, which receives the heat from the fire, with iron ribs projecting three or four inches beyond the surface of the case. These ribs greatly increase the surface, and thereby reduce the temperature, which is the great desideratum in these stoves.

(301.) The method of heating by cockle stoves is but little more than a modification of the ordinary hot-air stoves adapted to a larger scale. Mr. Strutt, of Belper, in Derbyshire, appears to have been the first person to introduce an improved method of heating by cockle stoves; and from the year 1792, when he first warmed his large cotton factories in this manner, various improvements have been made in these stoves, which, without at all altering their principles, have rendered their application more general.

The cockle stove consists of a very thick iron case, which forms the top and sides of the furnace. This case, or cockle, is enclosed in another case of brick or stone, placed so as to allow a space of three or four inches, or more, between them in every part; and appropriate openings are left for the admission of cold air at the bottom, and for the emission of the hot air at the top, which is from thence conveyed through channels or pipes to any place which is required to be warmed. A vast number of ingenious contrivances have been proposed for the improvement of this apparatus, and for many years it was the principal method of warming all the large buildings in England. It has been sometimes called the Belper stove, after the name of the residence of the inventor. The unwholesome effects of these stoves, however, have caused them to be now nearly superseded by the use of hot-water pipes, which, from their lower temperature, are free from the injurious tendency which the hot-air stoves have always been found to exhibit.*

(302.) A stove invented by Dr. Nott, of Philadelphia, has been found to produce very consider-

---

* The heating power of these cockle stoves is very considerable. Mr. Sylvester states, in his work "On Heating the Derby Infirmary," that a cockle which has 17 square feet of heating surface warmed 344,600 cubic feet of air 56° in twelve hours, with the consumption of 60 lbs. of coal. This is equal to 26,768 cubic feet raised 1° per minute by 17 square feet of

able effect with but small expenditure of fuel, and requiring very little attention. This stove is usually something of a pyramidical form. The lower part, which forms the firebox, is lined with firebricks, and the stove is divided vertically into two compartments. The fuel is put into this stove through an opening near the top, which forms a reservoir for the fuel, and occupies the front part of the stove, and the smoke passes downwards through a grating placed at the bottom, and then escapes through the back part of the stove into the chimney. The air for the support of the combustion enters the stove principally on the top of the fuel, and a small portion also enters below. As the fuel burns but slowly in consequence of the small quantity of air admitted, the firebox and reservoir, when once filled, will supply fuel for several hours, and give a very great heat. In fact, the heat is generally so considerable, and the air is thereby rendered so arid, that it is extremely unpleasant and unwholesome to those who are exposed to its influence.

(303.) One of the most economical stoves as regards the consumption of fuel, that has yet been invented, is that which has been introduced by Dr. Arnott. This invention is an improvement (in some respects) upon Dr. Nott's stove above described, the principal difference being the limiting the admission of air by which the combustion is regulated, and by separating the burning fuel more perfectly from actual contact with the heat-

heating surface, or 1574 cubic feet of air raised 1° per minute by each square foot of the surface of the cockle. This is just 7 times the effect produced by hot-water pipes (See Art. 99 and 105); but in economy of fuel it is inferior, being in the proportion of about 19 to 22. For a full description of the cockle stove, see chapter xi., part i.

ing surface of the stove, by which means the
excessive heat of Dr. Nott's stove is in a great
measure avoided. The plan used in these stoves
for regulating the admission of air was, long previous to its application by Dr. Arnott, employed
for limiting the intensity of the heat of furnaces;
for which invention Dr. Ure, some years since,
obtained a patent. Dr. Arnott's stove consists of
an external case of iron, of any shape that fancy
may dictate; within this case a fireclay box, to
contain the fuel, is placed, having a grating at the
bottom, and a space is left between the firebox and
the exterior case, so as to prevent, as much as
possible, the communication of too much heat to
the exterior case. The pedestal of the stove forms
the ashpit, and there is no communication between
the stove and the ashpit, except through the
grating at the bottom of the firebox. A small
external hole in the ashpit, covered by a valve,
admits the air to the fire, and according as this
valve is more or less open, the vividness of the
combustion is increased or diminished, and thence
the greater or less heat produced by the stove.
The quantity of air admitted by this valve is
governed by a self-regulating apparatus, either by
the expansion and contraction of air confined by
mercury in a tube, or by the unequal expansion of
two bars of different metals riveted together, on
the plan proposed by Dr. Ure.* The smoke escapes
through a pipe at the back of the stove: very little
smoke, however, is eliminated from these stoves,

* This ingenious and simple invention consists of two thin
narrow bars, one of brass or copper, and the other of iron,
about eighteen inches or more in length, riveted together at
each end. On subjecting this compound bar to heat, the
brass expands considerably more than the iron, and the
elongation of the bars causes them to open in the centre and
recede from each other (forming a bow), in consequence of

the fuel being always either coke or anthracite coal. By adjusting the regulator so as to admit only a small quantity of air, the temperature of the stove is kept within the required limits; and owing to the slow conducting power of the fireclay, of which the firebox is formed, the heat of the fuel is concentrated within the firebox, and the fuel burns with less air, and therefore more slowly than it would otherwise do. This slow combustion of the fuel, however, produces a large quantity of carbonic oxide, which is liable to escape into the room, and being a strong narcotic poison, is attended with considerable danger to those who breathe it. The escape of this gas from these stoves has been experimentally ascertained by Dr. Ure, by attaching to the ashpit of a stove a glass vessel containing a solution of sub-acetate of lead, which was speedily acted upon by the carbonic gas, and formed into the insoluble carbonate of lead.[*]   Carburetted hydrogen gas is also frequently formed in these stoves, and many dangerous explosions have in consequence occurred, and many calamitous fires have been produced by them. They require, therefore, to be used with the utmost caution. The endeavours to prevent these explosions have hitherto been unsuccessful; and so long as the principle of the extreme slow combustion and small admission of air is preserved, it will probably be impossible to prevent them occasionally taking place. They arise from the inflammable gases, generated during the combustion of the fuel, being detained in the

their being riveted together at both ends. The greater the heat, the more the centres of the bars recede from each other, or the greater the arc of the bow becomes; and by combining together a series of these bars, almost any extent of motion may be obtained for the purpose of opening or closing valves or air passages of furnaces or ovens.

  [*] Ure's "Dictionary of Arts," &c.; art. "Stove."

stove and in the chimney by the want of sufficient draught; and they are particularly liable to occur when the chimney is very large, or if it consists merely of an iron pipe and is much exposed to the cooling influence of the atmosphere. In such cases the heat which escapes into the chimney is insufficient to cause the ascent of the liberated gases; they therefore collect in the stove, or in the chimney, and the explosive effects occur whenever the air and the gases mix together in certain proportions. The compound thus formed is similar to the gas known in mines by the name of firedamp; and this mixture will explode whenever the carburetted hydrogen is not less than one-twelfth and not more than one-sixth of the whole mass. The explosion of carbonic oxide only takes place under particular circumstances; but red-hot charcoal will cause it to explode when mixed in the proportion of two measures of oxide to one of atmospheric air.[*] Dr. Dalton has remarked other circumstances under which this gas may explode when mixed with atmospheric air, the carbonic oxide being not less than one-fifth, and the oxygen not less than one-thirteenth of the whole mixture.[†] Many plans to remedy these evils have been proposed, principally by the admission of air above the fuel as well as through the ashpit in the usual manner; but occasionally this increases the evil, during certain stages of the combustion, though generally the effect is to increase the temperature of the stove so much as to render it extremely unpleasant, and liable to all the objections which exist against the old hot-air stoves. The immense number of serious accidents which have occurred from these stoves ought to render those persons who use them extremely careful. The great source of danger

[*] Dr. Henry's "Chemistry," vol. ii., p. 347.
[†] Dr. Dalton's "Chemical Philosophy," p. 373.

arises from their fancied security. This erroneous notion has caused them to be placed in situations not sufficiently protected; and either by explosions, or by the stoves from accidental causes becoming red-hot, many buildings have been set on fire, and most serious damage sustained.*

(304.) A very excellent contrivance for a stove, which burns coal, and at the same time consumes its own smoke, was invented by Dr. Franklin in the year 1771. Nearly a century previous to that time, M. Delesme, a French engineer, described an exceedingly rude contrivance on a similar principle, which was afterwards mentioned by Dr. Leutmann in a work on stoves, published by him in Germany, in 1723. Dr. Franklin expressly acknowledged that this stove of Delesme gave rise to his own invention. This stove is in the shape of a large vase, of which fig. 53 is a section, and it is fully described in Dr. Franklin's works.† Near the bottom of the body of the vase a grating c is placed, on which the fuel rests, and the top A B opens for the purpose of supplying the fuel : D is a square box forming the pedestal of the stove, at the bottom of which another grate is fixed to allow the ashes to fall into the box E. This latter box is open at the back, and communicates with the

* The author is in possession of an extensive list of accidents caused by these stoves, some involving total destruction of the buildings and many more of serious damage. In one of the former three lives were lost; and among buildings totally destroyed is to be reckoned Okehampton Church and (almost beyond a doubt) the Armoury at the Tower of London in 1841. Howth Castle near Dublin; Larkfield House, Sussex; and a vast number of other private and public buildings have had narrow escapes from destruction from this cause. Wherever these stoves are used the utmost caution ought to be observed to keep them insulated from everything of an inflammable nature.

† Franklin's "Works," vol. ii., pp. 296–300.

hot-air passages ff and gg, through which the heated gaseous matter given off from the fuel passes before it enters the chimney. It will be perceived that the draught of this stove is *downwards*; that is, the air enters at a small hole, about one and a half inch diameter, at the top of the stove, passes through the fuel in the vase, and escapes, together with the gaseous products of the fuel, through the pedestal D, the box E, and the hot-air passages ff and gg. That portion of the fuel which lies on the grate c is always red-hot; and the smoke

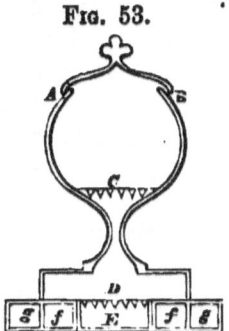

Fig. 53.

from any fresh fuel having to pass through this heated medium, is consumed, and the gases pass off in a clear and almost invisible state, so that no smoke lodges in the air-passages, or in the chimney. Like all stoves with a downward draught, the fire is troublesome to light in the first instance; but when once lighted, it can be made to burn for almost any length of time without attention by merely adjusting the opening at the top for the admission of air, and the consumption of fuel is extremely small.

This stove, with some slight alterations, has been brought before the public on more than one occasion since Franklin's time as a new invention, and within the last few years it has been again brought forward with arrangements very decidedly inferior to those proposed by Franklin. As designed by him, this stove is the best and most economical of its class, superior to Dr. Arnott's stove in its equality of temperature, economy of fuel, and permanence of action; though like the latter, and all others of the same general character, the want of ventilation, and the peculiar effects produced on atmospheric air by highly heated iron, render it undesirable for constant use in rooms for ordinary and domestic occupation.

(305.) The application of a method of burning carburetted hydrogen or coal gas, for the production of artificial heat for warming buildings, is an invention of rather recent date. The apparatus in which this is effected is very simple. A metallic ring, pierced on its upper side with a great number of holes of very small size, is connected with a pipe communicating with the gas main, and is placed within a double drum or cylinder of iron, raised an inch or two from the floor on small legs. This double drum is so made that there is a space between the inner and outer cylinder of about two inches; and in this space, near to the bottom, the ring pierced with holes is fixed. A stop-cock in the pipe connecting the pierced ring with the gas main shuts off the supply of gas when the stove is not in use. On opening this cock, and applying a light to the pierced ring, a brilliant ring of flame is immediately produced, which soon warms both the inner and outer case of the stove by the heat generated during the combustion of the carburetted hydrogen gas. The top of the drum is covered with a large open ventilator, by which the heated air that passes through the inner cylinder is allowed to escape into the room; but the products of the combustion, having no means of escape, pass downwards from the bottom of the stove into the room. This having been found exceedingly unwholesome, a plan has been contrived by which a considerable portion of the products of the combustion are carried off by a pipe inserted between the two cylinders, which conveys away the gaseous products into the open air.

A moderate sized stove of this description burns from 12 to 15 cubic feet of carburetted hydrogen gas per hour; and this is converted into two new compounds—water and carbonic acid gas. The quantity of water formed will be 2·6 cubic inches for each cubic foot of carburetted hydrogen, or about a pint to a pint and a quarter of water per hour; and from 12 to 15 cubic feet of carbonic acid gas, and *eight* times that quantity of nitrogen, will be the constant products of the combustion, and produce the most serious deterioration in the quality of the air. When these gaseous products are carried off by a chimney, the loss sustained will be very nearly one-half the total heat which is produced by the burning of the gas.

(306.) Dr. Dalton's experiments have proved that the combustion of one pound in weight of carburetted hydrogen generated sufficient heat to melt 85 lbs. of ice. A cubic foot of this gas weighs 292·89 grains; and a stove burning 15 cubic feet an hour, for fifteen hours a day, will consume 225 cubic feet, or 9·41 lbs. of gas, which therefore would melt 799 lbs. of ice ; and the cost of this quantity of gas at the present price of three shillings and sixpence per 1000 cubic feet, would be nine-pence halfpenny. The latent heat of water being 140°, the same quantity of heat that would melt 799 lbs. of ice would heat 179·2 cubic feet of water 10°. And as one cubic foot of water will raise the temperature of 2990 cubic feet of air as many degrees as the water loses (Art. 97), the combustion of 225 cubic feet of carburetted hydrogen gas would raise the temperature of 535,808 cubic feet of air 10°. This, then, will be the total effect of a stove of this kind. But we find (Art. 94), that 15·91 lbs. of coal will produce exactly the same quantity of heat, the cost of which, at 24 shillings per ton, will be only about twopence. It therefore appears that the cost of fuel for a gas stove without a flue will be about five times as much as a hot-air stove that burns coal, and about ten times as much as for coal, if the gas stove has a flue for carrying off the products of the combustion.* Sir John Robison, who did much to improve the gas stoves, and particularly in applying them to the purposes of cooking, in a paper read before the Society of Arts for Scotland, after describing some of his improvements, concluded by stating, "on the whole, it may be assumed that this mode of heating apartments is the most expensive, the least efficient, and, excepting that by Joyce's charcoal stove, the most insalubrious that can be resorted to." †

Since the early editions of this work were written, an improved method of burning gas, where heat and not light is required, has been introduced. It consists in mixing atmospheric air with the gas at some distance before it reaches the actual place of combustion. The quantity of air thus mixed is about twelve measures of air to one of gas, and by this means the expenditure of gas is reduced one-half when the same effect is produced. (See Art. 66, *ante.*) Scarcely any light is produced when gas is burnt in this way; but where heat only without light is required, the saving is very considerable, the cost of gas being then about three times the cost of coal.‡

---

* See Art. 66, *ante.*

† *Mechanics' Magazine*, vol. xxxii., p. 292.

‡ It may be doubted whether the most economical method of burning gas has yet been practised. The burning of "atmospheric gas" in the Bunsen burner will probably be further improved if the true principles of combustion be studied. There can be no true and perfect combustion until the gaseous products attain a temperature of between 800° and 900° of Fahrenheit (Art. 391 *et seq.*). If this heat, or any considerable portion of it, could be communicated to the atmospheric air, before it was mixed with the carburetted hydrogen, there is every reason to suppose greatly increased economy would be the result. The vast economy by the use of the "hot blast," in the manufacture of iron is a well known scientific fact: and further that the nearer the tem-

(307.) The stove invented by Mr. Joyce, for burning charcoal, has been so universally admitted as too unwholesome for use, that it would be almost unnecessary to describe it, were it not for the vast interest it excited on its first announcement to the public. The stove consists of a thin metal case, generally in the form of an urn or vase. Through the bottom of this, there is a small pipe which rises for two or three inches into the body of the stove, and terminates about the centre in a conical-shaped funnel, closed at the top, and pierced full of holes. This pipe conveys air to the fuel; and at the top of the stove there is a valve or regulator, by which the rate of combustion can be controlled; for no more air will enter the lower pipe than is sufficient to replace the volume of gas given off from the valve at the top of the stove. A small quantity of ignited charcoal being placed in the stove, the remaining space is filled up with charcoal not ignited; and the combustion is slowly carried on by the air, which enters at the lower pipe, and can be continued for a vast number of hours without any attention. The whole of the charcoal is converted into carbonic acid gas, the effects of which are fatal to animal life. Little or

perature of the air can be raised to the point of complete combustion (800° or 900°) the greater the economy that results from the operation. There is but little doubt that if the air admitted to the Bunsen burner could be heated to 400° or 500° Fahrenheit before it entered into combination with the carburetted hydrogen gas in the Bunsen burner, a greatly increased economy would be the result. It is by no means easy to effect this process of extrinsically heating the air to any considerable extent, as it involves many difficulties of a practical character (see Art. 217, &c.); but it is probable that any considerable improvement in the combustion of gas may be looked for in this direction, if greater economy is to be obtained, than that which results from the present Bunsen burner, and the employment of the "atmospheric gas."

no smell is emitted by these stoves, the charcoal being deprived of its usual pungent quality by reburning it thoroughly in a close oven, and quenching it while hot with an alkaline solution. Independent of all other considerations, these stoves are a most expensive mode of producing heat, owing to the high price which charcoal always bears in this country.

These stoves are remarkable chiefly for the extraordinary interest they excited on their first introduction to public notice. Before the specification for the patent was enrolled, one of these stoves was publicly exhibited for several weeks; and probably no invention ever excited so much attention in so short a time ; as the nature of the fuel and plan of combustion were strictly kept secret, and it was supposed a new era in the production of artificial heat was about to commence. No sooner did the stoves come into use, however, than their deleterious effects were apparent. Several persons were suffocated by using them ; and the high anticipations which at first were entertained of them were quickly dissipated, and their use reduced within very narrow limits.

(308.) An immense number of stoves have been brought before the public during the last few years, of which the novelty consists in nothing but their names. To describe them would be useless. Like other ephemeral productions they will mostly sink into oblivion, and probably be succeeded by others as devoid of originality. There is still room, however, for great improvements in the production of stove heat; but, until the salubrity of the open firegrate can be combined with the greater economy of the close stove, there does not appear much probability of satisfying the required conditions, as all which have hitherto been invented are deficient in one or other of these particulars.

(309.) The same objections also, to a greater or less extent, apply to all the methods of heating by flues. The extremely unequal temperature of the flues causes an insurmountable objection to their general adoption, even if their great expense and difficulty of adaptation to dwelling-houses or public buildings did not operate against them. One of the best systems of flues was proposed a few years since, by Mr. Alfred Beaumont, in which the principal novelty consisted in the furnaces being built *above* the flue and having a downward draught. By this means the smoke from the fire was nearly or quite consumed, and a very great evil of common flues thereby prevented, as the surface of the flues being by this means free from soot, which is a non-conductor of heat, a more free distribution of heat took place than by the old flues, and the fire was also much more under control. The principle, already described, for downward draughts applies equally to this plan. The furnace (in some cases) was merely a circular hole sunk in the floor, about two or three feet deep. The bottom, when in use, was a solid plate; which was movable, to take away the ashes. About three inches from the bottom there was an entrance into the flue; through this the flame and the products of combustion passed to the flues, and the fire being always bright at the bottom, the combustible gases from the fresh fuel, by passing through this, were converted into other products, which did not deposit smoke. In other cases, a stove made of fireclay was raised on the floor; and the flame and heated gases passed into the flues, which formed the stone flooring of the room to be heated. In all these cases the principle was the same; and several public buildings were heated on this plan, which is probably the best of any system of flues.

The Palmaise system of heating was, a few

years since, a great favourite with some persons
who anticipated from it results which a very
slight acquaintance with true science would have
shown to be utterly fallacious. This plan was in
reality nothing more than the old mode of flue-
heating, with a very slight modification in the
form and construction of the furnace. The plan
is now so completely abandoned, that it hardly
merits more than a very casual description. The
furnace was merely a plain square or oblong box
of firebrick, with a strong cast-iron plate forming
the entire top; and beyond this were the usual
flues, such as were ordinarily used in the old plan
of heating greenhouses. The only novelty, therefore, consisted in using a cast-iron plate to cover
the top of the furnace; and this plate becoming
intensely heated, the air was brought in a continued stream over it, thus imbibing heat in its
passage into the room or building to be heated.
It is useless to point out the defects of this plan;
they were numerous and palpable. It combined
all the defects of the old flue system, added to
those of the cockle stove, and others peculiar to
itself, arising from its primary construction, which
rendered it so liable to accident and derangement
that it was found utterly valueless for any practical purpose. Its name was derived from a small
town in Scotland, where this particular form of
flue-heating was first used; and its supposed
cheapness of construction obtained for it at first
many advocates, who, however, shortly found
that its want of durability rendered it unsuitable
for general use, while in other respects it was
decidedly inferior to the old system of heating by
flues. During its short-lived popularity it, like
many preceding inventions, was supposed to be
the perfection of economy and efficiency, which
belief was soon dissipated when the apparatus was
applied to general use.

CHAPTER III.

ON THE CHANGES PRODUCED IN ATMOSPHERIC AIR,
BY HEAT, COMBUSTION, AND RESPIRATION.

Necessity for Ventilation—Constitution of the Atmosphere—
Ventilation first proposed in Sixteenth Century—Subse-
quent Inventions for Ventilating—Effects of Contaminated
Air on the Human Frame—Effects of Climate on Health
and Longevity—Cause of Miasmata—Effects of Excessive
Moisture and Excessive Dryness of the Atmosphere—Re-
spiration: its Products and Effects—Impure Air—Carbonic
Acid Gas—Vapour from the Body—Effect of Diminished
Pressure—Electric Condition of the Air—Production of
Ozone—Decomposition of Extraneous Matter in the Air—
Effects of Hydrogen, Carburetted Hydrogen, and Carbonic
Oxide—Quantity of Air required for Ventilation—Impor-
tance of the Air to Animal and Vegetable Life.

(310.) PROBABLY no subject connected with the
health and vigour of the mind and body deserves
more and receives less attention than the condition
of the internal atmosphere of our houses and
apartments. Attempts are indeed occasionally
made to introduce some system of ventilation ·in
public buildings, but they are far more frequently
unsuccessful than otherwise, in consequence of the
arrangements not having formed any part of the
original plan, and being mere additions of a very
imperfect character. In private dwellings, ventilation appears never to be considered as at all
necessary; but if the contaminations and impurities that are frequently contained in the air, which

forms the *pabulum vitæ* of human beings, could be
seen by the eye, in the same way as contaminations
or impurities in ordinary alimentary food, the evil
would not be endured for even the smallest period
of time.

(311.) The real constitution of the atmosphere
has been known, comparatively, but a short time;
the experiments of Priestley, Scheele, and Lavoisier,
in the latter part of the last century, having first
made known its true nature. More than twenty
centuries previously, however, Hippocrates wrote
so justly on the immense importance of breathing
pure air, of the great influence which it exerted on
health and longevity, and laid down such excellent
rules upon the subject, that few writers, even of
the present day, appear to have more correct notions
of the vast and important effects which the air
produces on the human frame, than were possessed
by this great father of medicine.[*] Agricola, however, in the sixteenth century, appears to have been
the first writer on artificial ventilation; he having
recommended the ventilation of mines by producing
a current of air by fire, much in the same manner
as has been practised in the ordinary ventilation of
mines ever since his time. Nothing of any considerable importance occurred after this, until Desaguliers
in 1727 proposed a ventilating pump,[†] which some
years afterwards Dr. Hales[‡] applied in a better
manner, and to a considerable extent, in the
ventilation of ships, hospitals, prisons, and other
places, which were found to be unwholesome by
confined air. In 1734, Desaguliers invented a
centrifugal ventilating wheel, or fan; and in
1739 Sutton proposed a plan of ventilating ships,

---

[*] "Hippocrates," lib. "De Aere, Aquis, et Locis."
[†] *Philosophical Transactions*, 1727, vol. xxxv., p. 353.
[‡] *Ibid.*, 1743; and Dr. Hales "On Ventilators." London,
1743.

which consisted in drawing the air required for the combustion of the fuel in the ship's cooking apparatus through pipes leading from the hold, and other confined places,* precisely on the same plan that has been repeatedly adopted since his time, and has also been applied (in the year 1836) to the temporary Houses of Parliament. Since the time of Hales, many plans have been proposed for ventilating buildings, but they are mostly modifications of one or the other of the methods here described; and these methods, if properly applied, are amply sufficient to accomplish all that is required.

(312.) Few persons have any notion of the vast consequences which result from impure air; and how seriously the duration of human life is affected by want of proper attention to this important subject. Dr. James Johnson,† speaking of the effects of impure air, says, "that ague and fever, two of the most prominent features of the malarious influence, are *as a drop of water in the ocean*, when compared with the other less obtrusive but more dangerous maladies that silently disorganise the vital structure of the human fabric, under the influence of this deleterious and invisible poison;" and experience proves that multitudes shorten their lives by breathing impure air, and many more lay the foundation of diseases, accompanied by years of pain and sorrow, by neglecting to avail themselves of the bountiful provision of nature, which spontaneously affords to all who choose it, an unlimited supply of this important element, and requiring merely an unrestricted and free passage to diffuse itself abundantly in every direction.

* *Philosophical Transactions*, 1742, vol. xlii., p. 42.
† "Diary of a Philosopher."

(313.) The powerful effects which are produced on the animal functions by certain deleterious gases are very imperfectly known, as they are generally found in so diluted a state as to render their action slow and almost imperceptible. But to be aware of their real nature and influence, we should see their effects in a more concentrated form. Dr. Christison, in his work on Poisons, quotes from Hallé a description of the effects of the gases from the *fosse d'Aisance* of Paris on those who inhale them; and Dr. Kay gives the following account of them in his work on Asphyxia.* "Often the individual exposed to them perishes in a moment, his head and arms falling, and the trunk being doubled up from the instantaneous loss of muscular power. If death does not immediately occur, the victim, when he recovers from the first effects of this exposure, is affected with pains in the head, nausea, fainting fits, severe pains in the stomach and limbs, and constriction of the throat. Sometimes he utters involuntary cries, or lapses into delirium, accompanied with the sardonic laugh and convulsions, or tetanus ensues. The face is pale; the pupil dilated and motionless; the mouth filled with a white or bloody froth; respiration convulsive; the pulsation of the heart irregular; the skin cold; until at length complete asphyxia and death terminate the scene of suffering." The gases which produce these effects are combinations of ammonia, sulphuretted hydrogen, and nitrogen; and all these gases are occasionally to be found in the contaminated air of close ill-ventilated rooms, though of course in smaller quantities. The effects of climate are of the same kind. The impurity of the air in certain localities produces the most frightful results. Cretinism,

* Dr. Kay's "Physiology and Treatment of Asphyxia," p. 326.

although its cause is not positively ascertained, is ascribed by medical writers to confined air and other agencies;* and so frightful and revolting is this state of degeneracy of both body and mind, that Dr. James Johnson's description of the disease fully realises the character which he has given it that "Goitre (or the enlarged neck), on such a scale as we see in the Vallais is bad enough; but Cretinism is a cure for the pride of man, and may here be studied by the philosopher and physician on a large scale, and in its most frightful colours."† The picture is almost too frightful to copy, and is only second to his description of another scourge, Pellagra, arising from the same cause, which afflicts nearly one-seventh of the inhabitants of the Lombardo-Venetian Plains,‡ and the wretched victims of which rot away in a state so painful and disgusting, that the description absolutely sickens the reader, and prepares him for the announcement that multitudes of these wretched beings end their state of hopeless misery by committing suicide, which they generally do by drowning.

(314.) The statistical reports laid before Parliament by the War Office, on the sickness and mortality of the troops of the United Kingdom stationed in different parts of the world, prove most clearly the immense effect upon human life produced by small and almost inappreciable differences in the quality of the atmosphere. For on the same class of persons performing the same duties, and placed as nearly as possible in the same circumstances, the average mortality varies in different parts of the world from 1·37 per cent. per annum to 66·83 per cent. per annum; or the mortality is nearly forty-

* Dr. Hawkins's "Medical Statistics," p. 198.
† Dr. James Johnson's "Change of Air; or, the Pursuit of Health," &c., p. 56.
‡ *Ibid.*, p. 75.

Y

nine times as great in some localities as in others.* The morbific influence of certain gaseous emanations from the earth, in various parts of the globe, is well known. " The banks of the Nile about Sennaar," says Bruce, " resemble the pleasantest part of Holland in the summer season ; but soon after, when the rains cease, and the sun exerts his utmost influence, the dora begins to ripen, the leaves to turn yellow and to rot, the lakes to putrify, smell, and be full of vermin, and all this beauty suddenly disappears —bare-scorched Nubia returns; and all its terrors of poisonous winds and moving sands, glowing and ventilated with sultry blasts, which are followed by a troop of terrible attendants—epilepsies,

---

* These reports present the result of twenty years' observation, and were laid before Parliament in the years 1838, 1839, and 1840, being respectively for " The West Indies," "The United Kingdom, the Mediterranean, and British America;" and "Western Africa, the Mauritius, &c." The following list is extracted from these reports, and gives the mortality *per cent. per annum*, among white troops only, exclusive of the native troops :

	Per Cent.		Per Cent.
British Guiana	8·4	Sierra Leone	48·3
Trinidad	10·63	Cape Coast Castle	66·83
Tobago.	15·28	St. Helena .	3·3
Grenada	6·18	Cape of Good Hope	1·37
St. Vincent's	5·49	Cape of Good Hope,	
Barbadoes .	5·85	Frontiers.	0·98
St. Lucia	12·28	Mauritius .	2·74
Dominica	13·74	Ionian Islands .	2·52
Antigua, &c.	4·06	New Brunswick .	1·47
St. Kitts	7·10	United Kingdom.	1·4
Jamaica	12·13	Canada	1·61

These interesting reports give many particulars relating to the climate of each place. The principal characteristic of Sierra Leone and Cape Coast Castle is the extreme humidity of the atmosphere. In the year 1828 upwards of 313 inches of rain fell in three months at the former place; and in the following year the quantity was 144½ inches in six months; but owing to the registers being imperfect, the annual mean quantity has not been exactly ascertained.

apoplexies, violent fevers, obstinate agues, and linger-
ing and painful dysenteries, still more obstinate and
mortal."* So pestilential is this spot, that "no
horse, mule, ass, or any beast of burden will breed
or even live at Sennaar, or many miles round it.
Poultry does not live there. Neither dog nor cat,
sheep nor bullock, can be preserved a season there.
They must all go every half-year to the sands."†
Dr. James Johnson's graphic account of the Cam-
pagna di Roma‡ and its poisonous exhalations,
and Signor Gaetano Giorgini's account of some
other pestiferous localities,§ give a sufficient idea
of the effects of apparently small corruptions of the
atmosphere. And these accounts, to which vast
numbers of others might be added, all prove how
very small an alteration in the constitution of the
atmosphere materially affects the health of all who
expose themselves to its influence. Professor Daniell
has lately ascertained some facts, which render it
probable that many of the localities desolated by
malaria owe their unwholesomeness to small quantities of sulphuretted hydrogen, produced by decomposition of the sulphates contained in sea-water
by decayed vegetable matter.‖

(315.) We have seen the effect of climate on the
military, by the returns from the War Office, already
alluded to. The duration of life among the inhabitants of different countries is not less remarkable.
The average deaths annually, throughout England
and Wales, are in the proportion of one for every

---

* Bruce's "Travels," vol. vi., p. 387.
† Ibid., vol. vi., p. 381.
‡ Dr. James Johnson's "Change of Air," &c., pp. 117 and
219.
§ Ann. de Chimie, vol. xxix.; also, London and Edinburgh
Philosophical Magazine, vol. xix., p. 15.
‖ Daniell "On Sulphuretted Hydrogen," &c.; London and
Edinburgh Philosophical Magazine, vol. xix., pp. 1–19.

sixty inhabitants; and they vary in every country,* being nearly three times more numerous in some parts of Europe than in England. Very remarkable differences occur also in localities differing but very little from each other. M. Quetelet, in his celebrated work on Man, states, in reference to the effects of climate on the duration of life, that the deaths annually in the different localities are as follow :—

Department de l'Orne.	. 1 death in every	52·4	inhabitants.
„ de Finisterre	1	„	30·4	„
Province of Namur	. . 1	„	51·8	„
„ of Zealand (Netherlands)	} 1	„	28·5	„

This great excess of mortality in the last-named place, M. Quetelet attributes to the extreme and constant humidity of the atmosphere, which produces an immense number of fevers and other maladies.† The observations of M. Bossi also

* Dr. Hawkin's in his "Medical Statistics," pp. 30–74, gives the following average of the *annual* deaths in different localities :—

England	. . 1 in 60	Leghorn	. . 1 in 35
Pays de Vaud	. 1 in 49	London	. . 1 in 40
Sweden .	. . 1 in 48	Manchester .	. 1 in 74
Holland.	. . 1 in 48	Birmingham.	. 1 in 43
France .	. . 1 in 40	Paris and Lyons .	1 in 32
Prussia .	. . 1 in 35	Strasburg and Bar-	
Kingdom of Naples	1 in 35	celona .	. 1 in 32
Wurtemburg .	. 1 in 33	Berlin .	. . 1 in 34
Russia .	. . 1 in 41	Madrid..	. . 1 in 29
Venetian Provinces	1 in 28	Rome .	. . 1 in 25
United States.	. 1 in 40	Amsterdam .	. 1 in 24
Nice .	. . 1 in 31	Vienna.	. . 1 in 22½
Naples .	. . 1 in 28		

, This average for England, which is stated to extend to the year 1821, does not quite agree with more recent calculations, founded on the Report of the Registrar-General; and the difference probably arises from some variations in the mode of taking the averages.

† Quetelet "Sur l'Homme, et le Développement de ses Facultés," &c. ; and *London Statistical Journal*, vol. i., p. 176.

confirm this opinion; for, by dividing the *Departe-ment de l'Ain* into four districts, he found the deaths annually to be as follow :—*

In the Mountainous Parts	1 death in every	38·3	inhabitants.
On the Banks of the Rivers	1 "	26·6	"
On the level Parts sown with Corn	1 "	24·6	"
In Parts interspersed with Ponds and Marshes .	1 "	20·8	"

And it can be shown that, in England also, the rate of mortality follows nearly the same ratio, from the same causes:† and in the reports on the mortality among troops already alluded to, the excessive unwholesomeness of some particular districts is attributed entirely to the excessive moisture of the atmosphere, particularly when accompanied by high temperature. All physiologists have agreed as to the injurious effects of a heated atmosphere saturated with moisture. Hippocrates, by a comparison of the prevailing diseases with the state of the weather, and particularly as respected the moisture of the air, drew conclusions which the observations of succeeding ages have fully confirmed. Excessive dryness, however, not less than an extraordinary degree of moisture, equally destroys the salubrity of the air; though the diseases produced by these opposite states are, as might be imagined, of a very different character.‡ The monsoon, or rainy season of India, the campsin, or southerly wind of Egypt, and the simoon, or hot wind of the Asiatic continent,

---

* *London Statistical Journal*, vol. i., p. 177.

† Dr. Buckland, at a public meeting (March, 1844) stated that in the parish of St. Margaret's, Leicester, containing 22,000 inhabitants, one part was effectually drained, some parts partially so, and others not at all. In the latter the average duration of life is 13½ years, while in those parts partially drained the average duration is 22½ years.

‡ Dr. Arbuthnot, "On the Effects of Air on Human Bodies;" and Dr. Paris, "Pharmacologia," vol. i., p. 197, 325, &c.

more rapidly destructive than either of the others,
and which sometimes strikes down its victims with
instant death, are some among the numerous
instances which might be adduced to prove the
effects of various excessive degrees of heat, moisture,
and dryness of the air, upon the animal system.

(316.) The effects produced on the hygrometric
condition of the air by various modes of artificial
heat are of much importance. Dr. Ure has described*
the result of his examination into the effects produced
upon a great number of the gentlemen in the Long
Room of the Custom House, London, arising from
the use of the powerful hot-air or cockle stoves,
used for heating that establishment. He found
they were all affected with the same sensations and
complaints; tension or fulness of the head, flushings
of the countenance, frequent confusion of ideas,
remarkable coldness and languor in their extremities,
feeble pulse, and other sensations of an unpleasant
character. The stoves, he found, were frequently
red hot, and the air passing over them was sometimes heated to 110°, and was thereby rendered
intensely dry. The animal and vegetable matters
floating in the air were decomposed, and imparted
a disagreeable smell to the atmosphere. This
apparatus was found to be so pernicious that it was
removed, and a different mode of heating adopted.
But these effects, described by Dr. Ure, are by no
means uncommon. The author examined a school
heated in the same manner; and it was found to
be so pernicious to the health of the children, that
they occasionally dropped off their seats in fainting
fits; and when this did not occur, they suffered
so much by the debilitating effects of the intensely
heated atmosphere, that they constantly required
the relief of going for a few minutes into the fresh
air. The usher of the school, a strong, healthy

* *Philosophical Transactions*, June, 1836.

young man, also suffered in the same way; and on one occasion he fainted away in the school-room, and it was with difficulty that animation was restored. These pernicious effects, though generally in a somewhat less degree, always result from the use of intensely heated metallic surfaces. They are, however, much modified by tempering the air by the evaporation of water. In Russia and Sweden, the Apennines, and other places, where close stoves are used, an earthern vessel of water is always placed on the stove for this purpose, and greatly mitigates the oppressive effects which would otherwise be experienced. The desiccating power of the air increases with its temperature to a very great extent. Air at 32° contains, when saturated with moisture, $\frac{1}{160}$th of its weight of water; at 59° it contains $\frac{1}{80}$th; at 86° it contains $\frac{1}{40}$th; its capacity for moisture being doubled by each increase of 27° of Fahrenheit.* But when air is heated artificially, without being in contact with water, it is prevented from acquiring this additional quantity of vapour, and it then possesses a harsh and arid feel, the effects of which we have already seen. This extreme aridity of the air causes it rapidly to absorb moisture from the skin and lungs of persons exposed to its influence; and the evaporation, by its refrigerating effect, contracts the blood-vessels at the surface, while other parts, not being exposed to this influence, become in consequence surcharged with the fluids which are repelled from the extremities.†

(317.) A comparison of the dew-point of the air, made at those seasons of the year which are the most salubrious and agreeable, shows that, in rooms artificially heated, the most healthy state of the atmosphere will be obtained when the dew-point

---

* Leslie " On Heat and Moisture," p. 123.

† Ure " On Ventilation," &c.; *Philosophical Transactions*, 1836.

of the air is not less than 10°, nor more than 20°
Fahrenheit, lower than the temperature of the room.
When these limits are exceeded, the air will be
either too dry or too damp for healthy and agreeable respiration; and attention to the hygrometric
condition of the air would, perhaps, tend more to
the amelioration of that numerous class of pulmonary
complaints which so peculiarly distinguish the inhabitants of this country, than any other remedial
measure.

The quantity of vapour contained in the free
external atmosphere, is found to average throughout
the year 82 per cent. of the total quantity it can
contain. This, therefore, should be the quantity
that we should endeavour to maintain in artificially
heated rooms. In the extensive experiments made
by Mr. Goldsworthy Gurney in heating the House
of Lords,* it was found that when the temperature
of the House was 64° Fahrenheit, the most agreeable condition of the air was obtained when the
degree of moisture coincided with that shown by
Mason's Wet and Dry Bulb Hygrometer, when the
difference between the two bulbs varied from 3 to
9 degrees. This reading of the hygrometer would
therefore agree with a dew-point varying from 6 to
16 degrees below the temperature of the room.†
In large assemblages of people, such as the Houses
of Parliament, the Courts of Law, and other places
occupied for many hours consecutively, it is most
important that the dew-point of the air should be
properly regulated, as it exercises a most serious

---

* Parliamentary Report, August, 1857, "On Warming and
Ventilating Dwellings," p. 129.

† See Glaisher's Hygrometric Tables. Also *Quarterly
Journal of Science* (1820), vol. viii., p. 298, for Professor
Daniell's account of his new Hygrometer, and the method of
calculating from it the *quantity* of vapour in the air, corresponding with any given dew-point.

influence on the mental activity as well as the bodily powers of those exposed to its influence. Old people, and those of delicate constitution or impaired health, cannot bear so low a dew-point of the air— that is, air with so large a difference between the actual temperature and the dew-point—as those who are younger and more robust. When the dew-point is low, the air carries away both from the lungs and from the skin, a larger portion of moisture than is agreeable or healthy for an old person, or one in infirm health. For this reason a higher dew-point would probably be required in the House of Lords than in the House of Commons; and hence the rather high range of the dew-point noted in the House of Lords by Mr. Goldsworthy Gurney, in his experiments. (Art. 330.)

(318.) The decreased consumption of oxygen, when a highly heated atmosphere is breathed, is also another circumstance which exerts considerable influence on health. The experiments of Seguin, Crawford, and De la Roche,* on this subject, show that, under these circumstances, the blood is not so thoroughly decarbonized as when a colder atmosphere is breathed; and this, as we shall presently have occasion to show, quickly operates on the nervous system, affecting the animal functions as well as the mental faculties. In winter, fully one-eighth more oxygen is consumed, than in the summer during the same period.†

(319.) The contamination of the air, produced by respiration and by artificial heat and combustion is very considerable. Wide-spreading as are the consequences of the malarious influences of climate, the effects produced by the contaminations of the air, by the causes now to be described, are scarcely less extensive, though less capable of

* Murray's "Chemistry," vol. iv., p. 480.
† Liebig's "Animal Chemistry," pp. 16, 17.

being numerically determined; and as the progress of civilization causes larger numbers of persons to congregate together, the necessity for attending to ‚these effects becomes continually of greater importance.

(320.) Respiration is a never-ceasing and most extensive source of contamination of the air, and the theory of its operation has long engaged the attention and divided the opinions of physiologists. The opinion held by some of the most eminent is that *inspiration* is involuntary, being caused by the pressure of the air on the lungs; while *expiration* is the effort of the lungs to discharge the air after it has been changed in its nature, and become hurtful to them.* The effects produced by breathing over again a portion of this mephitic air depends upon the constitution of the individual who is exposed to its influence, and upon the amount of the contamination which the air has sustained. Considerable differences exist in the experiments which have been made upon the changes produced by respiration on atmospheric air: but this is unavoidable, on account of the very different capacity of the lungs of different individuals, and also in consequence of the mephitis itself differing in the same individual during the various stages of digestion, exertion, or repose, and according to the nature of the ingesta which he receives.

The contamination of the air is produced not only by respiration but by animal exhalations, which are given off by both the lungs and the skin. That which is produced by respiration is chiefly by the formation of carbonic acid gas, and by the vapour which is exhaled from the lungs. The proportion of both the vapour and the carbonic acid gas has been variously estimated by different

* Blumenbach's "Institutes of Physiology," by Eliotson, p. 84.

experimentalists. Dr. Prout* has collected together the results of the different experimenters on the quantity of carbonic acid given off in respiration. His own experiments give a mean of about three and a half per cent. as the quantity expired by himself, and four and a half per cent. when his experiments were made on another person. Sir Humphry Davy's estimate is about four per cent.; Menzies, five per cent.; Mr. Murray, six per cent.; Allen and Pepys, eight per cent.; Dr. Fife, eight and a half per cent.; Goodwin, and also M. Jurin, ten per cent. But Dr. Prout has shown that the quantity varies greatly at different periods of the day, and that the maximum quantity is given off about noon, up to which period it gradually increases from the beginning of twilight; and after noon it as gradually decreases till evening, and is at its minimum during the night.† Some recent and extensive experiments have confirmed these results generally; but it appears the times of maxima and minima depend principally on the state of digestion and the periods of taking food.‡

The quantity of vapour given off from the lungs has also been very variously estimated. Dr. Menzies calculated it at six ounces; Mr. Abernethy, nine ounces; Sanctorius, eight ounces; and Dr. Hales, twenty ounces, in twenty-four hours; § and the average is supposed to be about three grains per minute. The amount of vapour given off from the skin was found by Thenard to vary from nine to twenty-six grains per minute. Keil found it to amount to thirty-one ounces in twenty-four hours, or ten and a half grains per minute. Seguin ascer-

* *Annals of Philosophy*, vol. ii., p. 336.
† *Ibid.*, p. 330.
‡ *London and Edinburgh Philosophical Magazine*, vol. xiv., p. 401.
§ Dr. Paris's "Medical Chemistry," p. 316.

tained that it varied from eight to twenty-four grains
per minute; and it is generally considered that the
average quantity is about ten grains per minute,
which agrees with the experiments of Lavoisier.

(321.) It is also found that the quantity of
oxygen consumed in respiration varies with the
state of exertion or repose. Lavoisier found that by
a man, while engaged in strong muscular exertion,
compared with the same individual while in a
state of repose, the consumption of oxygen was as
32 to 14. It is also ascertained that the quantity
of oxygen consumed in respiration exceeds the
bulk of carbonic acid gas which is expired from
the lungs; the remainder unites with the hydrogen
derived from the food, and forms the vapour given
off from the lungs and skin: and these proportions
differ so greatly with the nature of the food, that
while some animals expire a quantity of carbonic
acid equal to that of the oxygen consumed, others
do not expire more than half as much carbonic
acid as the oxygen consumed would produce.*

(322.) Notwithstanding these differences of opinion among physiologists respecting various points
connected with respiration, there are certain fundamental facts which are agreed upon, that we shall
find amply sufficient for the illustration of the subject
before us.

(323.) The quantity of air admitted into the
lungs varies in different individuals, and even in
the same individual at different times, being
greatly influenced, as already stated, by the

* Liebig's "Animal Chemistry," p. 26. The production
of carbonic acid gas by respiration is stated to be as follows :—
By a man, 13·9 ounces; by a horse, 97·8 ounces; and by a
cow, 69·9 ounces in 24 hours. Dr. Carpenter states that air
once passed through the lungs contains 1·26 per cent. of
carbonic acid. For other changes produced in the blood by
respiration, see Liebig's "Animal Chemistry," pp. 12 and
272, &c.

relative amount of exertion or repose. Some experimentalists have estimated that as much as 800 cubic inches of air enters the lungs per minute; but it is more generally supposed to be about 330 cubic inches per minute under ordinary circumstances: * and this air, after passing through the lungs, returns charged with carbonic acid gas and vapour of water, as already stated.

(324.) Whenever the same portion of air is breathed a second time, a great sensation of uneasiness is experienced. This arises from several causes. The quantity of oxygen contained in the air, when in its natural state, varies from $20 \cdot 58$ to $21 \cdot 12$ per cent; † but it is found impossible to separate the whole of this oxygen by respiration, on account of the affinity which exists between the gases; and, however often the air is breathed, only about one-half of this quantity of oxygen can be separated from it. If, therefore, the process of respiration consumes (in forming the carbonic acid and the vapour) a quantity of oxygen, varying from six to eight per cent., by merely passing the air once into the lungs, it is evident that but very little can be afterwards abstracted by any further process of respiration.‡ Carbonic acid gas contains its own volume of oxygen; therefore, when any deficiency of oxygen occurs, the proper quan-

---

* Menzies and Goodwyn estimate the quantity of air taken into the lungs, at each inspiration, at 12 cubic inches; Jurin, at 20; Cuvier, at 16 or 17; Davy, at 15; and Thomson, at 33 —(Dr. Kay, "On Asphyxia," p. 123). The number of inspirations in a minute, Haller considers to be about 20; Menzies, 14; Davy, 26 or 27; Dr. Thompson, 19; and Majendie, 15. (Dr. Paris's "Medical Chemistry," p. 315.)

† Dalton "On the Atmosphere": *London and Edinburgh Philosophical Magazine*, vol. xii., p. 402.

‡ See the experiments of Messrs. Allen and Pepys, *Philosophical Transactions*, 1808; and *Nicholson's Journal*, vol. xxii., p. 204.

tity of carbon cannot be given off from the lungs
—a process which is absolutely indispensable for
the preservation of life.

(325.) The physiological effects of a deficiency
of oxygen are very remarkable. When the lungs
are not sufficiently supplied with oxygen, sangui-
fication ceases to be performed; and the arterial
blood retaining the dark colour of venous blood,
circulates in this state through the system. Bichât
proved by experiments,* that when venous or dark-
coloured blood is injected into the vessels of the
brain through the carotid artery, the functions of
the brain are immediately disturbed, and in a very
short time cease entirely; the heart instantly loses
its motion, and death speedily follows. Bichât
considers the effect of the venous blood circulating
through the brain to be similar to the action of
a narcotic poison; and this takes place when the
air is impure, and does not perfectly oxygenate
the blood. When the lungs receive impure air,
imperfectly oxygenated blood is circulated through
the brain, producing a cessation of the functions of
that organ, by which respiration is immediately
affected, and the heart ultimately ceases to act.
Sir B. Brodie also found by experiments made with
various active poisons,† that when the action of
the brain is impeded by other causes than that
produced by the blood, a similar result obtains: for
the instant the brain loses its action, respiration
stops, the heart gradually fails of its power of contracting and propelling forward the blood, and
death speedily ensues. But it has been found
that, after the brain has ceased to act, if an artificial rsepiration of pure air be produced, and continued for a short time, the functions of the brain
will be restored, and the animal ultimately recovers.

* Bichât "Recherches physiologiques sur la Vie et la Mort."
† *Philosophical Transactions*, 1811, pp. 36 and 178, *et seq.*

(326.) The effects so speedily experienced by some persons in close and ill-ventilated rooms are by these experiments easily accounted for. Headache is usually the first sensation of uneasiness which is experienced; and the succeeding symptoms of languor, uneasy respiration, faintness, and syncope, are all clearly referrible to the same cause as that which produced in Bichât's experiments the cessation of the vital functions. These effects always result from breathing air containing any considerable quantity of carbonic acid gas.* When the quantity of this gas is very considerable, it produces such a painful irritation of the epiglottis, as immediately causes it to close spasmodically on the glottis, and thus prevents the entrance of the gas into the lungs; but this also prevents the entrance of the atmospheric air, and thus produces immediate suffocation.

(327.) Physiologists are divided in opinion as to the precise nature of the action on the human frame exerted by carbonic acid gas. Sir Humphry Davy, Dr. Christison, Dr. Bird, and Dr. Paris, appear to consider that it acts as a strong narcotic poison. Dr. Thompson, and some others, entertain a different opinion, and think it only acts on the animal economy by preventing the proper quantity of oxygen from entering the lungs, and thence producing suffocation.

(328.) The greatest proportion of carbonic acid gas which may be breathed with impunity has not been exactly determined, and opinions respecting it are very various.† But it is evident

---

* Report of the British Scientific Association, 1839, p. 108. Experiments by Coathupe.

† In the Parliamentary Report, August, 1857, on *Warming and Ventilating Dwellings*, p. 127, it is stated that 1 per cent. of carbonic oxide is fatal to animal life, while 5 per cent. of carbonic acid may be breathed for a short time without fatal consequences.

that a given quantity of the gas produced by respiration or combustion will reduce the proportion of oxygen in atmospheric air to a much
greater extent than the same quantity of carbonic
acid gas added by simple mechanical mixture.
For, in the formation of a given quantity of carbonic acid gas, either by combustion or respiration,
exactly the like volume of oxygen is consumed—
carbonic acid gas being a compound of one volume
of gaseous carbon united to one volume of oxygen;
therefore, by these modes of forming carbonic acid
gas, we both reduce the quantity of oxygen, and
increase the quantity of carbonic acid; and it has
been estimated that two per cent. of carbonic acid
gas produced by combustion, deteriorates the air
as much as 10 per cent. added mechanically by
simple mixture.*   But in ill-ventilated rooms this
is by no means the only cause of the oppression
and inconvenience which are experienced.   The
vapour given off from the lungs and from the skin
forms a very important source of contamination;
and, being charged with animal effluvia, its effects
in contaminating the air are very considerable.
The baneful effects of mephitic exhalations from
animal respirations are not confined to the human
frame.   The glanders, of horses; the pip, of fowls;
and a peculiar disease to which sheep are subject,
all arise from the bad air generated by their being
too closely crowded together; † and in no case can
animals be confined in a vitiated air for any length
of time, without serious injury resulting, although
the effects may be shown in various ways.   In
ordinary cases the contamination is insufficient to
produce these violent and fatal effects; but the
slow and insidious effects of a less deteriorated
atmosphere are matters of far more importance

* The *Lancet*, for December, 1838.
† Paris, "Medical Chemistry," p. 309.

than most persons believe;* and we have already
seen by the unerring evidence of statistical facts,
how fearfully human life is shortened by even the
smallest conceivable differences in atmospheric and
climatic influences.

(329.) The quantity of vapour given off from
the body (which we have already seen averages
about 12 or 14 grains per minute) is greatly influenced, not only by the different degrees of
muscular exertion and repose, but also under the
ever-changing hygrometric condition of the atmosphere: for the greater the quantity of vapour
which the air contains, the less will the air be able
to carry off from the human body. For the air
possesses a desiccating power on the human body;
but, of course, that power is lessened in proportion
as it is nearer to the point of saturation.

(330.) The *hygrometric* condition of the atmosphere is ascertained by the *dew-point*.† The
lower the dew-point, the greater will be the
quantity of moisture carried off from the lungs
by the air in respiration; and therefore less will
be given off by perspiration from the skin, than
when the dew-point is higher. This is often the
case in very cold weather, when a large quantity
of vapour is carried off from the lungs, and but

---

* Some most excellent observations on this subject may be
found in Dr. Combe's " Principles of Physiology," 4th edition,
pp. 236-238, and 244-248.

† The dew-point is that thermometric temperature of the
atmosphere at which vapour is condensed. By exposing a
cold body to the air, a fine dew is deposited on its surface,
and by observing the temperature of this cold body, we know
the exact quantity of vapour contained in the air at that time.
Warm air contains a larger quantity of vapour than that
which is colder; for air has the property of taking up water
in solution in a quantity proportional to its temperature. The
Table II., Appendix, shows the quantity of vapour that the
air contains when the dew-point is obtained in this manner.

little by perspiration. When air is respired from the lungs it is nearly of the temperature of the blood, which is 98° Fahrenheit; and it is then charged with a large quantity of vapour. If we ascertain the quantity of vapour which the air contains when expired, and deduct what it possessed before it was inhaled, we shall learn the amount given off by the lungs; the quantity of air breathed per minute being known. Now, suppose the temperature of the air before it is inhaled to be 40°, and the dew-point 30°; as 330 cubic inches of air is the average quantity breathed per minute, $\frac{43}{100}$ of a grain* of vapour will be received into the lungs with the air per minute. But when the air is again expired, the temperature will be 98°, and the dew-point is always found to be 94°; it will then contain 3·07 grains of vapour in the 330 cubic inches; so that upwards of two and a half grains per minute are given off from the lungs under these circumstances. But if the dew-point of the air before it is breathed be 50°, which is frequently the case in damp or warm weather, then a less quantity of vapour will be given off in the same time. Dr. Dalton states that in the torrid zone, the dew-point sometimes rises to 80°, and that even in this country it occasionally reaches to 60°, while in winter it is sometimes below zero. This easily accounts for the variable quantity of moisture which is exhaled from the body and lungs at different times.

(331.) The atmosphere, during damp weather, when it is frequently nearly in a state of saturation, is unable to carry off the full quantity of vapour from the body. This causes the oppressive sensation that is so often experienced under such circumstances; and the slightest exertion causes the perspiration to condense upon the surface of the body, and a degree of heat is experienced much greater

* See Table II., Appendix.

than the simple thermometric temperature would occasion.

(332.) Although the carbonic acid gas given off from the lungs is rather more than 37 per cent. heavier than the oxygen which is consumed; still, in consequence of the dilatation of its volume by the increased heat, and the greater levity of the vapour given off from the lungs, the air is specifically lighter at the moment of its expiration than at its inspiration. For 800 cubic inches of pure air at the temperature of 60°, and the dew-point 40°, will weigh 243·395 grains; but 800 cubic inches of air at 95°, containing eight and a half per cent. of carbonic acid gas,* and 5·6 grains of vapour, with the dew-point 85°, will only weigh 232·450 grains; being nearly five per cent. lighter. Hence air, when expired from the lungs, always rises upwards, and will flow through ventilators in the ceiling, or the upper part of the walls of a room, if such be provided for its escape; but otherwise the vapour condenses, and the volume of the air collapses as it

---

* The quantity of carbon given off from the lungs being so considerable, we cannot wonder that the subject of its origin has been a deeply disputed question. Supposing 26 cubic inches of carbonic acid gas to be given off from the lungs per minute on an average, that quantity will contain 3·3 grains of pure carbon, which in 24 hours will amount to 11 ounces. Besides this, if the quantity of vapour from perspiration and pulmonary transpiration be taken at 10 grains per minute for the former, and three grains for the latter, they will amount to 42 ounces in 24 hours, making the vapour and carbon together amount to nearly 3·3 lbs., besides other excrementitious matter from the body. Some other source, then, besides the food, must exist for obtaining the matter which supports vitality, and this probably is the air. Liebig has shown (*Animal Chemistry*, p. 287) that a man takes in with his food about 13·9 ounces of solid carbon daily; and that the more food he consumes, the greater quantity of oxygen he inspires. It has been ascertained by Dr. Prout, that a vegetable diet diminishes the quantity of carbonic acid gas given off, and, of course, reduces the quantity of oxygen

cools; it then becomes heavier than the substrata of air, and sinks to the lower part of the room, contaminated with impurities.

(333.) The sensation of uneasiness produced by breathing impure air is an indication of the injurious effects that result from it which is too often neglected. When the air is not sufficiently pure to effect the complete decarbonization of the blood, we have already seen that the result is the circulation of venous blood through the brain; the respiration then becomes impeded, and the nervous system deranged; the extent of these effects, of course, varying with the amount of the exciting cause, and with the peculiar constitutions of the individuals exposed to their influence. Dr. Harwood remarks on this subject, "The want of wholesome air, however, does not manifest itself on the system so unequivocally, or imperatively; no urgent sensation being produced, like that of hunger, and hence the greater danger of mistaking

consumed; because carbonic acid gas contains exactly its own bulk of oxygen, united to the given weight of pure carbon. The accuracy of Dr. Prout's experiments has been confirmed by divers, and persons making use of the 'diving-bell. In all hot climates, also, where from the rarefied state of the air, less oxygen is received at each inspiration than in the higher latitudes, the inhabitants feel but little desire for animal food, and use principally a vegetable diet; while on the contrary, the inhabitants of the Arctic regions use animal food almost exclusively. Dr. Richardson, who accompanied Captain Franklin on his voyage of discovery to the Polar seas, says that himself and the other individuals who composed the expedition, never felt the slightest wish for vegetable diet, but desired the most stimulating animal food, and in much larger quantities than they had ever before been accustomed to. In such a climate, in consequence of the coldness and density of the atmosphere, the quantity of oxygen inhaled is much greater than in warmer regions, and therefore allows the larger quantity of carbon to be carried off, which the dieting on animal food produces. These results, therefore, accord with Dr. Prout's experiments.

its indications. The effects of its absence are only slowly and insidiously produced; and thus, too frequently, are overlooked until the constitution is generally impaired, and the body equally enfeebled." *

(334.) The diminished pressure of the air in inhabited rooms, caused by rarefaction, by chimney draughts, or by exhaustion of the air by mechanical means for the purpose of ventilation, has been supposed by some physiologists to produce considerable effects on the health of persons exposed to its influence. It cannot, indeed, be doubted that a diminished pressure of any considerable extent would be productive of great inconvenience, and cause considerable derangement of the animal economy. But the diminution of pressure from these causes seldom exceeds a quantity equal to about $\frac{1}{100}$ of an inch of the common mercurial barometer, and generally it does not reach $\frac{1}{200}$ of an inch. The ordinary fluctuations of the barometer, produced by meteorological causes, amounts to upwards of two inches and a half in the altitude of the mercurial column; and this difference is not found generally to produce any remarkable effects on the animal functions—a difference of pressure so very far exceeding anything which can occur in consequence of the rarefaction of the air in an ordinary room, that it cannot be conceived possible that this is the cause of the pathological effects which are experienced by persons much confined within doors.

The extremely small differences of pressure which occur by the rarefaction of the air in inhabited rooms by heat and ventilation, can only be detected by experiments with the differential barometer. But the natural differences of pressure

* Harwood's " Curative Influence of the Southern Coast of England," p. 282.

which occur in many parts of the world are very great. At Mont Louis, in Roussillon of the Pyrenees, one ·of the highest of the inhabited parts of Europe, the mean height of the barometer is only 24·65 inches,* which is more than five inches below the standard mean height in London; and many other places might be named where the pressure of the atmosphere is much reduced below our own standard. Some physiologists, indeed, have been of opinion that the inhabitants of places and districts, of which the height above the level of the sea is such as to cause a pressure considerably less than ours, are subject to peculiar diseases arising from this cause. Even the comparatively small differences which occur in the several parts of our own island, have been supposed to produce very marked pathological effects. Dr. Harwood has written at large on this subject; † and Drs. Wells, Darwin, Beddoes, and Cullen, and also Mr. Mansfield, have given similar opinions. From the facts they have collected, it appears that the inhabitants of high situations, where the atmospheric pressure is, of course, reduced, are more liable to pulmonary consumption than those residing in low, and even in marshy districts. But whether these effects are attributable or not to a diminished atmospheric pressure, the small difference which is caused by rarefaction and ventilation, wholly inappreciable as it is by the ordinary test of the barometer, can scarcely be supposed capable of producing any important results: while we possess the most ample evidence that in the vitiated atmosphere, and altered hygrometric condition of the air, there exists abundant cause for the languor and deficiency of vital energy which are so frequently experienced by persons exposed to their influence.

* Kirwan "On Temperature," p. 89.

† Harwood "On the Curative Influence of the Southern Coast of England."

(335.) Another cause, however, exists, which probably exerts far greater influence than that arising from altered pressure; a more accurate knowledge of which would enable us to account for frequent apparent anomalies in the sensations experienced in buildings which are artificially heated. The electric condition of the air, and its influence on the nervous system, are subjects apparently of the highest importance; but unfortunately our knowledge at the present time is so imperfect on this interesting and important branch of science, and almost daily experience of new discoveries convinces us we have still so much to learn respecting it, that it is difficult at present to assign to this agent its due place among the operative causes which undoubtedly combine to render many methods of artificial heat peculiarly prejudicial to the animal economy.

(336.) Experiments of a very extensive character have shown that the electric state of the atmosphere when no peculiar disturbance takes place, is always of that kind which is known by the name of *positive* or *vitreous* electricity. The quantity of electricity contained in the air appears to be far greater than is generally supposed, as the experiments of Mr. Crosse, Mr. Sturgeon, and Mr. Weekes, clearly show.* The quantity undergoes diurnal variations, there being two maxima and two minima in twenty-four hours;† the greatest quantities being a few hours after sunrise and after sunset; and the smallest quantities being just before sunrise, and again a few hours before sunset—or, as some suppose, about mid-day and midnight.

* Noad's " Electricity," pp. 89–102.

† *Ibid.*, p. 99: Daniell's " Chemical Philosophy," p. 253. Schubler's " Researches": *Quarterly Journal of Science*, vol. ii., p. 416.

(337.) The experiments which have been made on the electric state of the air of close and ill-ventilated rooms, have shown that in these cases the electric condition of the air is reversed, and that instead of being *positively* electric, the electricity of the air is of that kind which is called *negative* or *resinous* electricity. This subject has generally excited so little attention, that but few experiments have been made upon it; and the causes of interference are so numerous, that it requires much delicacy both in the instruments and in the manipulation to arrive at satisfactory conclusions. The similarity of the effects produced on the human frame under these circumstances, compared with the effects which are frequently experienced during the passage of thunder-clouds, when the air is generally negatively electric, or else in a constant state of oscillation between the positive and negative, also leads to the conclusion that the electric state of the air of close rooms performs a very important part in the production of those unpleasant effects which are generally experienced in such cases. In the natural state of the atmosphere, the earth is always found to be *negatively* electric with respect to the air (the air itself being *positively* electric), and of course all conducting bodies in communication with the earth are in a similar electric state. But when the air also becomes *negatively* electric, the earth still retains its original condition, and the human body thus becomes charged with electricity; the air in this condition being unable to carry off the electricity, as would be the case with a moderately moist atmosphere *positively* electric. The body in this case receives a charge of electricity similar to what is called the electric bath, produced by placing the person on a glass stool, and communicating a charge from an electric machine. The effects of the electric bath, when it is continued for a considerable time, are similar to those which result from the atmosphere of close rooms—headache and nervousness. Drowsiness, which is another frequent consequence of close rooms, is produced by a different cause, arising from the imperfect arterialization of the blood circulating through the brain, as we have already seen (Art. 325). Whatever therefore increases the *positive* electricity of the air, also relieves to a considerable extent the unpleasant effects of close, ill-ventilated and highly-heated rooms. Evaporation of water, and, at certain periods of the day, vegetation, both tend to relieve the oppressive effects of close rooms ; and both are known to produce *positive* electricity of the air.* So greatly does evaporation affect the electric condition of the air, that the diurnal variation in the quantity of electricity follows nearly the same course as the exhalation of moisture, and evaporation is considered to be the principal source of atmospheric electricity.† But evaporation, by adding to the moisture of the air, also renders it a good conductor of electricity; while dry air, on the contrary, is an extremely bad conductor ; and hence it follows, that evaporation of water in a highly-heated and desiccated atmosphere must produce salutary effects in relieving the unpleasant effects experienced in close rooms. It must not, however, be understood, that adding moisture to the air is at all times desirable. Sometimes the air may be deteriorated by any additional moisture, and the proper quantity of moisture is a question of great importance on health, as the fluids from the body cannot be carried off if the air be too moist; while on the

* Thomson, " On Heat and Electricity," pp. 440 and 502.
† Noad's " Electricity," pp. 92 and 100.

other hand, the electric condition of the body will
be disturbed if the air be too dry.

(338.) The disturbance of the electric condition
of the air certainly adds another link to the chain
of causes which produce the unwholesome and
depressing effects of ill-ventilated rooms. Dr.
Faraday * considers that the deterioration of the
air under these circumstances " depends as much
or more on matters communicated to the air by
the living system, as by any direct injury to the
air due to the deficiency of oxygen or presence
of carbonic acid; " and as electricity is produced
by every change of state, both respiration and
the vaporization of the fluids of the body must
produce considerable effect upon the electric condition of the atmosphere of rooms where large
numbers of persons are congregated together. It
is also probable that other causes of deterioration
of the air may be discovered ; for various reasons
appear to point to the existence of a distinct substance being found mixed with atmospheric air
when analytically searched for under these circumstances. Dr. Faraday has on more than one
occasion alluded to the probable discovery, that
some new substance may be formed in atmospheric air under the circumstances we are considering ; and all that is known respecting the
formation of *ozone* renders it highly probable
that to this newly-discovered agent in chemistry
many of the injurious effects of close rooms and
a contaminated atmosphere are referrible; and
that most important results to the animal system
are produced by it, even when it exists in the
smallest possible quantities.† And we may look

* " Report of Select Committee on Ventilation of Houses
of Parliament," p. 21.

† Faraday's " Lectures on Non-Metallic Elements," delivered at the Royal Institution ; with Notes, &c., by Scroffern
(1853), pp. 105–116.

to the further discoveries in *ozone* or *ozonized oxygen* for an elucidation of many of the effects which have hitherto baffled philosophers in accounting for various phenomena produced by gaseous influences on the animal functions.

(339.) The preceding remarks will enable us to estimate the comparative wholesomeness of various methods of distributing artificial heat. Many of these methods are highly injurious to the animal economy, and cannot be persevered in without permanent derangement of the health of those who are exposed to their influence.

(340.) There are always suspended in the air myriads of particles of animal and vegetable matter; but these almost unheeded atoms possess a high philosophical importance, however they may generally be disregarded. They are the evidences of the unceasing changes which the material world is continually undergoing—the irrefragable proofs that the visible matter of the universe is slowly and almost imperceptibly passing through a series of transmutations, which affect both organic and inorganic nature. Many of these particles are easily decomposed by heat, and are then resolved into the various gases, either in their elementary or mixed state. Hence many of the methods of producing artificial heat are materially affected, as regards their wholesomeness, by the fact of their being able or not able to decompose or chemically alter these floating particles of matter. To this cause is mainly attributable the unpleasant smell produced by several modes of warming buildings by highly-heated metallic surfaces; and we have already seen that the hygrometric and electric condition of the air is also altered by the same means. All the different descriptions of hot-air stoves are more or less liable to these objections; as also the high-pressure system of

hot-water apparatus, and still more the cockle
or hot-air furnaces. Dr. Nott's stoves, and also
the Russian and German stoves, are subject to
this inconvenience; and asphyxia is frequently
produced in Russia by the use of these stoves.*
The cockle or hot-air furnace is particularly liable
to these objections; for not only will it act power-
fully in decomposing the floating particles of extraneous matter contained in the air, resolving
them into sulphuretted, phosphuretted, and carburetted hydrogen, with various compounds of
nitrogen and carbon, but it will likewise decompose a portion of the vapour contained in the
air, absorbing the oxygen and liberating the
hydrogen.

(341.) These various gases thus exhaled into
the air cannot be breathed without considerable
inconvenience. Signor Cardone made some experiments on breathing hydrogen gas. He inhaled
30 cubic inches, which is about one-ninth part
of the total quantity of air contained in the lungs;
and the almost immediate effects he experienced
were an oppressive difficulty of breathing, and
painful constriction at the superior orifice of the
stomach, followed by abundant perspiration, tre-
mor of the body, heat, nausea, and violent headache; his vision became indistinct, and a deep
murmur confused his hearing. Some of these
symptoms lasted a considerable time, and were
with difficulty got rid of.† Sir Humphry Davy
tried the effect of inhaling carburetted hydrogen.
He made three inspirations of the gas. "The
first inspiration produced a sort of numbness and
loss of feeling in the chest, and about the pectoral
muscles. After the second inspiration, he lost all
power of perceiving external things, and had no

---

* Dr. Kay " On Asphyxia," p. 344.
† *Annals of Science*, vol. xxviii., p. 149.

distinct sensation, except a terrible oppression on the chest. During the third inspiration this feeling disappeared, and he seemed sinking into annihilation, and had just power enough to drop the mouthpiece from his unclosed lips." The effects of this experiment lasted for several hours, producing excessive pain, extreme weakness, nausea, loss of memory, and deficient sensation.* Carbonic oxide is still more prejudicial in its action on the animal system. Sir Humphry Davy, on trying the effects of inhaling a small quantity of it, was seized with a temporary loss of sensation, succeeded by giddiness, sickness, and acute pains in different parts of his body, and it was some days before he entirely recovered ; but Mr. Witter, of Dublin, who tried to repeat the experiments, was immediately affected with apoplexy, and was restored with difficulty.†

This last-mentioned gas is generated by all stoves which are constructed so as to burn with a very slow draught : and Dr. Arnott's stove has been found peculiarly liable to produce this deleterious gas, which escapes into the room through the ventilator in the ashpit, and is extremely unwholesome in small close rooms. The carburetted hydrogen is abundantly produced by the gas stoves, in consequence of a portion of the gas escaping unburned from the stove ; and this unburned gas, when combined with the large quantity of vapour which is produced by the combustion of carburetted hydrogen, as already described (Art. 305), renders these stoves peculiarly unwholesome. All these causes of deterioration of the air affect different persons in very different

---

* Davy's " Researches on Nitrous Oxide"; also Paris, "On Diet," p. 296.
† Ure's "Dictionary of Chemistry," art. "Carbonic Oxide."

degrees; but wherever the causes exist, the result
will necessarily be derangement of the animal
system, however robust the persons may be who
are exposed to their influence; but, of course, the
sensations will be soonest experienced by the delicate
and the valetudinarian.

(342.) It remains to estimate the quantity of
air which is required to be changed by ventilation,
in order to preserve its purity when deteriorated
by respiration and the exhalations from the human
body.

(343.) The quantity of air necessary for respiration is very much less than is required to absorb
the vapour given off from the skin and from the
lungs. The amount of vapour from this cause, we
have already seen, is about 12 grains per minute,
when the individual is not making any particular
muscular exertions. If the temperature of the room
be 60°, the air, when quite dry, will absorb 5·8 grains
of vapour per cubic foot; but the average dew-point
being about 45°, the air will previously contain
3·5 grains; so that each cubic foot of air will be
able to absorb only 2¼ grains of vapour. Under
these circumstances, the perspiration from the body
will saturate 5¼ cubic feet of air per minute. But,
in estimating the quantity of air which is to be
warmed, in order to allow of sufficient ventilation,
this amount may be considerably reduced; because,
as 45° is the average dew-point for the whole year,*
it will be much lower in winter and higher in
summer, and probably will not exceed 20° or 25° on
an average during the time that artificial heat is

* This is for the neighbourhood of London. It varies of
course in different places, and is much influenced by the prevailing winds. An easterly wind travelling to us from the
Continent of Europe, and across the dry and arid countries
of the Asiatic Continent, must necessarily part with much of
its moisture, acquired from the Pacific Ocean, before it reaches

required. Every cubic foot of air will then absorb an additional quantity of about $3\frac{1}{2}$ to 4 grains of vapour; and we may therefore estimate the quantity of air which is requisite to carry off the insensible perspiration at $3\frac{1}{4}$ cubic feet, and for the pulmonary supply, a quarter of a cubic foot per minute, for each individual.

This calculation is sufficient for estimating the quantity of air which in winter is required to be warmed per minute, as explained (Art. 105). But for the purpose of summer ventilation, a larger allowance should be made. As the dew-point is much higher in summer, the air will absorb less moisture from the body, while at the same time the exhalations from the body are considerably greater in summer than in winter. For summer ventilation, therefore, from five to ten cubic feet of air per minute, for each person, is the smallest quantity which ought to be changed, in order to maintain the purity of the room.

(344.) Other causes of deterioration of the quality of the air exist, such as the consumption of oxygen and the elimination of extraneous gases by the burning of fires, candles, lamps, &c.; but as all gases are capable of absorbing equal quantities of vapour, it follows that, when air has been deteriorated by these causes, so as to be less fit for respiration, it is still just as capable of carrying off the vapour from the surface of the body as pure air; and therefore no special allowance need be made for these causes of vitiation.

---

us, and therefore it will be to us a dry wind; while, on the contrary, a westerly wind is always charged with a large quantity of moisture, absorbed during its passage from the American Continent, across the Atlantic. Its passage over this ocean—a distance of 3000 miles—occupies a period varying from 3 to 10 days, during which time it is constantly imbibing moisture from the ocean.

(345.) Some persons have calculated the quantity of air which is required for ventilation at a much greater amount than is here stated. Dr. Reid considers that ten feet per minute ought to be allowed for each individual ; and in some of his experiments at the House of Commons, he states that 30 cubic feet and sometimes even 60 cubic feet, per minute, has been allowed.* Such cases of extreme ventilation are not absolutely necessary ;† and although in the House of Commons, where these experiments were made, a larger proportionate amount of ventilation is requisite than in almost any other case, it is very difficult, if not impossible, to avoid the effect of draughts when such excessive ventilation is produced. In all ordinary cases also, the expense of providing such a great amount of ventilation would be an insuperable objection; though in such buildings as the Houses of Parliament, the expense is perfectly unimportant, provided the desired object of an improved atmosphere be obtained.

(346.) The important changes in the chemical constitution of the atmosphere which we have seen result from respiration, heat, and combustion, imperatively demand a constant change in the air, in order to maintain its purity and to enable it to support organic life. It would be difficult to describe this fact in more forcible language than that which formed the concluding lecture, at the *Ecole de Medecine,* by M. Dumas, on "The Chemical Statics of Organized Beings." From this most interesting lecture some very short extracts can alone be given; but those who have the means of

---

* *Reports of the British Scientific Association,* vol. vii., p. 131.

† Wyman "On Ventilation," p. 296, considers that in hospitals 20 cubic feet per minute is necessary to maintain proper purity of the air.

consulting the document at large will derive both pleasure and instruction from its perusal.*

"We have proved in fact," says M. Dumas, "that animals constitute, in a chemical point of view, a real apparatus for combustion, by means of which burnt carbon incessantly returns to the atmosphere under the form of carbonic acid ; in which hydrogen, burnt without ceasing, on its part continually engenders water, whence, in fine, free azote is in cessantly exhaled by respiration, and azote in the state of oxide of ammonium by the urine. Thus from the animal kingdom, considered collectively, constantly escape carbonic acid, water in the state of vapour, azote, and oxide of ammonium ; simple substances, and few in number, the formation of which is strictly connected with the history of the air itself. Have we not, on the other hand, proved that plants, in their normal life, decompose carbonic acid for the purposes of fixing its carbon, and of disengaging its oxygen ; that they decompose water to combine with its hydrogen, and to disengage also its oxygen ; that, in fine, they sometimes borrow azote directly from the air, and sometimes indirectly from the oxide of ammonium or from nitric acid, thus working, in every case, in a manner the inverse of that which is peculiar to animals ? If the animal kingdom constitutes an immense apparatus for combustion, the vegetable kingdom in its turn constitutes an immense apparatus for reduction ; in which reduced carbonic acid yields its carbon, reduced water its hydrogen, and in which also reduced oxide of ammonium and nitric acid yield their ammonium or their azote.

"If animals, then, continually produce carbonic acid, water, azote, oxide of ammonium—plants incessantly consume oxide of ammonium, azote,

* Translated in the *London and Edinburgh Philosophical Magazine*, vol. xix., p. 338, *et seq.*

water, carbonic acid. What the one class of
beings gives to the air, the others take back from
it; so that, to take these facts at the loftiest point
of view of terrestrial physics, we must say that *as
to their truly organic elements,* plants and animals
spring from air—are nothing but condensed air."

These views have been beautifully investigated
and explained in the lecture of M. Dumas, here
alluded to, and also by M. Liebig, in his excellent
works on animal and vegetable chemistry. The
effects of animal respiration we have already examined at some length. The respiration of plants
is a subject of much interest to the horticulturist,
although several points connected with it are still
undecided, and remain matters of dispute even
among men of the first rank in science.

(347.) The effects of a factitious atmosphere are
less injurious to vegetable than to animal life.
Vegetables appear to have a power of accom-
modating their functions, in some degree, to the
nature of the gaseous elements by which they are
surrounded. That solar light has a powerful
effect on vegetables has long been acknowledged;
and under this influence they exhale large portions of oxygen and moisture. Dr. Daubeny has
ascertained that the same action is produced in
the absorption of moisture by the roots, and the
exhalation of it by the leaves of plants, whether
they are exposed to a strong light, or with a
smaller degree of light they receive a considerable portion of radiant heat.* So powerful,
indeed, is the action of light, that M. Condolle
has found that plants, during the day, and when
exposed to the light, are wholly uninjured by the
action of gases which quickly destroy them at
night; and even the application of chlorine and
other deleterious substances to the roots of plants

* *Reports of the British Scientific Association,* vol. iv., p. 73.

is innocuous during the day, though they are
presently destroyed by similar treatment at night.
Sulphuretted hydrogen, nitrous acid gas, muriatic
acid gas, and chlorine, were severally tried in this
manner, with similar results in each case.*

It has been generally supposed that plants
exhale carbonic acid gas during the night; but
this, by Dr. Dalton's experiments, appears doubtful: for he states that, by numerous analyses of
the air of hothouses, he has always found it to
contain during the day, as well as during the
night, the same proportions of carbonic acid gas.†

The action of fruits on the air has been stated by
M. Berard in his essay, which received the prize
from the French Academy of Sciences, to produce
a constant elimination of carbon, under all circumstances.‡ This opinion has been controverted, and,
as it is supposed, successfully, by M. de Saussure,
who states that green fruit has the same influence
on the air as the leaves have—the action of the
former being rather less intense; but, in proportion
as the fruit ripens, its power to decompose carbonic
acid gas becomes feebler.§

Although it would appear that the purity of the
air is not of so much importance to vegetation as
it is to the animal economy, still, as many of the
gases which are innocuous to plants during the day
are deleterious to them in the night, it is necessary
to prevent any considerable deterioration of the
air, in order to preserve them in a healthy state.
Hence it becomes an important matter, when it is
an object to obtain fruits and flowers of the finest

* *London and Edinburgh Philosophical Magazine*, vol. iv.,
p. 316.
† *Reports of the British Scientific Association*, vol. vi., p. 58.
‡ *Annales de Chimie*, vol. xvi., p. 152; and *Quarterly Journal
of Science*, vol. xi., p. 395.
§ *Quarterly Journal of Science*, vol. xiii., p. 152.

descriptions, to employ only such means of producing artificial heat as do not eliminate extraneous
gases to the air; and experience has proved that
since the general introduction of the plan of heating
buildings by hot water, horticulturists have found
their plants to be more healthy and productive, than
with the old methods of warming buildings.

The general principles of physiology, which have
here been discussed, will enable a correct opinion to
be drawn as to the effects on organic life of any
method of producing artificial heat. It is not to
any particular invention that these remarks apply
—they equally affect all; and when any new plans
for heating buildings are brought under the public
notice, it will be well for those who value their
health, to test the merits of these inventions by the
general principles which have here been explained.
That a vast amount of ill-health, and great consequent physical suffering, is produced solely by
ignorantly using noxious modes of artificial warming, is certain ; and it is really lamentable sometimes
to see people slowly and surely undermining their
health, by persisting in the use of some of these
modern inventions, and thus sacrificing the greatest
blessing which Heaven bestows on man, by wilful
ignorance of, and even setting at defiance, those
laws which bind man to the whole physical universe,
and of which not even the smallest can be violated
with impunity.

CHAPTER IV.

ON THE VARIOUS METHODS OF PRODUCING VENTILATION.

Spontaneous and Mechanical Methods of Ventilation—Cause of Motion in Spontaneous Ventilation—Velocity of Discharge—Effects of unequal height of Discharge Pipes—Defects in Ventilation of Churches—Proportions of Induction and Abduction Tubes—Quantity of Air discharged through Ventilators—Ventilation by Heat and by Chimney Draughts—Mechanical Ventilation by Fans, Bellows, and Pumps—Quantity of Air discharged by these Means—Calculation of Power expended—Cowls used for promoting Ventilation—Effect of different Forms of Cowls.

(348.) THE different modes of producing ventilation may all be classed under two general heads—the natural and the mechanical. All the methods of spontaneous effusion, produced by the unequal density of two columns of air, whether caused by chimney draughts or otherwise, belong to the former class; while the various methods of ventilating by fans, bellows, pumps, and other similar contrivances, belong to the latter class. Of these different modes, the mechanical is the most effective; the natural generally, but not always, the most economical.

(349.) The primary force which produces motion in spontaneous ventilation is the difference of specific gravity of the two columns of air. If a column of air contained in a tube or chimney be heated, it expands according to an ascertained

law, applicable to all gaseous bodies, namely, that
the expansion is equal to $\frac{1}{480}$ of its volume for each
degree of Fahrenheit that the temperature is raised
from 32° to 212°. If this column of air be 10 feet
high, and have its temperature raised 20°, then it
will expand $\frac{20}{480}$, or $\frac{1}{24}$ of its bulk ; so that its specific
gravity would be diminished, and it would require
a column of air 10 feet 5 inches high to balance
a column of the external air 10 feet high, when the
temperature of the latter is 20° lower than that of
the former. But as the height of the heated column
is limited by the height of the tube or chimney,
which we suppose to be only 10 feet high, the
colder column presses it upwards with a force
proportionate to this difference in weight, and with
a velocity equal to that acquired by a body falling
through a space equal to the difference in height
that two columns of equal *weight* would occupy,
which in this case is 5 inches. Now the law of
gravitation is this : that the velocity of descent is
relatively as the square root of the distance through
which the body falls ; and as the body falls $16\frac{1}{12}$ feet
in a second (or 16 feet, neglecting the fraction),
the velocity will be, agreeable to the well-known
law of gravitation, equal to eight times the square
root of the height of descent, in decimals of a foot ;
or $2\sqrt{g \cdot h}$, where g is the distance through which a
falling body descends in one second of time, namely,
16·09 feet, and h the height of the descent.*

In the case we have supposed, five inches is the
height of the effective descent of the heavy column
of air. This fall of five inches is equal to ·416 of
a foot; therefore, by the rule, $2\sqrt{16\cdot09 \times \cdot416}$
$= 5\cdot174$ feet per second, or 310 feet per minute,
will be the velocity with which the heated column
of air would be forced through the tube or chimney

---

* See Chapter V., Part II.

under the circumstances we have supposed.. If, therefore, the tube were one foot square, there would pass out 310 cubic feet of air per minute. This rate of efflux, however, is subject to certain corrections, on account of the contaminations which increase the specific gravity of the escaping air, and also in consequence of friction, arising from various causes, but more particularly in consequence of angular deviations in the tubes. In straight tubes, the friction is found to be in all cases directly as the length of the tube, and inversely as the diameter. In general practice, a deduction of from one-fourth to one-third of the initial velocity is necessary to compensate for these several effects, and to represent the true rate of efflux. The velocity of discharge per second through ventilating tubes, or chimneys, will therefore be found (after the difference in height of the two columns of air has been calculated in the manner already stated) to be equal to *eight times the square root of the difference in height of the two columns of air in decimals of a foot*; this number reduced one-fourth, to allow for friction, and the remainder multiplied by 60, will give the true velocity of efflux *per minute*; and the area of the tube in feet, or decimals of a foot, multiplied by this latter number, will give the number of cubic feet of air discharged per minute.

(350.) In calculating the rate of efflux of the air from any room or building, it is not merely the height of the room which must be considered, but the total height of the column of heated air. Thus, if the ventilating tube passes through another room or loft over the room to be ventilated before it discharges the vitiated air into the atmosphere, the total vertical height from the floor of the room to the top of the tube is the effective height of the column of heated air. If the tube in its course passes horizontally, this additional length may be neglected in the calculation in all ordinary cases, as it makes no other difference in the result except that of increasing the friction by so much additional length. The vertical height is that which alone governs the rate of the discharge, and the horizontal length of the tube is merely one of the fortuitous circumstances which slightly modify the result.*

(351.) As the vertical height of the column of heated air governs the velocity of discharge in the ratio of the square root of the height of the column, it is necessary if more than one ventilating tube or opening for the escape of the heated and vitiated air be made, that they shall all be similar in height, otherwise the highest vent will prevent the efficient acting of the lower one, and the discharge may even be less through the two tubes than it would be with the upper one alone. The cause of this apparently paradoxical effect is not difficult of explanation. If we suppose a tube, open at both ends, to be filled with heated air, it is evident the velocity of the ascent will be proportional to the height of the tube and the excess of temperature of the air which it contains, the weight of the external air pressing the lighter column upwards, as already explained. But if another opening be made at the side of the tube, at one-half the total height, then this opening at the side will not emit any portion of the heated air, but will, on the contrary, admit a quantity of cold air ; and the velocity of its admission will be, like that of the cold air at the bottom of the tube, in proportion to the height of the heated column of air in the tube above the opening. Now, as the

---

* This remark must be taken with its proper limitation, for cases may arise where the friction caused by the horizontal tube may become a very important element in reducing the velocity of the discharge when the horizontal length is great in proportion to the vertical height.

column of air above this opening is only one-half the height of that above the former opening at the bottom of the tube, the velocity with which the air enters it will be, compared to the velocity with which it enters the opening at the bottom, as the square root of the height of the heated column of air above the respective openings. Both these openings will therefore admit cold air at the same time; but, by the admission of the cold air at the middle of the tube, the temperature of the superincumbent atmosphere above the lower opening will be reduced, and the velocity with which the air enters the lower opening will therefore be diminished; the excess of temperature of the air in the tube being the *primum mobile* of the efflux. The interference of the different currents will likewise reduce the quantity of air discharged; and the total result will be that somewhat less air would be discharged under these circumstances than if the whole of the air had entered at the bottom.

(352.) Precisely the same effects as here described take place in the ventilating of rooms by openings at any height above the level of the floor. The highest opening alone will act as the abduction tube, and all openings below this will act as induction tube, reducing the discharge by lowering the temperature of the air of the upper part of the room, and also by causing in it countercurrents. Some modifications of this result will, however, occasionally occur, as, for instance, when the abduction tube is too small; in which case, the next lowest opening will also act in carrying off the heated air. On the other hand, when the openings for the admission of cold air are too small in proportion to those for the egress of the hot air, then the current of cold air will descend through part of the hot-air tube, and the hot air will ascend through the other part of the same tube.

These effects are frequently very sensibly felt in churches and other buildings, where part of the ventilation is effected by means of the windows. The cold air entering at these windows generally descends upon the heads of those who are placed near them. The effect of this entering current is to lower the temperature of the vitiated air, which parts with a portion of its heat to the fresh air entering the building, and the vitiated air being heavier than fresh air of the same temperature, it falls by its greater specific gravity, and is again breathed by the persons assembled, instead of the pure air which they would have received, had the openings for the admission of the fresh air been at or near the floor of the building.

No plan of ventilation can be worse than that just described, which, however, is the method adopted in a very great majority of churches and other large buildings. Notwithstanding this plan has obtained such extensive adoption, it is certain that it is opposed to every sound principle of science, and has had its rise in the most perfect ignorance of the physical laws; and no better proof than this need be adduced to show how very little the true principles of ventilation have been studied, and how erroneous any conclusions on this subject are likely to prove that are not based on the known laws which govern the motion of fluids.

(353.) In all the methods of ventilation, it is advisable to make the aggregate area of the openings that admit the fresh air larger than the aggregate openings for the efflux of the vitiated air. This becomes necessary notwithstanding the increase of volume which takes place in the heated and vitiated air. If the opposite course be adopted, and the abduction tubes be larger than the eduction, then a counter-current takes place in the hot-air or ventilating tubes, and the cold air descends through them ; but by making the induction tubes numerous, and of a large total area, the velocity of the entering current is reduced, and unpleasant draughts are avoided. It is also expedient to divide the entering current as much as possible. By so doing it prevents the dangerous effects of cold draughts, when the entering current is colder than the air of the room ; and when it is hotter than the air of the room, it prevents the air from rising too rapidly towards the ceiling, and therefore distributes it more equally throughout the apartment. Provided the aggregate openings for the admission of cold air be not less in size than those for the emission of the heated air, the quantity of air which enters a room depends less upon the size or number of the openings which admit the fresh air than upon the size of those by which the vitiated air is carried off. This will be evident when it is considered that, the room being always absolutely full of air, no more air can enter until a portion of that already in the room be removed. But as soon as a portion of the air which previously occupied the room is removed, a similar quantity of fresh air rushes in to supply its place, the quantity entering being exactly equal to that which escapes.* The only exception which occurs to this rule is the slow interchange among the particles of air which takes place, according to the laws of gaseous diffusion, through the lower as well as the upper openings of the room, and which

* This description, perhaps, scarcely gives the exact circumstances of the case ; for in spontaneous ventilation a small portion of the cold air will enter before any discharge takes place from the room. The *compression* produced by the fresh air entering causes the heated air to flow out as described Art. 350.

continues so long as any inequality exists either in the specific gravity or in the composition of the gaseous matter. This diffusion among the particles of different gases is known as the laws of *endosmose and exosmose*, and the effects are remarkable in many cases.

(354.) The following Table will show the discharge *per minute* through a ventilator one foot square, for various heights and differences of temperature, the allowance which has already been stated (Art. 349) having here been made. The discharge through a ventilator of any other size

TABLE XXVIII.

Table of the Quantity of Air, in Cubic Feet, discharged per Minute through a Ventilator of which the Area is one Square Foot.

Height of Ventilator, in Feet.	Excess of Temperature of Room above the External Air.					
	5°	10°	15°	20°	25°	30°
10	116	164	200	235	260	284
15	142	202	245	284	318	348
20	164	232	285	330	368	404
25	184	260	318	268	410	450
30	201	284	347	403	450	493
35	218	306	376	436	486	531
40	235	329	403	465	518	570
45	248	348	427	493	551	605
50	260	367	450	518	579	635

*** The above Table shows the discharge through a ventilator of any height, and for any difference of temperature. Thus, suppose the height of the ventilator, from the floor of the room to the extreme point of discharge, to be 30 feet, and the difference between the temperature of the room and of the external air to be 15°, then the discharge through a ventilator one foot square will be 347 cubic feet *per minute*. If the height be 40 feet, and the difference of temperature 20°, then the discharge will be 465 cubic feet *per minute*.

may easily be calculated, because, as the area is
here 144 square inches, we have only to multiply
the number of feet found by the Table, by the
number of square inches in the area of the proposed
ventilator, and then, by dividing that number by
144, the quotient will be the quantity sought,
which will represent the number of cubic feet of
air that will be discharged per minute by the proposed ventilator.

(355.) As the discharge through any given
height and size of ventilator is less in proportion as
the difference between the external and internal
temperature is smaller, it follows that it will be
most difficult to obtain ventilation in hot weather.
In summer, either the number or the dimensions of
the ventilators should be increased; otherwise a
room which is well ventilated in winter will be
extremely uncomfortable in summer. The increase
in size can be effected by having movable ventilators, which can be contracted at pleasure; and the
actual size of the trunk or channel which conveys
the air away should be sufficiently large to carry
off the greatest quantity of air required for summer
ventilation.

(356.) The method of spontaneous ventilation
which has been described requires, in every case,
that the air of the room to be ventilated shall be
of a higher temperature than the external air. In
very hot weather and with crowded assemblies, this
method is generally insufficient to secure a wholesome and comfortable state of the atmosphere.
But artificial means have long been in use for
increasing this effect. This is accomplished in
two ways: either by heating the air in the upper
part of the ventilating tube, which causes it to
ascend with greater rapidity, and thereby to draw
it out of the room or building; or by causing the
air of the building to pass through a furnace,

from which all other supply of air is excluded. Both these plans have been extensively used, and both answer the intended purpose. The principal theatres in London are ventilated by the former method, advantage being taken of the heat of a large chandelier placed near the ceiling in the centre of the house. The heat of this chandelier causes a great rarefaction of the air, and increases the draught; and it thence passes out through tubes into the open atmosphere, the buildings being supplied with fresh air from below. The method of ventilating by causing the vitiated air to pass through a furnace has also been long and extensively employed.* In many manufactories, this method is economically applied, where fire-heat is used either for steam-engines or other purposes. All that is necessary is to conduct the air from the rooms requiring ventilation, through tubes, into the ashpit of the furnace, all other supply of air to the ashpit being prevented; and the draught of the fire causes a rapid abstraction of air from the building, which is immediately supplied by fresh air, and produces a thorough ventilation.†

(357.) The most extensive application yet made of this principle was that employed by Dr. Reid, for the ventilation of the temporary Houses of Parliament in the year 1836. For this purpose a large

---

* In 1739, Mr. Sutton proposed this plan of ventilation; but Dr. Desaguliers, in 1723, appears to have adopted a plan somewhat similar for ventilating the House of Commons (Desagulier's *Experimental Philosophy*, vol. ii., p. 560).; though in reality the invention of applying fire-heat to produce a draught of air was long prior to either of these dates, and was first proposed by Agricola, in the sixteenth century. (See Chapter III., Part II., Art. 311.)

† The quantity of air which would be thus withdrawn from a building by these means may be estimated by what is stated in the Note, Art. 397.

chimney was erected, with a furnace of proportionate size, and the air was drawn off from the ceilings of the buildings with great rapidity, passing downwards through a tunnel into the bottom of the furnace, and thence through the fire. The quantity of air thus withdrawn is governed by the force of the fire. In this case it was very great, for by means of this furnace the whole air in the House of Commons could be changed in a very few minutes.

(358.) The ventilation of the old House of Commons was, for many years, a subject of complaint, and it engaged the attention of many practical and scientific men. The apparatus which was erected for this purpose in the temporary House of Commons in the year 1836, appears, however, to have been a most expensive contrivance for accomplishing the object.

Dr. Ure has written a useful memoir on the subject of ventilation,* in which he compares the advantages of these two methods; and he estimates the relative cost of ventilating by a fan, compared with that by chimney draught, as about 1 to 38. In his calculations on this subject, however, he has apparently been led into an error. His experiments on the consumption of fuel to produce a given effect by chimney draught were all made on furnaces used either for steam-boilers or for brewers' coppers. But, as it could only be the residual heat of the furnace which became available in his experiments, after the principal part of the heat given off by the coal had been absorbed by the boiler, it is certain that any calculation founded on the effect produced in this manner must be considerably below the truth. But although the relative cost of fuel will not be so greatly different as Dr. Ure supposes, under

* *Proceedings of Royal Society*, 16th June, 1836.

any circumstances the difference between the two methods must be very considerable.

(359.) The efficiency of the mechanical method of ventilation by a fan turned by machinery, has been proved so extensively in some of the largest buildings and manufactories in the kingdom, that it might perhaps appear singular Dr. Reid should have adopted so expensive a method of ventilation as that obtained by chimney draught. But the requirements of such buildings as the Houses of Parliament are quite different to all other buildings : and the successive experiments made in the ventilation of these buildings for now nearly forty years, appear to result in the conclusion that ventilating by chimney draught, though very much the most expensive, possesses advantages for these particular buildings, hardly obtainable by other means.

These experiments, which have been carried on at a cost, as it is supposed, of half-a-million of money, commenced in the year 1836, under Dr. Reid. By his plan large coke fires, burning under an immense chimney, were employed to draw out the foul air from the house. This was afterwards combined with revolving fans, 30 feet in diameter, driven by steam, for forcing in fresh air, either heated or cooled, according to the season of the year, for supplying the place of the air extracted by the chimney draught. On the retirement of Dr. Reid, about the year 1848, various schemes for the warming and ventilating these buildings were proposed and partially carried out, without any very definite results. In 1851, Mr. Goldsworthy Gurney was consulted, and a year or two later he received the appointment for arranging and superintending all the warming and ventilating arrangements. Great alterations were then made. The steam jet was now applied for the ventilation, and after considerable trials was given

up as unsuitable for the purpose. The warming
arrangements were all reconstructed. The ventilation was changed from an upward to a downward
draught, and the windows of the House were made
to open which had before been closed. These
plans underwent several alterations. The ventilation by chimney draught was again introduced.
Fans for giving motion to the air were again
employed; and the heating was effected by high-
pressure steam in conjunction with an arrangement called "steam batteries," constructed on the
same principle as Sylvester's stoves, described
Art. 208. After various alterations the present plan
was adopted, by which the powerful furnaces with
an upcast shaft were used for extracting the foul
air from the ceiling. The fresh air, either heated
by steam batteries in winter, or cooled by ice in
summer, was made to enter through the cast-iron
perforated floor covered with a peculiar open
matting of whipcord, and also partly forced in by
air pumps or bellows worked by steam. The
arrangements of Dr. Reid for washing the air by
passing it through canvas wetted by fine spray
jets, were improved and extended, and the arrangements may now be considered as a conbination of
all the previous plans which had been separately
tried, condemned, and approved, but which in
combination appear to accomplish all that can
reasonably be expected in such a building. By the
present arrangement any degree of warmth; any
amount of ventilation; any degree of moisture or
dryness of the air can be produced. Any one of
these conditions can be altered in a few moments
to almost any extent, and with absolute certainty;
and could the various members of the two Houses
only agree as to the temperature, the quantity of
air and its degree of moisture, which they prefer,
every one of these conditions could be fulfilled by

2 B

the present apparatus with instant and unfailing
precision.   This agreement of opinion is never
likely to be attained.   The different physical conditions of the occupants of these buildings, by age,
constitution, and habit, and the ever-varying
feeling of excitement, or repose, or exhaustion
which they experience, renders it impossible to
meet the desires of individuals so variously affected.
If satisfaction in this matter could be looked for
under such circumstances, it might be found under
the present very efficient arrangements carried on
under the superintendence of Dr. Percy, and those
with whom he is associated.

(360.) The Marquis de Chabannes, about the
year 1816, extensively applied the other mode of
ventilation which has been alluded to, by artificially heating the air by means of stoves, *after
it has passed through openings in the ceiling.*   By
these means the draught is greatly increased.   This
plan was applied to several very large buildings,
and among others to the old Houses of Parliament.
It has been fully described by the Marquis, in a
pamphlet published by him in 1818 ; but it appears
now to have fallen into disuse.   More recently the
same principle has been applied by using hot-water
apparatus fixed in the roof to heat the air, instead
of stoves, and by these means producing a rapid
and continuous ventilation.   For moderate-sized
buildings, where a fan moved by steam power
would be unsuitable, this plan of ventilation offers
many advantages, and has been used in several
very important buildings.

The proper dimensions of such an apparatus
would be thus calculated :—Suppose a room to be
ventilated which holds 500 people.   By Art. 345
we estimate the quantity of air for each person at
ten cubic feet per minute.   Five thousand feet of
air per minute must thus be abstracted through the

ventilators. If we raise the temperature of this air 20° in the heating chamber in the roof by the hot-water apparatus, we shall find by Table IV., Art. 106, that to raise 1000 cubic feet per minute from 50° to 70° (the assumed temperature for the purpose of calculation of the room and of the heating chamber) it will require 86 square feet of heating surface. Therefore 430 square feet of hot-water piping would be required to heat the 5000 cubic feet of air. By Table XXVIII., Art. 354, it appears that a velocity of 235 feet per minute will be acquired by a difference of 20° Fahrenheit; therefore an opening of 21 square feet will be required to provide for the escape of 5000 cubic feet of air per minute under the circumstances here assumed.

(361.) The mechanical method of ventilation appears to possess many advantages. It is of course only suitable for extensive buildings, on account of the cost of erection and maintenance of the apparatus being too great in any case, except where a large quantity of air is required to be withdrawn. A rotary fan was used as long ago as the year 1734, by Dr. Desaguliers, for ventilating hospitals, prisons, and other buildings; * though the plan he recommended appears to have been but little used for that purpose. In fact, although the principle was good, it failed in consequence of the trouble attending its use. About the year 1741, Dr. Hales introduced his method of ventilation by bellows; and it was applied in many cases with great success. Several of the prisons, hospitals, and other buildings of the metropolis, as well as nearly all the ships in the navy, were successfully ventilated by this apparatus, which was extremely simple in its construction and operation.† It con-

* *Philosophical Transactions*, 1735, vol. xxxix., p. 41.
† Dr. Hales, " On Ventilators." London, 1743.

sisted of a large box with valves opening inwards, and other valves opening outwards, which alternately admitted and discharged the air, when an internal diaphragm, fixed by leather hinges to the centre of the box, was moved up and down by a handle passing through the upper part of the box. One defect, however, was common both to this apparatus and that of Dr. Desaguliers; they, were both made dependent for their operation on manual labour, and therefore their use was limited both in duration and extent. For, however efficacious the operation might be, the trouble attending it, when the whole effect was produced by manual exertion, rendered it inconvenient, expensive, and uncertain. The extensive introduction of machinery throughout every department of manufactures has again, after a lapse of many years, brought into use both these methods of ventilating buildings. The ventilating fan of Dr. Desaguliers, with some improvements suggested by modern discoveries and experiments, is now extensively used in the manufacturing districts of England; the fan being turned by the steam power employed in the manufactories. The bellows of Dr. Hales, slightly altered in their form, have also again been recently brought into use. The late Mr. Oldham, the engineer to the Bank of England, applied this principle of ventilation, in the year 1838, to a part of that establishment with complete success. This apparatus differs, however, slightly from that of Dr. Hales, particularly in being fitted with a piston instead of a movable diaphragm, which gives it more of the character of a pump, though the difference in its construction is very inconsiderable. In this case also the motive power is steam, the powerful machinery constantly in use at the Bank being employed to work the pump which ventilates the building. Both these methods of ventilating, when thus applied, are unerring in their operation, and appear to accomplish all that is necessary or desirable on this important subject; for the quantity of air discharged can, by both methods, be either increased or diminished at pleasure, by the mere shifting of the band which drives the pulley.

(362.) By these methods of ventilation, the rarefaction or diminished pressure of the air can be effectually prevented. In the method adopted at the Bank of England, instead of the vitiated air being drawn from the building by the apparatus, the operation consists in forcing in fresh air, which in cold weather is warmed by passing through a steam chamber; and the vitiated air escapes from the room in consequence of its greater levity; the quantity which escapes being equal to that which is forced in by the ·pump. By this means no diminution of pressure can arise; but, from what has already been stated (Art. 334), it may be questioned whether the small diminution of pressure which occurs under ordinary circumstances is a matter of any importance.

(363.) The fans that were used by Dr. Desaguliers were seven feet in diameter, and one foot wide. They revolved in a concentric case, closed in every part, except an opening at the centre, which communicated by a pipe with the room to be ventilated, and another pipe at the circumference, by which the foul air that was drawn in at the centre was thrown out with considerable force by the rotating leaves of the fan. This construction, however, has been found objectionable. Considerable loss of power accrues from employing a fan moving concentric with the case, on account of a large quantity of air being carried round by the leaves of the fan, instead of passing out through the discharge-pipe at the circumference. The most advantageous form is when the case is eccentric to the revolving leaves

of the fan, the discharge-pipe being placed at that
part of the circumference of the case where the
eccentricity is the greatest; the air being admitted
at the centre in the same manner as before stated.
In this form there is comparatively little loss of
power; but owing to the inertia of the air, some
loss must always occur between the calculated
and the actual discharge, the difference being
always greater in proportion to the greater speed
with which the fan revolves.

(364.) The mode of calculating the quantity of
air discharged by any mechanical method, as also
the power expended in discharging it, is necessary
to be known, in order to apportion an apparatus of
a proper size to any particular building. Both
these subjects have been investigated by Dr. Ure,*
who has made various experiments connected with
this branch of inquiry.   The mean velocity of the
portion of the vanes of the fan by which the air is
discharged is about seven-eighths of the velocity of
the extremities of the leaves; but, owing to the
inertia of the air, there will be a further loss in the
velocity of the issuing current, increasing with the
greater velocity of the vanes ; so that, under ordinary
circumstances, the current will be discharged with
a velocity equal to about three-fourths of the velocity
of the extremities of the leaves.   This velocity, in
feet per second multiplied by the area of the
discharge-pipe in square feet, will give the number
of cubic feet of air discharged per second.   To
estimate the force necessary to cause the rotation
of the fan, the following method of calculation,
founded on the ordinary mode of estimating steam

* *Philosophical Transactions*, 1836.   See also Péclet, " Traité
de la Chaleur," 3rd edition, p. 100 *et seq.*, for some useful
calculations on the relative cost of various methods of discharging air for ventilating purposes. Also Wyman "On
Ventilation," p. 169.

power, will be found sufficiently accurate. Suppose the effective velocity of the vanes of the fan to be 70 feet per second, and the sectional area of the eduction tube to be 3 square feet, then $70 \times 3 = 210$ cubic feet will be the quantity of air discharged per second; and this number, multiplied by 60, will give the quantity per minute. As a cubic foot of air weighs 527 grains, there will be about 13 cubic feet of air to a pound; therefore $\frac{210 \times 60}{13} = 969$ lbs. is the weight of air put in motion per minute with a velocity of 70 feet per second. The height from which a gravitating body must fall in order to acquire a velocity of 70 feet per second is $\frac{70^2}{82} = 76 \cdot 5$ feet, which, multiplied by the number of pounds weight moved per minute, will give the power necessary to be expended in order to discharge this quantity of air at the stated velocity; and this product divided by 33,000 (the number of pounds weight that one horse will raise one foot high per minute) will give the amount of steam power required. Therefore $\frac{76 \cdot 5 \times 969}{33,000} = 2 \cdot 24$, or nearly $2\frac{1}{4}$ horses' power, will be necessary to discharge the given quantity of air at the velocity stated.

The quantity of air discharged by bellows is easily calculated. The cubic contents of the box (or that portion of it which is filled and emptied at each alternation of the handle), multiplied by the number of strokes per minute, will, of course, give the quantity of air discharged; making such deduction from this amount as may be necessary for imperfect fitting of the diaphragm. The ventilating pump differs from the bellows simply in making the whole diaphragm move up and down, instead of one end being fixed. The force requisite for discharging the same quantity of air by either of these methods is the same as with a fan. For, suppose a ventilating pump three feet square and five feet high, and that the piston makes 25 double strokes per minute, each four and a half feet long; in this case 2025 cubic feet of air per minute will be discharged; and if the valve for its discharge be 10 inches square, the velocity of its discharge will be equal to $48\cdot6$ feet per second. This quantity of air reduced into weight will be $\frac{2025}{13} = 155$ lbs., put into motion every minute at the rate 'of $48\cdot6$ feet per second; and therefore we shall have $\frac{48\cdot6^2}{8^2} = 36\cdot9$ feet, as the height from which a gravitating body must fall to obtain the velocity of $48\cdot6$ feet per second; and $\frac{155 \times 36\cdot9}{33,000} = \cdot17$, or one-sixth of a horse's power, as the necessary force to discharge this quantity of air at the stated velocity.

(365.) The Archimedean screw has repeatedly been proposed for the ventilation of buildings, and its application to this purpose has been more than once the subject of a patent. In comparison with a fan, it appears to be every way inferior; for neither in quantity nor in velocity can it at all approach the performance of the fan in discharging air from buildings. By the fan almost any velocity of the discharged current may be obtained; but not so by the Archimedean screw. The friction between the air and the threads or spirals of the revolving screw must be the power which produces the discharge; but so soon as the pressure or condensation of the air between the spirals equals the friction existing between the air and the surface of the spirals, any further increase in the amount of the discharge ceases; and the screw might revolve with any increased velocity without increasing the quantity of air discharged, and a loss of power in turning the screw would then necessarily occur. The plan, however, has been used with some success in large manufactories in Leeds and elsewhere,

where an abundance of steam power is always in operation.

(366*.) In the year 1873, a patent was obtained by Mr. Tobin of Leeds, for a mode of ventilation which excited much attention, and a Joint Stock Company was formed to carry out the invention on an extensive scale. The plan was to introduce *cold* fresh air through tubes, tunnels, or drains, under the floors of rooms, and thence by means of upright or *vertical pipes*, placed at the corners of the room, or elsewhere, to give the air an upward direction, through open pipes, of from four to six feet in height. It was assumed that this method would produce ventilation without draught, but this assumption was a perfect fallacy. It was, moreover, discovered that this same invention had been extensively tried more than 90 years previously, and a full description of it published in 1794, taken from the papers of Mr. John Whitehurst, F.R.S., who died in 1788, who several years before his death had applied the plan in question to various buildings. The patent of Mr. Tobin is described almost in the identical words of Mr. Whitehurst; and both the legal validity of the patent and the plan itself appear to be equally worthless. In cold windy weather the draughts become intolerable; and in warm and still weather, when ventilation is most needed, no effect whatever is produced. The principle, as applied by Mr. Tobin, is founded on a fallacy, which, however excusable in 1780, when first used by Mr. Whitehurst, is utterly unsuited either to the requirements or to the knowledge of the present day. The fallacy of the plan consists in supposing that perfectly cold external air can be introduced into rooms in this way without causing cold currents; but when the air is heated before admission into the room to be ventilated, this mode of introducing fresh air possesses advantages in many cases.

(367*.) More than twenty years before Mr. Tobin revived this invention of Mr. Whitehurst, a much more judicious plan for bringing in *warmed* air through upright ducts or channels above the heads of the occupants of a room, had been successfully employed. In one of the courts of the Old Bailey Sessions House in London, the ventilation was obtained by setting the panelling of the court about six inches off from the wall for about seven or eight feet in height. Through this large channel, extending almost round the court, a large volume of air was forced into the court, and having an upward motion given to it, by the position of the channel, no inconvenient draught was experienced.

(368*.) This plan of ventilating, though possessing much to recommend it, is not absolutely perfect. It requires, in addition, that a further supply of air be brought in at a lower level. When this is not done the air nearer the floor is not in so fresh a state as it ought to be; and it requires, therefore, for perfect ventilation, that a portion of air should enter near the floor as well as at the higher level already described.

By heating the air before it enters the room to be ventilated, the motive power of the wind is almost entirely lost. Some other motive power is therefore necessary when any considerable quantity of air is required. In the Old Bailey Court House, the air is set in motion by a large fan. And either by mechanical means, or by exhaustion by heat by some of the methods previously described, a sufficient motive power must be applied to produce a current varying from sixty to one hundred feet per minute. A greater velocity than this produces a very sensible and unpleasant draught.

(369*.) Very extensive use has at various times been made of ventilating cowls, similar in principle to those used for increasing the draught of smoky chimneys. These cowls are extremely useful in many cases, but they are limited in their effects; and they depend altogether on the action of the wind, their effects being scarcely appreciable in still and warm weather when ventilation is most required.

(370*.) Cowls of almost every imaginable variety of form have been invented. They all owe their efficacy to one general law belonging to all aeriform fluids,—that a current moving in any direction at a high velocity, readily imparts its own velocity to another current moving at a lower velocity in the same or nearly the same direction. This peculiar property of fluids, both liquid and aeriform, to communicate motion to other similar bodies, not in immediate contact with them, may be illustrated as follows :—Suppose a pipe several feet in length, and of considerable diameter—say for example twelve or fourteen inches—and open at both ends. Let there be also a small pipe, of about one inch diameter, inserted into one end of the large pipe for a short distance, and let there be a current of air or steam forced through the small pipe with considerable velocity. The action of this current of air or steam will be such, that it will continue its course after leaving the small pipe for a very considerable distance along the large pipe, at its original velocity, and with scarcely any lateral expansion ; and it will communicate its own velocity to a very great body of air in the large pipe, which will thus have a current produced in it of considerable intensity. The distance to which the current passing along the small pipe will proceed without losing its velocity and mixing with the aeriform fluid contained in the large pipe, will depend upon the initial velocity which is given to it ; but this distance is considerable, and the quantity of air is very great which may be put in motion in the large

pipe, by an extremely small jet of air in the small
pipe. The currents and eddies of many rivers and
lakes have their origin in this cause. The draught
caused in the chimneys of locomotive engines by a
jet of steam, the action of all chimney cowls, and
other of the phenomena which occasionally present
themselves in practical science, are due to this fact.
In the case of chimney cowls the effect is very remarkable. Whatever efficacy any chimney cowl
really possesses arises from this cause; though
very few of the contrivers of these machines are
aware of the fact. Some interesting experiments
on the action of cowls, by Mr. Ewbank, are instructive on this point.* The figures 54 to 62 are
some of the forms on which he experimented. His
models were made of glass, and a strong blast of
air was made to blow equally on them all. The
lower ends of the models were all made to dip into
a trough of water, and the height to which the
water rose in the stem of the model, showed the
relative effect produced by each different form.
The annexed cuts show the models in the pro-
gressive order of their efficiency, namely—

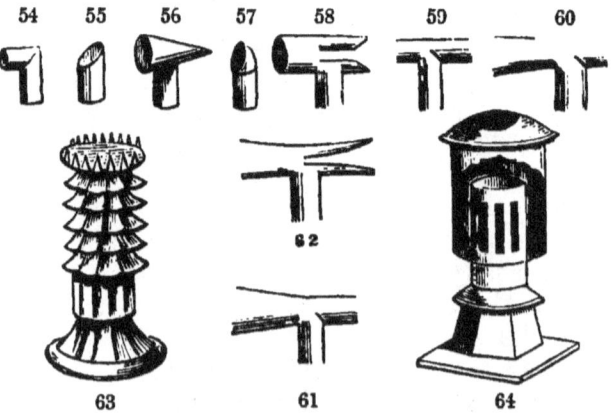

* See *Ewbank's Hydraulics*, New York, 1854; *Franklin
Journal of Pennsylvania*; *Mechanic's Magazine*, vol. xxxvii.,
p. 372; and *Wyman on Ventilation*, Boston, 1846.

No. of Figure........	54	55	56	57	58	59	60	61	62
Height to which the water rose, in inches.	$2\frac{1}{4}$	$2\frac{1}{2}$	$3\frac{1}{2}$	$4\frac{1}{2}$	$4\frac{1}{2}$	$4\frac{1}{2}$	$8\frac{1}{2}$	15	18

In the case of fig. 54, the blast was made to act in direct line with the outlet; but when the blast acted at an angle of 45 degrees with this direction, the water rose $3\frac{1}{4}$ inches; and when it acted at an angle of 90 degrees, the water rose only $2\frac{1}{2}$ inches. The other forms, acted on by the same blast, caused the water to rise to the heights stated against each. This height of water may fairly be assumed as the relative exhausting power of these different forms of cowls, when acted upon by a strong blast. It is, however, considered that in such forms as those of Nos. 58 and 62, the effect is best secured by bringing the small blow-pipe tube beyond the vertical junction of the main tube, as otherwise it is liable to blow downwards.

(371*.) There are many other forms of ventilating and smoke cowls which possess considerable advantages, some being more suited for smoke and some being only fit for ventilation, by reason of the difficulty of freeing them from deposits of soot. Except for this latter objection, whatever form is best for one purpose would be equally advantageous for the other. Hawkesley's Patent Cowl, fig. 63, is an extremely useful apparatus, and so also is the Funnel, or Cowl, fig. 64, known as the Himalaya funnel. Both these owe their effective action to the well-known principle that the wind is deflected at the same angle at which it strikes any object. In both these last shown figures, this deflection is produced, *upwards*, by the wind striking against surfaces fixed at angles of about 45° from a horizontal line. In figure 63 the striking surfaces are

slightly curved, which is supposed to give a still
further advantage to the upward current.

Another patented invention for this purpose is
Carson's Ventilator,* which has been extensively
used. It is precisely similar in principle to the
" Blow Pipe Ventilators," shown figures 58 and 62.
It is almost one of the best forms of those ventilators which turn round on a centre with the wind.
In all forms of revolving cowls increased steadiness
and certainty of action is produced by making the
acting portion of the vane double, and so placed
that the heavy end of the vane or director shall
form a figure like the capital letter Y, the diver-
gence between the two members being about 30°.

In a somewhat recently patented invention by
Boyle and Son, of Glasgow, a ventilator is employed
which is divided vertically into four parts. The
action arises from the wind striking on a number
of inclined surfaces fixed vertically and forming an
external circumference of many narrow inclined
surfaces surrounding the centre tube which carries
off the air from the room to be ventilated. These
inclined surfaces stand off from the main ventilating
tube some two or three or more inches. The
effective action in this ventilator is due to the
wind being driven against the inclined surfaces of
one of the sides of the ventilator; and as the
pressure of the wind will then be a *plenum* on this
side, while one of the other sides will be a *vacuum*,
the foul air of the room will be withdrawn through
the inner tube with a corresponding velocity to
that of the external current. The action of this
ventilator is precisely the same as that of the
Himalaya funnel, or of Carson's, Hawkesley's, and
other ventilators already described. They are all
subject to the same laws, and are all dependent on
the action of the wind to produce any effect what-

* See *Repertory of Arts*, vol. xv. (1841), p. 71.

ever. In still weather no effect is produced by
any of these ventilators beyond what would result
from merely cutting a hole through the roof or
ceiling of equal size.

(372*.) Such modes of ventilating as have been
here described may be useful where only comparatively small quantities of air are required to be
changed. They are all of them utterly inefficient
in calm and warm weather. Where very large
numbers of persons are congregated together, and
consequently a large supply of fresh air becomes
absolutely indispensable, no plan of spontaneous
ventilation is sufficient. The several plans described in the early part of this chapter are alone
to be relied on for this purpose. They may be
classed under three heads. Either by mechanically
forcing in, or sucking out, air by means of fans,
bellows, or pumps; or by heating the air in a false
roof or chamber above the room to be ventilated ;
or by drawing the foul air either upwards or
downwards through a large furnace, in the manner
used by Dr. Reid in the Houses of Parliament.
Which of these plans may be most suitable must
depend upon the character and size of the building
in which it is to be applied.

(366.) Considering the great importance of ventilation, it is much to be regretted that so little
attention has generally been devoted to the subject by scientific men. The treatises which have
been written upon it are extremely few, and those
generally but too indefinite and inconclusive. In
the year 1835, a Committee of the House of
Commons was appointed to report upon the best
plan of ventilating the Houses of Parliament; but
the various scientific men who were then examined were unable to point out any public
building of which the ventilation was at all
deserving of the consideration of the Committee,

as a model for adoption. In fact, it is not merely the method but the amount of ventilation which is desirable, that appears to have been hitherto unascertained. When Sir Humphry Davy ventilated the House of Lords, in 1811, he used a pipe only one foot diameter to convey away the foul air. In 1813, this was increased to three feet diameter; an alteration which allowed *nine* times the quantity of air to escape; but even this was found quite insufficient for the purpose. And the various experiments of Watt, Rumford, Davy, Chabannes, and others, in ventilating various public buildings, prove that their methods were inadequate for the purpose, notwithstanding their plans might be considered as improvements upon preceding arrangements. The subject of ventilation has now, however, attracted more of public attention; and we may therefore hope that this important means of improving the public health will henceforth be more fully considered, and that the time may come when architects will consider it as great a defect to neglect providing the means for the admission and discharge of the air required for ventilation, as they would to omit the doors and windows of the buildings they are called upon to design and erect. The vast importance of ventilation was most forcibly demonstrated by the evidence taken before the Committee of the House of Commons on the Health of Towns, in the year 1840. Scrofulous diseases are stated by the medical witnesses to be the result of bad ventilation; and that in the case of silk weavers, who pass their lives in a more close and confined air than almost any other class of persons, their children are peculiarly subject to scrofula and softening of the bones.* Most of the witnesses state that a deterioration of the race undoubtedly occurs among

* " Report on the Health of Towns," pp. 18 and 201.

those classes most exposed to defective ventilation; and they consider that bad air deadens both the mental and bodily energies.* The statement of some of the diseases produced by bad air is absolutely sickening; and presents the consequences of violating the physical laws, in a point of view, which will scarcely find a parallel.† These are undoubtedly extreme cases; but although the ordinary effects of defective ventilation are less marked, it is certain that no violation of the physical and organic laws, however slight, can possibly be allowed without the consequences becoming apparent in the deteriorated health of those who violate them; and it will be well when this fact is as universally acted upon, as it is generally assented to.

* *Ibid.*, pp. 54 and 201.
† *Ibid.*, Mr. Walker's Evidence, p. 211, *et seq.*

CHAPTER V.

ON THE THEORY OF GASEOUS EFFLUX.[*]

(367.) In the preceding chapter the theoretical determination of the flowing of air and other gases through apertures has been given, and its utility in calculating the proper size of ventilators pointed out: it will here be desirable to show the grounds for believing that this theory truly represents the case with the accuracy required for determining a physical law.

(368.) The theoretical determination of the velocity with which gaseous fluids are discharged through tubes and apertures under pressure has often been submitted to mathematical investigation; and the subject being of importance in various branches of practical science, it is to be regretted that considerable differences exist in the results of the several formulæ which have been propounded for its elucidation. Dr. Papin,[†] in 1686, first showed that the efflux of all fluids follows a general law; and that the velocities are

---

[*] This chapter contains the paper by the author which was read before the Institution of Civil Engineers, May 1840, "On the Efflux of Gaseous Fluids under Pressure." Since it was written, anemometers for registering the velocity of aerial currents have been greatly improved, and as now constructed they afford an easy mode of measuring the velocity of aeriform fluids with very considerable accuracy.

[†] *Philosophical Transactions*, 1686.

inversely as the square roots of the specific gravities. Dr. Gregory has likewise given various formulæ for calculating the velocities of air in motion under different circumstances; and Mr. Davies Gilbert, Mr. Sylvester, Mr. Tredgold, and many other writers of equal authority, have also investigated the subject.

(369.) The hydrodynamic law of spouting fluids has by all writers been applied in the calculations for the determination of this question. This law, it is well known, is the same as that of the accelerating velocity of falling bodies; and is proportional to the square root of the height of the superincumbent column of homogeneous fluid. But, although the various writers all agree in this fundamental principle, they differ materially in the mode of applying it, and the several corrections introduced in their theorems; and the results they have arrived at are of a very contradictory character.

(370.) Dr. Gregory's formulæ for calculating the velocity with which air of the natural density will rush into a place containing rarer air, is based upon the velocity with which air flows into a vacuum. This is equal to the velocity a heavy body would acquire by falling freely from a height equal to that which a homogeneous atmosphere would have whose weight is equal to thirty inches of mercury. The height of this homogeneous atmosphere is 27,818 feet: and the velocity which a body would acquire by falling from this height (and consequently the velocity with which air will flow into a vacuum) is $\sqrt{(27{,}818 \times 64 \cdot 36)} = 1339$ feet per second. The density of the rarefied air divided by the density of the natural atmosphere, and this number subtracted from unity, represents the force which produces motion; and the square root of this number multiplied by 1339 feet (the velocity with

2 c 2

which air rushes into a vacuum) is the velocity
with which the atmosphere will rush into any
place containing rarer air.*

(371.) The method employed by Mr. Davies
Gilbert is also based upon the velocity with which
air rushes into a vacuum, when pressed by a homogeneous atmosphere equal to the weight of the
natural atmosphere at the earth's surface. This
supposed homogeneous atmosphere is, according to
Mr. Davies Gilbert's calculation, 26,058 feet: and
the velocity with which air would rush into a
vacuum when pressed by this weight, will be
$\sqrt{(26,058)} \times 8 = 1295$ feet per second. When this
calculation is applied to two columns of air of unequal density—as for instance, the discharge of
air through a chimney-shaft—the height of the
heated column of air divided by the height of this
homogeneous atmosphere, and the square root of
this number multiplied by the velocity with which
air flows into a vacuum, and this product again
multiplied by the square root of the number representing the expansion of the heated air, will give
the velocity in feet per second. The expansion of
air when heated is found (by Mr. Gilbert's method)
by raising the decimal $1 \cdot 002083$ (which represents
a volume of air expanded by one degree of
Fahrenheit) to the power whose index is the
number of degrees which the temperature of the
air is raised; or it is equal to the fraction $\frac{481}{480}^{n}$
n being the number of degrees of Fahrenheit which
the temperature of the ascending column exceeds
that of the external atmosphere.†

(372.) Mr. Sylvester's method of calculation proceeds upon the supposition that the respective
columns of light and heavy air represent two un-

---

* Gregory's " Mechanics," vol. i., p. 515.
† *Quarterly Journal of Science*, vol. xiii., p. 113.

equal weights suspended by a cord hanging over a pulley; and this mode of calculation gives a result very much less than by any other method.

The unequal weight of two columns of air is found by Mr. Sylvester nearly in the same manner as by Mr. Gilbert. The volume of air expanded by one degree of heat is equal to $1 \cdot 00208$; and this number, when raised to the power whose index is the excess of temperature of the heated column, gives the expanded volume of the air; and assuming the atmospheric density to be unity, we have $1 - \frac{1}{(1 \cdot 00208)^e} = d$; e being the excess of temperature of the heated column, and d the difference of density between the two columns. This difference of density multiplied by eight times the square root of the height of the tube or shaft containing the heated air, gives the velocity in feet per second.[*]

(373.) In Mr. Tredgold's theorem for calculating the efflux of air, the force which produces motion is assumed to be the difference in weight of a column of external and one of internal air, when the bases and heights are the same. The difference of temperature of the two columns by Fahrenheit's scale, divided by the constant number 450 plus the temperature of the heated column, and this quotient, multiplied by the height of the tube or shaft, gives the difference in weight. Then by the common theorem for falling bodies, eight times the square root of this number will give the velocity in feet per second; or, accurately, $V = \sqrt{\frac{64\frac{1}{3} h (t - x)}{450 + t}}$; h being the height of the tube, t the temperature of the internal, and x the temperature of the external air.[†]

(374.) The method of calculation proposed by Montgolfier appears, however, by recent experiments to be the most accurate, as it is also the most simple

* *Annals of Philosophy*, vol. xix., p. 408.
† Tredgold, " On Warming Buildings," p. 76.

of all the modes of determining this question. The difference in height must be ascertained which two columns of air would assume when the one is heated to the given temperature, the other being the temperature of the external air ; and the rate of efflux is equal to the velocity that a heavy body would acquire by falling freely through this difference of height.

The space which a gravitating body will pass through in one second, we know to be 16·09 feet ; but, by the principle of accelerating forces, the velocity of a falling body at the end of any given time is equal to twice the space through which it has passed in that time ; or the velocity is equal to the square root of the height of the fall, multiplied by the square root of 64·36 feet ; or, again, to the spuare root of the number obtained by multiplying 64·36 feet by the height of the fall in feet

When the *vis viva* is the difference in weight between two columns of air, caused by the expansion of one of these columns by heat, the decimal ·00208, which represents the expansion of air by one degree of Fahrenheit, must be multiplied by the number of degrees the temperature is raised, and this product again by the height of the heated column. Thus, if the height of the column is 50 feet, and the increase of temperature 20°, we shall have $20 \times ·00208 \times 50 = 2·08$ feet ; or 52·08 feet of hot air will balance 50 feet of the cold air, and the velocity of efflux of the heated column when pressed by the greater weight of the colder column will be equal to $\sqrt{(2·08 \times 64)} = 11·55$ feet per second.*

---

* This mode of calculation supposes an equal expansion of air by equal increments of temperature, which is generally assumed to be true at all moderate differences of temperature. There can, however, be but little doubt that air expands more, proportionately, at high temperatures than at low ones, for equal increments of heat ; but, as all other bodies expand even

The efflux of air under any given pressure can also be calculated by the same means; for the pressure being known, it is only necessary to calculate the height of a column of air which would be equal in weight to this pressure. Thus, if the pressure be equal to one inch of mercury, water is 827 times the weight of air, and mercury 13·5 times the weight of water; therefore, 827 × 13·5 = 11164 inches, or 930·3 feet; and according to the preceding formula $\sqrt{(930\cdot3 \times 64)} = 244$ feet per second for the velocity of efflux under this pressure of one inch of mercury.

(375.) In all these cases the velocity thus ascertained is independent of any loss by friction; a certain deduction must be made for this loss, which will vary greatly, according to the nature and size of the tube or shaft through which the air passes, as well as with the velocity of the air. Like all other fluids, the retardation of the air by friction, in passing through straight tubes of any kind, will be *directly* as the length of the tube and the square of the velocity, and *inversely* as the diameter. This question, however, becomes very complicated under these circumstances, and particularly so when there are angular turns in the tube through which the air passes. The present state of our knowledge on this subject does not allow of any very accurate determination of the amount which

more irregularly than air, we possess no means of measuring this deviation from regular expansion. Mr. Davies Gilbert's and Mr. Sylvester's mode of calculating the expansion of air already given, supposes a very considerable increase in the rate of expansion, and the following formula is used by Dr. Gregory: The expansion of air for 180° is ·376; therefore, any other temperature will be $(1\cdot376)^{\frac{1}{180}} \times (1\cdot376)^x = (1\cdot0018) \times (1\cdot376)^x = V$; x being the temperature required, and V the volume of the air at this increased temperature (Gregory's "Mechanics," vol. i., p. 486). This mode of calculation gives a less expansion than that of Mr. Gilbert.

ought to be deducted for friction from the initial velocity obtained by calculation; and it is only by empirical means we can arrive at an estimate of its amount.

(376.) We shall proceed now to ascertain how far these theoretical calculations agree with the results obtained by experiments.

In some new furnaces which Sir John Guest has lately added to his extensive ironworks at Dowlais, some experiments have been made on the quantity of blast injected into the furnaces. In these experiments, the machinery employed being new and of the best construction, the loss occasioned by the escape of air through imperfections of the apparatus, was, perhaps, as small as possible. The engine for blowing the furnaces made, at the time of the experiments, 18 double strokes per minute. The diameter of the blowing cylinder was 100 inches, and the effective length of the stroke seven feet six inches. From these dimensions, therefore, it appears that 14,726 cubic feet of air was taken into the blowing cylinder per minute; and the tubes through which it was discharged from the receiver were six of four inches diameter, and six of one and a quarter inch diameter; the area of all these tubes was therefore ·5747 of a square foot, and the pressure of the blast, measured by a mercurial gauge, was equal to four and a half inches of mercury. Calculating by the formula already given, we shall have $\sqrt{(827 \times 13\cdot58 \times 4\cdot5 \div 12 \times 64)} = 519\cdot2$ feet, which is the velocity per second; and this number multiplied by 60, and then by the area of the tubes, will give $519\cdot2 \times 60 \times \cdot5747 = 17,903$ cubic feet of air discharged per minute. From this amount some deduction must be made for friction. The velocity of the discharged air is 354 miles per hour; and with this immense velocity, and through such small pipes, the friction is no doubt considerable. By deducting 18 per cent. from the calculated amount of 17,903 cubic feet, we shall have 14,681 cubic feet, which agrees within a fraction (namely, 45 feet) with the quantity obtained by measurement.

(377.) In other experiments made at the same place, the following were the results: The quantity of air which entered the blowing cylinder was the same as before, namely, 14,726 cubic feet; the total area of the tubes which discharged the blast was ·5502 of a square foot, and the pressure of the blast was equal to four inches of mercury. The calculation, therefore, will be $\sqrt{(827 \times 13 \cdot 58 \times 4 \div 12 \times 64)} = 489 \cdot 5$ feet per second : and therefore $489 \cdot 5 \times 60 \times \cdot 5502 = 16,159$ cubic feet discharged per minute. The velocity of the blast in this case was 333 miles per hour, and if we deduct for friction nine per cent. from the calculated amount, the remainder is exactly the quantity of air which is ascertained by experiments to be discharged through the tubes.

(378.) In a work published in 1834 by Mr. Dufrenoy, being a Report to the Director-General of Mines in France on the use of the Hot-Blast in the Manufacture of Iron in England, the results are given of many similar experiments to the above; but with two exceptions, the details are not sufficiently ample to found any calculations upon. The two exceptions named are the furnaces at the Clyde, and at the Butterley Iron Works, when they were blown with cold air. Both these blowing machines are described as having been in use for several years ; and it is therefore natural to suppose the various parts were more worn and fitted less accurately than in those experiments already described. The experiments were also made with less care. They show a different result

to those already detailed ; as in these the calculated
quantity of air appears to be less than the quantity
which entered the blowing cylinders, in about the
same proportion as it exceeded it in the former
cases. This difference, no doubt, arises from the
imperfect fitting of the piston of the blowing cylinder, which by allowing a portion of air to escape,
would diminish the apparent pressure on the mercurial gauge placed at the further extremity of the
apparatus, and hence the calculated rate of efflux
would of course be diminished.

(379.) In the experiments at the Clyde works
the quantity of air which was discharged into the
furnaces, when estimated by the quantity that
entered the blowing cylinder, was 2827 cubic feet
per minute. The pressure of the blast was equal
to six inches of mercury, and the area of the tubes
·0681 of a cubic foot. Calculating the discharge
of air under this pressure, it amounts to 2450 cubic
feet, being 13 per cent. less than the measured
amount, supposing no loss to occur by imperfect
fitting of the apparatus.

(380.) At the Butterley works, the quantity of
air discharged into the furnace, estimated by the
contents of the cylinder, was 2500 cubic feet per
minute. The pressure of the blast was equal to
five inches of mercury, and the area of the tubes
·0681 of a cubic foot. The quantity, by calculation, appears to be 2235 cubic feet, being less by
$10\frac{1}{2}$ per cent. than that shown by experiment. In
both these last cases, however, there is but little
doubt that the loss of air from the cylinder caused
the pressure on the mercurial gauge to be less than
it would have been had the apparatus been perfectly tight; and a very small diminution in the
observed height of the mercury would account for
a much greater difference in the velocity of efflux
than is here shown.

We are fully warranted in the conclusion from these experiments, that this method of calculation is as accurate as any theoretical determination of such a question can be; but, from the results so obtained, an allowance must always be made for friction, which will necessarily vary with the peculiar circumstances of each case.

The following Table will exhibit the results of the preceding experiments at one view :—

TABLE XXIX.

Place and Number of Experiment.	Pressure of Blast, in Inches of Mercury.	Area of Tubes, in Square Feet.	Velocity of Blast, Miles per Hour.	Quantity of Air by Experiment. Cubic Feet.	Quantity of Air by Calculation. Cubic Feet.	Difference in Quantity. Per Cent.
Dowlais, No. 1	4·5	·5747	354	14726	17903	+ 18
„ No. 2	4·0	·5502	333	14726	16159	+ 9
Clyde, No. 3	6·0	·0681	408	2827	2450	− 13
Butterley, No. 4	5·0	·0681	372	2500	2235	− 10·5

In order to show the results of the several modes of calculation which different mathematicians have adopted, the following Table has been calculated from the data given in experiment the second of the preceding Table; and it shows how far the several modes differ from each other in their results :—

TABLE XXX.

Place of Experiment.	Pressure of Blast, in Inches of Mercury.	Area of Tubes, in Square Feet.	Quantity of Air (by Experiment).	Quantity of Air discharged (by Calculation).				
				Mont-golfier.	Gre-gory.	Gilbert.	Sylvester.	Tredgold.
Dowlais	4	·5502	14726	16159	15152	14855	5017	15555

Considering the amount of friction which must result from the discharge of air at the immense velocity which was ·obtained in this experiment, namely 333 miles per hour, and also that some of the tubes were only 1¼ inch diameter, it will probably be considered that the highest of these calculations is the nearest the truth, as it only allows of a deduction of nine per cent. being made for friction, to reduce the calculated amount to the quantity obtained by experiment. It may, therefore, be concluded that the method which gives this result is the most accurate, as it is also the most simple for general use.

397

CHAPTER VI.

ON THE CHEMICAL CONSTITUTION OF COAL, AND THE COMBUSTION OF SMOKE.*

Early Use of Coal in England—Chemical Composition of Coal —Analyses of Coal—Combustion of Coal—Loss by Imperfect Combustion—Loss from the Escape of Smoke—Loss by Carbonic Oxide—Causes of Imperfect Combustion— Theory of Combustion—Temperature required for Combustion—Effects of Rarefaction of the Air—Effects of Hot Air—Quantity of Air required for Combustion—Methods of admitting Air—Combustion of Anthracite Coal—Description of various Plans for consuming Smoke—Artificial Fuels.

(381.) THE value and importance of coal, whether considered in its commercial or its physical character, or in the effects it has produced upon civilization, are sufficient to render all investigations concerning it singularly interesting. The changes which have been effected in the social condition of man by the instrumentality of this valuable substance are as instructive as they are interesting; and great as are the other mineral riches of England, they would be comparatively valueless without the possession of that substance which, under the generic name of coal, is so extensively distributed throughout its various districts.

* This chapter contains the principal part of a Paper read before the Institution of Civil Engineers, May 10th, 1840, and also the evidence given by the author before a Committee of the House of Commons " On the Prevention of Smoke."

Important, however, as coal has now become, its value has been fully recognised for comparatively but a few years; and the apathy and indifference with which it was formerly regarded, contrast most singularly with the present importance which it assumes.

(382.) The ancient Britons were acquainted with the use of coals before the arrival of the Romans. The Anglo-Saxons also knew and partly used them ; but they are not mentioned by any of the writers in the times either of the Danes or Normans, till the reign of Henry III., who, in 1234, granted a charter to the inhabitants of Newcastle to work them. In 1306, they were prohibited at London as a nuisance; but in 1321 they were used at the Palace, and became soon afterwards an important article of commerce. They, however, gradually fell into disuse except among the poor ; and even in the seventeenth century their use was confined to the lower orders, except for the working of metals.* Such a prejudice existed against them in the middle of the sixteenth century, that, in the reign of Elizabeth, an attempt was made again to prohibit their use in London, even for manufacturing purposes, during the sessions of Parliament; it being supposed that the air was rendered unwholesome by the smoke which they produced.

Previous to the commencement of the seventeenth century the smelting of iron and other metallic ores was performed with charcoal; but at that period the large consumption of wood for making charcoal had so thinned the country of timber, that a probability appeared that many ironworks would be stopped for the want of fuel. An important era now commenced. In 1619 a patent was obtained by Dudley, for smelting iron with the coke made from bituminous coal, which in-

* Fosbroke's "Archæology," vol. i., p. 72.

vention was applied on a limited scale for several years. Shortly previous to this, other persons had attempted the same process without success. Sturtevant in 1612 and Ravenson in 1613, both obtained patents for methods of using coal in blast furnaces for smelting iron ; but both failed of success. Many years, however, elapsed before the use of coke, for this purpose, became general; but, from the period of the middle of the seventeenth century the use of coal became common, both for domestic and manufacturing purposes, in all cases where heat was required to be produced ; and its consumption has steadily increased from that period.

These few historical remarks on the use of coal show, in a striking point of view, the neglect with which this valuable substance was formerly treated ; and it contrasts most singularly with the vast importance which is now attached to it. And when we consider the gigantic effects produced by its agency in our own times, and the still greater effects of which it yet gives promise, we can scarcely credit the apathy with which it was regarded at a period so little remote from our own age.

(383.) The several properties of coal are the result of two distinct substances, which are combined in different proportions in the specimens obtained from different localities. Carbon and bitumen are the two substances which give to coal its distinctive characters, though other substances exist in it in a greater or less degree; some being extraneous and the result of mere local situation, and others partaking of a more generic character. In discussing this subject we shall proceed to consider—first, the chemical character and composition of coal; secondly, its properties as a combustible ; and, lastly, the methods of consuming the smoke given off during its combustion, and of rendering it available in increasing the calorific effects of the fuel.

(384.) Some chemists have objected to the opinion that coal is a compound of carbon and bitumen. No process has yet been devised by which it has been possible, in the analysis of coal, to resolve it entirely into these two substances; but, together with them, there is always obtained a quantity of gaseous matter. Some portion of this gaseous matter readily distils from coal with comparatively a very slight increase of temperature; while other portions, evidently resulting from the decomposition of the bitumen, require for their elimination a much more intense heat. No specimens of coal yet discovered are entirely free from these gaseous products. The only kinds which may be supposed to be exempt from them are the anthracites; but these all contain a small portion of volatile matter, considered by some to be merely water, or its elements; but by others, on more accurate analysis, decided to be similar in composition, though far less in quantity, to the products obtained from bituminous coal.

Considerable discrepancies exist in the analyses by different chemists of some varieties of coal. Enough, however, is known to afford very valuable information in the successful application of coal to the various useful purposes of the arts; though it must be confessed that hitherto the operations of the practitioner have availed far more than the theories of philosophy.

(385.) No accurate analyses of coal were made previous to those of Dr. Thompson, in 1819. The method previously employed, by ascertaining the quantity of residuary carbon left after the volatile matter was driven off, was necessarily very inaccurate; because the gaseous products of the coal carried off a considerable portion of carbon in the form of carburetted hydrogen and carbonic oxide, by which the quantity of carbon in any given specimen appeared to be considerably less than really existed. Such a mode of analysis, however, is not without its use; because it shows, more accurately than any other, the quantity of coke which can practically be obtained from any given specimen of coal, and which by any other mode can only be found by calculation.

The earliest analyses of coal by this process

TABLE XXXI.

Name of Coal.	Carbon.	Ashes.	Volatile Matter.	Specific Gravity of the Coal.	Specific Gravity of the Coke.	Authority.
Welsh Furnace Coal ..	88·068	3·432	8·500	1·337	1·0	
Alfreton Furnace Coal	52·456	2·044	45·50	1·235		
Butterley Furnace Coal	52·882	4·288	42·830	1·264	1·100	
Welsh Stone Coal ..	89·700	2·300	8·000	1·368	1·393	
Welsh Slaty Coal ..	84·175	6·725	9·100	1·409		
Derbyshire CannelCoal	48·362	4·638	47·000	1·278		Mushet.*
Kilkenny Coal	92·877	2·873	4·250	1·602	1·656	
Do. Slaty Coal ..	80·475	6·525	13·000	1·445		
Scotch Cannel Coal ..	39·430	4·000	56·570			
Boolavoonen Coal ..	82·960	3·240	13·800	1·436	1·596	
Corgee Coal	87·491	3·409	9·100	1·403	1·656	
Queen's County Coal	86·560	3·140	10·300	1·403	1·621	
Stone Wood, Giant's Causeway ..	54·697	11·933	33·370	1·150		
Caking Coal	75·900	1·500	22·600	1·269		
Splint Coal..	55·230	9·500	35·270	1·290		Dr. Thomson.†
Cherry Coal	42·246	10·000	47·754	1·265		
Cannel Coal	29·000	11·000	60·000	1·272		
Kilkenny Coal	89·3	4·0	6·7	1·435		
Anthracite, Lehigh	90·1	3·3	6·6			
Do. Rhode Island	90·03	5·07	4·9			Vanuxem.‡
Do. Do.	77·7	15·6	6·7			
Anthracite, Llanelly ..	41·62	39·98	18·4	1·571		
Do. Milford LowerVein	88·74	2·46	8·8	1·374		Daniells.
Do. American	85·98	4·62	9·4	1·518		

* *Philosophical Magazine*, vol. xxxii., p. 140. The same author has recently increased this list by some hundreds of analyses of different coals, and published the results in his "Papers on Iron and Steel."

† *Annals of Philosophy*, vol. xiv., p. 81; and vol. xv., p. 394.

‡ *Ibid*, vol. xxvii., p. 104. In this analysis the volatile matter is considered to be entirely composed of water.

2 D

which have been published, with the exception of those by Kirwan, were made by Mr. Mushet: they are given in the preceding Table, together with several other more recent analyses by different experimentalists.

These analyses show the very great difference which exists in the composition of the various descriptions of coal. This difference not only exists between the several species of coal, but likewise in

TABLE XXXII.

Name of Coal.	Carbon.	Hydrogen.	Azote.	Oxygen.	Specific Gravity.	Authority.
Caking Coal	75·28	4·18	15·96	4·58	1·269	
Splint Coal..	75·00	6·25	6·25	12·50	1·290	
Cherry Coal	74·45	12·40	10·22	2·93	1·265	} Dr. Thomson.*
Cannel Coal	64·72	21·56	13·72	0·0	1·272	
Splint Coal..	70·90	4·30	0·0	24·80	1·266	} Dr. Ure.†
Cannel Coal	72·22	3·93	2·8	21·05	1·228	

Name of Coal.	Carbon.	Hydrogen.	Oxygen & Azote.	Ashes.	Authority.
Splint Coal, Wylam ..	74·823	6·180	5·085	13·912	
Ditto, Glasgow	82·924	5·491	10·457	1·128	
Cannel Coal, Lancas.	83·753	5·660	8·039	2·548	
Ditto, Edinburgh ..	67·597	5·405	12·432	14·566	} Richardson.‡
Cherry Coal, Newcas.	84·846	5·048	8·430	1·676	
Ditto, Glasgow	81·204	5·452	11·923	1·421	
Caking Coal, Newcas.	87·952	5·239	5·416	1·393	
Ditto, Durham	83·274	5·171	9·036	2·519	
Anthracite, Wales ..	92·56	3·33	2·53	1·58	} Regnault.§
Ditto, Pennsylvania ..	90·45	2·43	2·45	4·67	
Ditto, Meyenn	91·98	3·92	3·16	0·94	
Ditto, Roldue	91·45	4·18	2·12	2·25	
Ditto, Wales	89·43	3·56	3·95	1·70	Jacquelain.‖
Ditto, Pembrokeshire	92·43	3·37	2·49	1·73	Schafhaeutl.¶

* *Annals of Philosophy*, vol. xiv., p. 95.
† "Chemical Dictionary," Art. "Coal."
‡ *London and Edinburgh Philosophical Magazine*, vol. xiii., p. 131.
§ *Annales de Chimie*, vol. lxvi., p. 337.
‖ *London and Edinburgh Philosophical Magazine*, vol. xvii., p. 213. In this analysis the coal contained 1·36 per cent. of water.
¶ *Ibid.*, p. 215. In the ashes of this analysis ·12 per cent. is sulphur.

the different specimens of the same species obtained from different localities. This is particularly the case with the anthracite coal, which passes through every stage of difference, from nearly a pure carbon, down to the state of ordinary bituminous coal.

(386.) A more intimate knowledge of the nature of coal was obtained by Dr. Thomson's analyses, before alluded to, by which the exact constituents of the volatile matter were ascertained. The preceding Table gives these analyses, together with the results obtained by other chemists in determining the nature of the gaseous products obtained from coal.

By comparing together the results of Dr. Thomson's analyses, given in the preceding Tables, we shall see how greatly the nature of the volatile matter contained in any specimen of coal affects the resulting quantity of coke. By Table XXXII. it appears that . the aggregate quantity of the gaseous products of caking, splint, and cherry coal, are very nearly similar; while by Table XXXI. we perceive that the quantity of coke obtainable from these several species varies more than 45 per cent. This, however, can readily be accounted for, when we ascertain the nature of the gas which predominates in each species: for, where hydrogen and oxygen abound, a large quantity of carburetted hydrogen and carbonic oxide is formed, at the expense of a certain proportion of the carbon; while in such specimens as contain azote in the largest proportion, a far smaller loss of the carbonaceous portion of the coal is sustained.

(387.) Of all the volatile constituents of coal, the azote is that which quits it with the greatest difficulty. Professor Proust* considers that coal always contains azote, even when reduced to coke; for when coke is treated with potass, a prussic

* *Nicholson's Journal,* vol. xviii., pp. 166 and 173.

lixivium is always obtained ; and the same he even found to be the case with anthracite coal. The prussic radical is a compound of azote and carbon ; and it may be a question whether any part of the difference which is known to exist in the heating power of "oven coke" and "retort coke," is owing in the former case to the presence of a larger portion of nitrogen. Some of the prussic compounds are very inflammable, and may therefore be supposed to produce some calorific effect by their presence.

Sulphur, which is another substance retained with the greatest pertinacity by coal, exists in nearly all the species in a greater or less degree. It is, perhaps, the only one of the constituents which, according to our present knowledge, is wholly valueless. Its injurious tendency is not more remarkable than the tenacity with which the coal retains it. Generally it exists in combination with iron, in the form of pyrites. In the process of coking a portion of the sulphur escapes in the state of sulphuretted hydrogen gas ; but no degree of heat is sufficient to drive off the whole of the sulphur; and many beds of coal are rendered almost useless in consequence of the large quantity of sulphur which the coal contains, preventing its use in metallurgy and many other processes in the arts. The presence of sulphur, indeed, is generally more perceptible in coke than in coal : the mass of volatile matter which escapes from the latter disguises the presence of the sulphur in a great degree ; while, with coke, the fumes of sulphurous acid are generally very perceptible. No mode, practically applicable, has yet been discovered for freeing coal from the sulphur it contains. By treating it with nitric acid the pyrites are dissolved, and the sulphur and the iron may be washed out; but the coal is converted by the operation into a bulky coke, and is entirely changed in its character, and ceases to afford the same gaseous products as before.* It is probable that some kinds of anthracite coal are nearly free from the presence of sulphur; and, indeed, several kinds afford no evidence of its existence.

(388.) The application of coal to the purposes of fuel depends like that of all other combustible bodies, on the chemical change which it undergoes in uniting by the agency of heat with some body for which it possesses a powerful affinity. In all ordinary cases this effect is produced by its union with oxygen; and we shall therefore inquire into the modes of effecting this in the best manner.

When coal is entirely consumed, the carbon is wholly converted into carbonic acid gas, and the hydrogen into water—the latter being in the state of vapour. The air supplies the necessary oxygen for this purpose; and in this state the products of the combustion are nearly or quite invisible, both the products being colourless fluids. *Smoke, therefore, is always the result of imperfect combustion.*

(389.) It has generally been considered that when coal is perfectly coked, the residuary coke will produce as much heat when applied as a fuel as the original quantity of coal would have done from which it was produced. This of course can only be taken in a general sense; because much must depend upon the method of coking and the prevention of waste, as well as the extent to which the process of coking has been carried. Many experiments confirmatory of this general view of the relative values of coal and coke have been made; the most recent are those of Mr. Apsley Pellat, Mr. Parkes, and the Count de Pambour. But it should be observed that in the residuary coke from the process of gas-making, the peculiar mode of carbonization lessens its heating power to

* *Nicholson's Journal,* vol. xviii., p. 170.

a considerable extent; and it is principally to what
is known as "oven-made coke" that this remark
will therefore apply.

We have the clearest evidence from the fact of
the great difference in the heating powers of *equal
weights* of coal and of coke, that the waste must be
very great in the usual modes of burning coal. We
know that a large proportion of the gaseous products of coal—which we have already seen constitute on a rough average about one-fourth of its
total weight—consists of matter which is capable
of producing the most intense heat: and yet we
find, practically, that its effect in furnaces is absolutely negative. This can arise only from some
imperfection in our methods of combustion; and
we may obtain a tolerably accurate notion of the
extent of the loss thus sustained, by a reference
to the analyses of coal already given. Let us take
as an example the caking coal, according to Dr.
Thomson's analysis in Table XXXII.—which is the
Newcastle coal so generally used. We find that
in every 100 lbs. of coal there are contained 4·18 lbs.
of hydrogen, and 4·58 lbs. of oxygen. When these
gaseous products are driven off by heat, they will
both combine with a portion of carbon. The quantity of carbon which combines with the hydrogen
is very variable; differing with the degree of
heat to which it is exposed. When the temperature is very high the hydrogen will combine
with three times its weight of carbon, forming
the true carburetted hydrogen; but, from a coke
oven, a large portion of the hydrogen escapes
nearly in an uncombined state, and therefore the
quantity of carbon thus abstracted will be only
about one-half the quantity which would constitute true carburetted hydrogen; or about 6 lbs.
of carbon may be assumed as the quantity carried
off in the latter case. Dr. Dalton ascertained that

the combustion of 1 lb. of hydrogen would melt 320 lbs. of ice ; therefore, 4·18 lbs. would melt 1337 lbs. of ice; and the heat produced by 6 lbs. of carbon will be sufficient to melt 376 lbs. of ice, according to the average of the experiments of Watt, Rumford, and Black.* The 4·58 lbs. of oxygen contained in the coal will combine with 3·5 lbs. of carbon, and form 8·08 lbs. of carbonic oxide. According to Dr. Dalton, 1 lb. of this gas will melt 25 lbs. of ice ; therefore 8·08 lbs. will melt 202 lbs. of ice. These results amount together to 1915 lbs. of ice melted by the heat obtainable from these several substances ; which number, multiplied by 140 degrees, the latent heat of ice, and this product divided by 7020, the number of pounds weight of water which can be raised 1 degree by the combustion of 1 lb. of coal, we shall find the total heat of these several gaseous products are equal to the calorific effects of 38 lbs. of coal. It is not known whether the azote which the coal contains produces any heating effect ; nor will the heat necessary for its expulsion from the coal cause any loss which is appreciable, even if the whole of it be driven off by heat, which, however, we have already seen, is not the case. By the experiments of Berard and Delaroche, on the specific heat of gases, we find that to raise the temperature of the 15·96 lbs. of azote to the temperature of 500 degrees Fahrenheit, will only require 4·9 ounces of coal—a quantity too small to be taken into account. The loss, therefore, by the escape of these gaseous products of the coal, amounts by these calculations

* The average of the experiments of Watt, Rumford, and Black, gives 39 lbs. of water raised 180 degrees, by the combustion of 1 lb. of coal; or 7020 lbs. of water raised 1 degree. The latent heat of ice being 140 degrees, this will be equal to melting 50·14 lbs. of ice with 1 lb. of coal; and the heating power of coke, compared with coal, being as 10 to 8, 1 lb. of coke will melt 62·7 lbs. of ice.

to 38 per cent.; and the coke which remains will be
of the description known by the name of "oven
coke," and will produce the same calorific effect
as the original quantity of coal would have done,
provided the coal were burned in the usual manner.

Another method may be employed for ascertaining the loss sustained by the escape of the volatile
matter of coal, by calculating the heating power of
the various products obtained from coal in the process of gas-making. The quantity of carburetted
hydrogen gas obtainable from 100 lbs. of caking
coal, of good average quality, may be stated at about
450 cubic feet; of which the specific gravity is
about ·50 to ·55. Twenty-four cubic feet of this gas
will weigh 1 lb.; and as 1 lb. weight of this gas
will melt 85 lbs. of ice, according to Dr. Dalton's
experiments, we shall have $\frac{450}{24} \times 85 = 1593$ lbs. of
ice melted by the combustion of the gas obtainable
from 100 lbs. of coal. Reducing this, as in the
former case, to the mean of the results obtained by
Watt, Rumford, and Black, we shall find that it is
equal to the total effect of 31·76 lbs. of coal. In
addition to this there will be about 8 lbs. of tar
obtained from 100 lbs. of coal. This, when decomposed by heat, yields about 100 cubic feet of an
impure hydro-carburet, mixed with about one-third
by weight of carbonic oxide. Reckoning the heat
of this by the data already given, it will be equal
to the effect of 10·4 lbs. of coal. The other product
of the distillation is ammoniacal liquor. Of this
about 7½ lbs. will be obtained from 100 lbs. of coal;
consisting of 5¾ lbs. of water, and 1¾ lbs. of ammonia; the former containing in its composition
·64 lb. of hydrogen, and the latter ·29 lb., making
together ·93 lb. of hydrogen; and the heat obtainable from this quantity of hydrogen will be equal
to 5·93 lbs. of coal. These several results amount
together to a loss equivalent to 43·09 lbs. of coal,

but as the residuary product of the distillation will
be only "retort coke," which is inferior to "oven
coke" in its heating power to the extent of $12\frac{1}{2}$ per
cent., according to the experiments of M. de Pambour, we must deduct from the above amount the
difference between the heating power of this coke
and that which was supposed to be obtained by the
former mode of calculation—the quantity being
considered the same in both cases. We shall therefore find the statement will stand as follows:—

			lbs. of Coal.
Heat obtainable from 450 cubic feet of carburetted hydrogen	=	31·76
Do.	8 lbs. of tar.	=	10·40
Do.	·93 lbs. of hydrogen contained in the ammoniacal liquor	=	5·93
			48·09
Deduct difference in heating power of residuary coke, viz. 75 lbs., at $12\frac{1}{2}$ per cent.	=	9·37
	Total loss	38·72

By this method of calculation, the loss occasioned
by the non-combustion of the volatile products of
the coal amounts to $38\frac{1}{2}$ per cent., which is an extremely near approximation to the result obtained
by the former method. We cannot, however, consider that the whole of this amount is always lost
by the escape of smoke in the combustion of coal.
With open fires, no doubt, this is the case, as well
as other sources of loss peculiar to this method of
combustion. But in furnaces, however imperfectly
they are constructed, some portion of the smoke is
always consumed ; and by that amount, whatever
it may be, the loss is diminished. The smaller the
quantity of volatile matter which the coal contains,
the less will be the loss in this way ; but the kind
of coal selected for these calculations is a quality
which may be considered to afford a fair average.

(390.) But, in addition to the loss of calorific
effect which is here shown by the escape of uncon-

sumed smoke, there is another source of loss which always exists in a greater or less degree, arising from the formation of carbonic oxide. It is of importance to understand correctly the theory of the formation of this carbonic oxide, as it materially affects the question of economy in the combustion of fuel.

(391.) When atmospheric air comes in contact with coal or coke at a very high temperature, the combination of the oxygen of the air with the carbon of the fuel always forms carbonic acid gas, and produces the phenomenon of combustion.* Such is the effect of atmospheric air entering through the grate-bars of a furnace, on a lower stratum of fuel, lying immediately on the bars. But while this carbonic acid gas passes upwards through the upper strata of the heated fuel, a further portion of carbon combines with it, and it becomes converted into carbonic oxide—carbonic *acid* consisting of two volumes of oxygen and one volume of carbon; while carbonic *oxide* is composed of equal volumes of oxygen and carbon.

The result of this conversion of carbonic acid gas into carbonic oxide, by passing the former through highly-heated carbon, is a considerable loss of heat when the carbonic oxide escapes from the furnace in this state; for a given weight of carbon, converted into carbonic oxide, only produces half the heat which it would do were it converted into carbonic acid gas. The combustion of fuel, therefore, cannot be perfect where any considerable quantity of carbonic oxide escapes undecomposed;

* The lowest temperature at which this combination of oxygen and carbon takes place, is a little above the boiling point of mercury. At this temperature carbonic acid gas is formed without any luminous appearance; at higher temperatures true combustion occurs, and carbonic acid is produced with great rapidity. (Davy's "Experiments on Flame," *Philosophical Magazine*, vol. 50, p. 10.)

although it may often happen that no smoke is visible even when there is a large escape of carbonic oxide. This is particularly the case when coke is used as fuel. There is in this case no smoke; but if there is a deficiency of atmospheric air to supply the necessary amount of oxygen to convert the product of the combustion wholly into carbonic acid gas, a large quantity of carbonic oxide is formed, which not only causes a very great loss of fuel, but it is probably even more unwholesome than the most dense smoke. No system of combustion can therefore be perfect which does not provide a supply of atmospheric air above the fuel, sufficient to reconvert the carbonic oxide (which has been formed by passing the carbonic acid through the mass of burning fuel) into carbonic acid; and in this process it absorbs just its own volume of oxygen, and it gives out another measure of heat exactly equal to that already produced by it in the furnace, the heat from a given volume of carbonic acid being just double what is obtained from the same volume of carbonic oxide. The combustion of the furnace is therefore produced by the oxygen of the air forming carbonic acid with the lower stratum of fuel lying on the furnace-bars; this is changed into carbonic oxide as it passes through the super-stratum of heated fuel; and this carbonic oxide is again reconverted back into carbonic acid, by bringing the oxide into union with a further quantity of atmospheric air.

(392.) In the case of bituminous coal, the economy of fuel must necessarily consist both in consuming the smoke and in preventing the escape of undecomposed carbonic oxide. Smoke always arises from one of two causes—deficiency of air, or an insufficient degree of heat to cause the chemical union between the constituents of the fuel and the oxygen derived from the air; and sometimes it

arises from both these causes combined. The loss
arising from the carbonic oxide is almost entirely
owing to deficiency of oxygen. In open fireplaces
the smoke is caused by deficiency of heat; in close
furnaces it is generally caused by deficiency of air;
and all the different methods which have been
proposed for *consuming smoke* in close furnaces,
however variously these plans may be applied, are
all based on the principle of supplying additional
air to the burning fuel. Two or three plans,
indeed, for *destroying smoke* have been proposed,
which will presently be mentioned; but we shall
first inquire into the methods of beneficially applying the combustion of the gaseous products of coal
to the ordinary purposes of fuel.

(393.) A vast deal of misconception upon a very
simple subject has occurred from parties interested
in particular inventions discussing the general
question of the combustion of smoke in the way
best calculated to recommend their own inventions, and to depreciate those of others. The
inquiries as to whether hot air or cold air is most
advantageous for consuming smoke, or whether
smoke is really capable of being *consumed* after it
is once formed, or whether the only remedy for it
is to prevent its formation, are entirely of this
kind. But that which will be found to be the
fact is, that smoke is as capable of being consumed
as any other combustible, and that there are many
methods of accomplishing this both by hot and by
cold air.

The combustion of smoke, and indeed of any
other substance, is not to be supposed to involve
its total annihilation; for matter of all kinds, so
far as our knowledge extends, is indestructible.
But the combustion of smoke is that change of
state produced by chemical union with other substances, which entirely alters its character and
appearance.

(394.) The constituents of smoke can be accurately judged of from a knowledge of the chemical composition of the coal which produces it. Nitrogen, oxygen, hydrogen, and carbon, with the various combinations of these bodies, namely, carbonic acid, carbonic oxide, carburetted hydrogen, ammonia, and vapour of water, together with minute portions of various resins, salts, earthy matters, and volatile inflammable vapours, must necessarily constitute the substance known under the general name of smoke. All these substances, except the carbonic acid, are capable of further combinations with atmospheric air, by means of a high temperature; and practically they do all undergo a change, except that the nitrogen exists in too large a quantity to enable any considerable proportion of it to combine chemically with the other substances. Thus, then, there is nothing to prevent a " true combustion of smoke " from taking place; by which means chemical combinations are produced, the principle being that the uncombined carbon which gives the black colour to smoke unites with oxygen derived from the air, and becomes converted into the colourless carbonic acid gas; while the carbonic oxide is also changed into the same chemical compound—namely, carbonic acid.

As regards the actual destruction of the black colour of smoke, it matters but little whether hot or cold air be admitted into the furnace; for so long as the furnace is sufficiently hot, and the quantity of air is sufficiently abundant, the combustion will take place. But before the air can enter into combustion, it is necessary that it be raised to a high temperature; in most cases, about 800° or 900° of Fahrenheit being required for this purpose.* When the air is not heated previous to

* See note to Art. 391, *ante.*

its entrance into the furnace, this heat which is necessary for its combination is obtained from the bodies with which it combines; their temperature is therefore necessarily lowered, by parting with the requisite heat to raise the temperature of the air to that high degree at which it will enter into chemical union with the gaseous and solid matter of the fuel.

(395.) The experiments of Sir Humphry Davy on combustion,[*] clearly show the necessity of a high temperature before active combustion can take place, and the advantages that must therefore result from extrinsically heating the air which supports combustion. They also entirely refute the assertions that the rarefaction of the air by heat is injurious to complete combustion, his experiments having in fact been undertaken with the view of testing the accuracy of a theory to this effect, propounded by M. de Grotthus and others, and which he found to be erroneous.

When Sir Humphry Davy caused a jet of hydrogen gas, one-sixth of an inch in height, to burn in the receiver of an air-pump, the flame enlarged as the receiver was exhausted by the pump, and was at its maximum when the pressure of the air was between four and five times less than that of the atmosphere; and when a larger jet was used, the same phenomenon occurred even when the air was rarefied ten times. This effect, from a larger jet, was found to arise from the increased heat produced : and the conclusion drawn from all the experiments was, "that among combustible bodies, those which require least heat for their combustion, burn in more rarefied air than those that require more heat; and those that produce much

---

[*] "Researches on Flame," by Sir Humphry Davy. *Philosophical Transactions*, Part I., for 1817; and *Philosophical Magazine*, vol. 50, p. 1, *et seq.*

heat in their combustion, burn, other circumstances
being the same, in more rarefied air than those that
produce little heat." The experiments also proved
that "by preserving heat in rarefied air, or giving
heat to a mixture, inflammation may be continued
when, under common circumstances, it would be
extinguished." When these mixtures were heated
before combustion, Sir Humphry Davy found "that
expansion by heat, instead of diminishing the com-
bustibility of gases, on the contrary enables them to
explode apparently at a lower temperature; which
seems perfectly reasonable, as a part of the heat
communicated by any ignited body must be lost in
gradually raising the temperature." It was also
found that "the cooling power of mixtures of elastic
fluids in preventing combustion increases with their
condensation and diminishes with their rarefaction; at the same time the quantity of matter entering into combustion in given spaces is relatively
increased and diminished. The experiments on
flame in rarefied atmospherical air show that the
quantity of heat in combustion is very slowly
diminished by rarefaction, the diminution of the
cooling power of the azote being apparently in a
higher ratio than the diminution of the heating
powers of the burning bodies." When the rarefaction of the air, however, is produced by heat, not
only is there no loss whatever in the available heat
produced by combustion, but the extensive application of heated air by means of the " hot blast " to
the smelting of iron, proves that there is an enormous increase in the effect, both on the solid matter
of the fuel as well as on its gaseous products. And
the same result will necessarily occur with respect
to all kinds of furnaces for the combustion of fuel.
The heated air, when carefully kept from imbibing
moisture, will always enter into combustion more
readily than cold air; will cause a much greater heat

in the furnace ; and will produce more perfect combustion of the fuel. Sir Humphry Davy not only proved "that the combustion of all gaseous mixtures is increased by rarefaction by heat," but he ascertained by his experiments, that a general law obtained "that at high temperatures gases not concerned in combustion will have less power of preventing that operation, and likewise, that steam and vapours, which require a considerable heat for their formation, will have less effect in preventing combustion (particularly of those bodies requiring low temperatures) than gases at the common heat of the atmosphere." The well-known effect of cold frosty air, in causing fires to burn clear and bright, in no way militates against these conclusions. The effect produced by cold air arises from the decreased quantity of moisture which it then contains, and not from the greater density of the air : for Sir Humphry Davy found that even with atmospheric air condensed to *five times* its natural density, scarcely any appreciable difference could be perceived in its effect on combustion. Neither can these effects be different, whether the combustible be a solid or a gaseous body ; except that the latter would be more easily lowered in its temperature and reduced below the temperature requisite for its accension.

(396.) The expansion of air by heat, previous to its entrance into the furnace, cannot at all reduce the quantity of oxygen which combines with the fuel. For as air will not support combustion until it be raised to a very high temperature (about 800° or 900°), its expansion must necessarily be the same, whether this heat be communicated to it within or without the furnace.

(397.) The quantity of atmospheric air required for the combustion of coal is very great. Taking Richardson's analysis of Newcastle coal (Table

XXXII., Art. 386), it appears that 355,376 cubic feet of air, of ordinary density, would be required for the combustion of one ton of this coal. If heated air be used, the number of cubic feet must be increased according as the density of the air is diminished; so that sometimes when the air is very highly heated, twice, or even three times this number of cubic feet of air may be necessary for the perfect combustion of the coal.

The actual quantity of air which enters into a furnace, where complete combustion takes place, must be sufficient to convert the carbon into carbonic acid, and the hydrogen into water. The former requires 2·66 times, and the latter eight times its weight of oxygen to make these combinations; and the oxygen of the air being one-fifth of its total volume, or nearly one-fourth of its weight, we can thus calculate the quantity of atmospheric air required for the combustion of any particular kind of coal.*

But this quantity of atmospheric air, large as appears its amount, will not be sufficient for perfect combustion; for this calculation supposes that the whole of the oxygen is abstracted from the air in the process of combustion, a result which experience proves is never practically produced. Some interesting experiments on this subject have been made by Mr. Hunt,† on the furnaces of the principal engines in Cornwall; and the average of the analysis of the air, taken from the chimneys of the furnaces, after it has performed its office in the combustion of the fuel, shows that the mixed gases passing off through the chimney contain

---

* The constituents of atmospheric air by weight are oxygen 22·22, and nitrogen 77·77; but by volume the constituents are oxygen 21, and nitrogen 79, or very nearly these proportions.

† *Transactions of the Cornwall Polytechnic Society*, 1843; and *Glasgow Engineers' Magazine*, vol. iii., p. 93.

2 E

one-tenth of their volume of free oxygen. The amount of carbonic acid was also found to be, on an average, one-ninth of the total volume of the gaseous matter passing through the chimney. It appears, therefore, that but little more than one-half the oxygen of the air is abstracted in the process of combustion; and these experiments prove that practically it requires double the quantity of air to produce complete combustion in furnaces that theoretical calculations would give, when based on the assumption of the entire abstraction of the oxygen from the air.*

(398.) Whenever the gases eliminated from the combustion of coal are made to unite with the proper quantity of oxygen, and the temperature of the mixture is sufficiently high, the smoke will be consumed, in whatever part of the furnace or flues the admixture takes place. This fact has been disputed, but without any grounds for so doing. It has also been asserted that more atmospheric air is required to produce combustion of the smoke after it is mixed with the carbonic acid formed in the furnace, than would be required previous to this intermixture. This is true, theoretically, as the experiments of Sir Humphry Davy proved; for he ascertained that carbonic acid gas has rather a greater power of preventing the firing of explosive mixtures than azote would have.† Perhaps, therefore, the most advantageous place to introduce atmospheric air would be at the front of the furnace; but, in all probability, the difference in this respect is very small, as we have already

---

* These experiments made by actual working on a very large scale, show that 312 cubic feet of air, weighing 23½ lbs., is required for the combustion of each pound weight of coal of the quality used in those experiments.

† *Philosophical Transactions*, Part II., 1815; and *Philosophical Magazine*, vol. xlvi., p. 449.

seen that, in ordinary cases of combustion, only about one-half the oxygen of the air combines in the process of combustion, arising, no doubt, from the difficulty of sufficiently mixing the gases together during their passage through the furnace. But the longer these gases are in contact, and the more they are agitated and mixed together by passing through the different obstructions of a furnace, the more likely is the oxygen of the air to be abstracted, and chemical combination to take place.* And, contrary to the opinion that smoke after it is once formed cannot be burned, a recent patent has been obtained for burning the smoke of furnaces by passing it over a second fire, at a considerable distance from the principal fire, with a fresh supply of atmospheric air; and, however distant this second fire may be from the primary one, the combustion of the smoke is complete, and an immense heat is derived from these gaseous products, which, under ordinary circumstances, would only produce the black smoke of common furnaces.†

(399.) The actual quantity of heat produced by different qualities of coal does not exactly depend upon the quantity of oxygen with which they combine. Dr. Ure made experiments ‡ on the actual amount of heat given out by several qualities of coal, when consumed in a calorimeter of very perfect construction; by which, as nearly as possible, the entire heat from the coals was obtained; and, taking the number of pounds of water raised 1° by

---

* See Parliamentary Report on " Smoke Prevention," 1843, for much useful information on this subject. Also three Reports on Coals suited to the Royal Steam Navy, published by the Museum of Practical Geology.

† See Collier's " Patent," Art. 456.

‡ *Reports of the British Scientific Association*, vol. viii. (1839), p. 20.

the combustion of 1 lb. of coal as the standard
of comparison,* the proportions were—

Lambton's Wall's End	.	.	. 7,500
Llangennech Coal	.	.	. 9,000
Anthracite Coal	.	.	. 12,000

The cause of the less degree of heat by the
combustion of coals containing large quantities of
hydrogen, Dr. Ure considers to arise from the great
amount of heat rendered latent by the formation
of steam and carburetted hydrogen gas; though
the experiments of Dalton, Davy, Lavoisier, and
Crawford all proved that the heat produced by the
combustion of hydrogen is greater than from any
other substance.†   All experiments, however, agree
in proving the great heat which is derivable from
the combustion of anthracite coal.   Considerable
difficulty attends its combustion, on account of this
kind of coal always breaking up in the furnace into
small pieces, except it be very gradually heated; and
unless this precaution be adopted, the draught of
the fire is wholly stopped.   The breakage of the
coal arises from its slow-conducting power, which
causes the outside surfaces to expand more than
the inner parts, when exposed to a high temperature; and this expansion causes the exterior parts
continually to separate from the interior, until the
whole substance is broken up.   The elasticity
which bitumen gives to coal prevents this result
with the ordinary qualities of bituminous coal.   To
remedy this inconvenience with anthracite coal,
various plans have been proposed for supplying it
with vapour of water, in order to render it less
brittle.   The advantages of this operation, how-

---

* These numbers represent the number of lbs. weight of
water heated 1° by 1 lb. of coal.
   † Ure's " Dictionary of Chemistry," Art. " Combustion."

ever, are very questionable. For, in America, where large quantities of anthracite are burned, experiments have been made in order to ascertain the cause of the corrosion which sometimes occurs to boilers, iron chimneys, and stove pipes, by the combustion of anthracite coal: and from a report of the Franklin Institute of Pennsylvania,* it appears that in these cases the ashy deposit has been found to contain muriate and sulphate of ammonia, sometimes as much as three-fourths of the deposit consisting of these salts. Where moisture is present, the action of these salts must be much increased in activity; and it therefore deserves serious inquiry whether by the addition of vapour of water, all descriptions of anthracite become in some degree corrosive, or whether the effect is peculiar to the coal of certain districts.† It is probable, that the destructive effect which is sometimes produced on thin copper boilers by particular kinds of fuel may arise from something of this kind. Where the boiler is of such a form that the fire acts particularly on the sharp edges which form the connection between the bottom and sides of the boiler, instances have repeatedly occurred of the corrosion being so active, that the bottom has separated from the sides as though it had been cut with a chisel; and in other cases, the surface of the boiler, when very thin, has been so corroded as to become full of holes in the course of a few months' wear (Art. 75).

---

* *Mechanics' Magazine*, vol. xxxvi., p. 439.

† Experiments have led the author to conclude that this effect is not peculiar to anthracite coal, but that coke when burned with moisture produces the same results. The circumstance is interesting in a chemical point of view; and if more careful and extensive experiments show this opinion to be correct, they may give rise to important inquiries concerning the compound nature of certain (so-called) simple substances.

The inventions which have been brought forward for consuming and for preventing smoke are very numerous. The following list will give a tolerable idea of the plans which have been proposed, and of the general methods by which the object is sought to be obtained. Those of which a description is known to have been published have a reference to such description; but the list is not given as a perfectly accurate account of all the inventions for this purpose, as no doubt there are others which have escaped the author's notice.[*]

(400.) The first attempt at consuming smoke appears to have been made by M. Delesme, some time prior to 1669, by means of a stove with a downward draught; but it was not at all suitable for furnaces. (*Philosophical Transactions*, 1686.)

(401.) Dr. Papin (1695) proposed a plan for forcing air down a shaft upon the fuel, in order to burn the smoke of furnaces. (*Philosophical Transactions*, 1697).

(402.) James Watt (1785), patent for consuming smoke by admitting air through openings in the front of the furnace door, and also by gradually coking the coals. (*Repertory of Arts*, vol. iv. (1796), p. 226.)

(403.) C. W. Ward (1792), patent for condensing smoke by drawing it, by means of an air-pump or bellows, through cold water. (*Repertory of Arts*, vol. i. (1794), p. 373.)

(404.) W. Thompson (1796) proposed a furnace

---

[*] In the following descriptions of the various inventions, there are several which profess to accomplish results utterly unattainable by the means proposed. The statements are chiefly taken from the published accounts of the various plans, and several of them are totally contrary to the principles of science. The description given must, therefore, not be taken as the author's explanation of the operations or the effects, but as those of the several inventors.

to burn smoke, by letting air in *behind* the bridge. (*Repertory of Arts*, vol. iv. (1796), p. 316.)

(405.) Roberton, of Glasgow (1800), patent for admission of heated air in thin streams over the fire-door; for coking the coals in front of the furnace; and also a hopper to supply the fuel. (*Repertory of Arts*, vol. xvi. (1802), p. 364.)

(406.) M. de Prony (1809 or 1810), Report on apparatus erected at the Royal Mint, Paris, for consuming smoke, by two pipes passing from the front of the furnace door and delivering hot air at the bridge. He also states that this plan had previously been used by MM. Clement and Desormes, and others. (*Annales de Chimie* and *Retrospect of Science*, vol. v., p. 439.)

(407.) Wm. Sheffield (1812), patent for hollow or split bridge, which delivered the air in a *horizontal* stream towards the front of the furnace. (*Gill's Technical Repository*, vol. i., pp. 16 and 42.)

(408.) J. Wakefield (Manchester), a similar patent to the above, and of subsequent date. (*Ibid.*)

(409.) Wm. Johnson (Salford), a similar and subsequent patent. (*Ibid.*)

(410.) Losh (1815), patent for dividing the furnace into two parts, lengthways; the two compartments being supplied with fuel alternately, by which the smoke from the fresh fuel passed over the more perfectly ignited fuel, and was thus consumed.

(411.) Brunton (1816), patent for revolving grate and feeding hopper; by which the fuel was equally distributed over the furnace, and in small quantities at a time, thereby preventing all dense smoke. (*Mechanics' Magazine*, vol. i., p. 121.)

(412.) John Gregson (1816), patent for bringing air down a shaft, near the bridge, to promote combustion. Mechanical means were used to supply the fuel by a snail wheel moved by springs. (*Quarterly Journal of Science*, vol. iii., p. 348.)

(413.) Josiah Parkes (1820), a patent for split bridge, precisely similar to Sheffield's patent of 1812. (*Mechanics' Magazine*, vol. ii., p. 250.)

(414.) W. Pritchard (1820), a patent for a regulating weight to close gradually the furnace door, so as to vary the quantity of air passing into the furnace. (*Repertory of Arts*, vol. xxxix. (1821), p. 140.)

(415.) Mr. Marsh (1824) consumed the smoke of furnaces by leaving two openings in the flue at the back of the furnace. (*Gill's Technical Repository*, vol. vi., p. 213.)

(416.) Chapman (1824) described a plan for hollow furnace bars, that conveyed heated air into the furnace through a split bridge which projected the heated air horizontally. Also a hopper to supply the fuel. (*Transactions of the Society of Arts*, vol. xlii., p. 32; and *Quarterly Journal of Science*, vol. xix., p. 138.)

(417.) James Nevill, patent for a fan fixed in the flue, which produced a rapid draught up the chimney.

(418.) Stanley's patent for a feeder for furnaces which consisted of a hopper, and grooved rollers to crush the coal. Two revolving pans scattered the coal over the fire in small quantities at a time, and thus prevented the dense smoke.

(419.) Jeffries (1824), patent for *destroying* smoke by a shower of water. Two chimneys are used, and the smoke passes up one and down the other, in which latter a shower of water falls from a colander, and carries the smoke in the form of soot into a drain. (*Mechanics' Magazine*, vol. xxxiv., p. 198.)

(420.) Mr. Oldham, of the Bank of England, employed a plan of rocking bars, moved by a small eccentric on a shaft worked by the engine.

(421.) G. Chapman, of Whitby, plan for consuming smoke by the use of hollow furnace bars delivering heated air at the bridge of the furnace, 1825. (*Repertory of Arts,* vol. xlvi., p. 360.)

(422.) J. Gilbertson (1828), patent for making the sides of the furnace of hollow plates, through which the air passes, and is delivered in a heated state at the back of the furnace. (*Repertory of Arts,* vol. vii. (1829), p. 65.)

(423.) Wm. Taylor (1830), patent for consuming smoke, by forcing it through the fire mixed with atmospheric air by means of an air pump. Also for a mode of passing the smoke through red-hot pipes placed in the furnace among the fuel, which pipes form the only outlet to the chimney. (*Repertory of Arts,* vol. i. (1834), p. 282.)

(424.) J. C. Douglass (1833), patent for two or more sets of bars. The smoke from the first set passes downwards below a bridge, and then upwards through a second set of bars, on which burning fuel is placed. (*Repertory of Arts,* vol. v. (1836), p. 346.)

(425.) J. G. Bodmer (1834), patent for traversing bars, moved by machinery, which receive the fuel from a feeder, and discharge the ashes at the further end of the furnace.

(426.) Richard Coad (1835), patent for heating the air by passing it through pipes placed in the flues, and delivering the air at the bridge of the furnace. (*Mechanics' Magazine,* vol. xxvii., p. 375.)

(427.) T. Hedley's patent for *purifying* smoke by four or six flues, of which one half ascend and the other half descend. A shower of water falls through the descending flues and washes the smoke, depositing the carbon in the form of lamp-black.

(428.) William Richard, of Leeds, patent for gasometer applied to furnaces, so that on opening the furnace door to feed the fire a passage is opened from the gasometer, containing condensed air, to a number of holes at the back of the bridge, and the supply of air gradually diminishes as the fire burns clear.

(429.) Samuel Hall (1836), patent for cast-iron pipes placed upright in the flue at the back of the furnace, and then passing towards the front. The air is thus heated to about 300°, and the gases inflame in front of the furnace. (*Mechanics' Magazine*, vol. xxviii., p. 226.)

(430.) John Hopkins (1836), patent for a curved bridge, by which the smoke and gases are thrown back again upon the burning fuel. (*Repertory of Arts*, vol. vii. (1837), p. 252.)

(431.) Jacob Perkins (1836), patent for two sets of bars and two ashpits. The second ashpit is closed, and supplied with air by a fan, so as to give more air to that part of the furnace, and thus burn the smoke from the fuel on both sets of bars. (*Repertory of Arts*, vol. viii. (1837), p. 268.)

(432.) Joseph Chanter (1837, &c.,) several patents for smoke burning, principally by inclined bars—double sets of bars—air supplied through tubes placed under the first set of bars—and hot air supplied at the bridge.

(433.) James Drew, Manchester, patent for two sets of bars. The coal is coked on the first set, and passed on to the second set, which is then raised by rack-work as near the boiler as possible.

(434.) Paul Chappe, patent for injecting small jets of boiling water over the fire, in front of the bridge.

(435.) Ivison and Bell (1838), patent for injecting small jets of steam into the furnace, and also for heating the air by passing it through tubes. (*Mechanics' Magazine*, vol. xxviii., p. 221, and vol. xxx., pp. 69 and 107.)

(436.) Rodda's patent for a furnace divided across in two parts. The fuel is first put into the compartment nearest the door, and afterwards thrown backwards to the further compartment. The smoke is burned by passing over the clear fire of the second compartment. (*Mechanics' Magazine*, vol. xxxi., p. 386.)

(437.) Cheetham and Bayley, patent for a fan which catches the smoke and forces it, mixed with fresh air, through the ashpit and furnace bars, the ashpit being made air-tight.

(438.) Thomas Hall (Leeds), patent for dividing the furnace into two compartments lengthways, which are supplied with fuel alternately, and the smoke thereby passes over red-hot fuel.

(439.) James Nevill (1837), patent for two sets of hollow bars containing water. A downward draught is produced by the chimney drawing only from the lower set of bars and the smoke burned by passing through the hot fuel in contact with the bars. (*Repertory of Arts*, vol. xii. (1839), p. 220.)

(440.) John Juckes (1838), patent for heating the fuel by passing it through highly-heated pipes or other surfaces, by which it is coked before it passes into the fire, which is effected by mechanical means. (*Repertory of Arts*, vol. xiii. (1840), p. 122.)

(441.) J. A. Caldwell (1839), patent for a rotary fan, by which air is forced into a closed ashpit; the furnace bars are placed very close together, and a movable damper is applied in the chimney, by which the velocity of the smoke escaping is retarded, and the heated gases retained longer in the furnace. (*Repertory of Arts*, vol. xiii. (1840), p. 83.)

(442.) William Miller (1839), patent for rocking bars, by which each alternate bar is made to move lengthways in opposite directions, backwards and forwards; and thus preventing clinkers, and consuming the smoke by allowing a free passage for the air through the bars. (*Repertory of Arts*, vol. xvii. (1842), p. 143.)

(443.) C. W. Williams (1839), patent for supplying air in jets to the furnace, principally behind the bridge, by a diffusion box. The air is supplied cold.

(444.) André Kurtz (1840), patent for three sets of bars, those at each end inclined, and the middle set lower than the others. Hollow bearing bars which convey heated air into the furnace. (*Mechanics' Magazine*, vol. xxxiv., p. 397.)

(445.) Junius Smith (1840), patent for a double fan or blower, which passes heated air with the smoke through the furnace bars a second time. The heavy gases are allowed to fall by their gravity below · the fan, which then forces them down, and filters them through gravel or sand. (*Repertory of Arts*, vol. xvi. (1841), p. 81.)

(446.) Baron Von Rathen (1840), patent for hollow firebars, resting upon bearers with steps, forming two sides of a triangle, which allows more air to pass into the furnace. Also, a coal-feeder placed over the dead-plate, which supplies fuel without opening the door. (*Mechanics' Magazine*, vol. xxxv., p. 27.)

(447.) Godson and Foard (Jan., 1841), patent for a box placed below the furnace bars with a movable bottom. The box being filled with fuel, the bottom is gradually raised by a lever, and supplies fuel from below, the smoke from which is consumed by passing through the red-hot fuel. (*Repertory of Arts*, vol. xviii. (1842), p. 129.)

(448.) M. Coupland (Sept., 1841), a patent for movable centre bars which pass downwards into a box; nearly similar to Godson's. (*Repertory of Arts*, vol. xviii., p. 207.)

(449.) F. Heindruckx (1841), patent for a furnacc without bars. The sides of the furnacc are inclined, and a narrow opeuing is left at the bottom and whole length of the furnace, through which the air enters, and is regulated by a longitudinal valve. (*Mechanics' Magazine*, vol. xxxv., p. 366.)

(450.) J. C. March (1841), patent for causing air to be blown in streams on the upper surfacc of the fuel without passing through the fire. No furnace bars are used by this plan. (*Mechanics' Magazine*, vol. xxxv., p. 492.) See also a somewhat similar plan, Art. 401.

(451.) John Juckes (1841), patent for a furnace grating passing over rollers at each end of the furnace like an endless chain. The bars revolve by machinery, receiving fuel from a feeder placed ncar the door, and deliver the ashes at the further end of the furnace. (*Repertory of Arts*, vol. xvii., (1842), p. 210.)

(452.) J. Prosser (1842), patent for a furnace bridge, with square holes for the admission of air. The bridge is fixed close against the bottom of the boilcr.

(453.) Kymer and Leighton (1843), patent for diagonal bars resting in small longitudinal troughs of water. A closed ashpit is used, and a fan forces air through the bars and also over the fuel. The plan used principally for anthracite coal.

(454.) Schofield, of Leeds (1842), proposcd the use of very narrow furnace bars, a quarter of an inch wide at top, and as thin as possible at bottom, and two inches deep. These bars admit a larger quantity of air than usual, and thus consume the smoke.

(455.) E. Billingsley, of Bradford (1842), proposed a focal bridge beyond the furnace bars, and a sliding rack or grating in front of the furnace to admit air. (*Mining Journal*, Jan., 1843.)

(456.) E. II. Collier (1843), patent for the use

of a second fire at a distance from the usual fire. The smoke, after passing through the ordinary flues, is carried over this second fire, and there mixed with an additional quantity of air, when it inflames, and is carried through a second set of flues before passing into the chimney.

(457.) Butler's (1845) registered plan for movable bridge, which opens to regulate supply of air. (*Mechanics' Magazine*, vol. xlii., p. 50.)

(458.) Whiteley's (1845) registered plan for feeding apparatus and admitting fresh air. (*Mechanics' Magazine*, vol. xlii., p. 147.)

(459.) Blackwell's (1848) patent for a double-chambered furnace for coking the fuel. (*Mechanics' Magazine*, vol. xlix., p. 122.)

(460.) Acock's (1848) patent for a feeding hopper and other improvements. (*Mechanics' Magazine*, vol. xlix., p. 554.)

(461.) Burrow's (1848) patent for a regulating feed-roller to furnaces. (*Mechanics' Magazine*, vol., l., p. 425.)

(462.) James Robertson (1848), patent for perforated tubes to convey air into the furnace. (*Mechanics' Magazine*, vol. l., p. 429.)

(463.) Grist's (1849) patent for revolving furnace bars. (*Mechanics' Magazine*, vol. l., p. 108.)

(464.) Newcombe's (1849) reciprocating furnace bars, moved by a cam. (*Mechanics' Magazine*, vol. li., p. 67.)

(465.) Samuel Hall (1849), reciprocating furnace bars moved by eccentrics. (*Mechanics' Magazine*, vol. li., p. 286.)

(466.) Elijah Galloway (1849), oscillating furnace bars. (*Mechanics' Magazine*, vol. lii., p. 239.)

(467.) Joseph Johnson (1849), patent for heated air delivered through pipes, at or near the bridge. (*Mechanics' Magazine*, vol. lii., p. 318.)

(468.) William Hargreaves (1850), patent for

oscillating bars combined with a feeding-bed which forces the fuel gradually forward in the furnace. (*Mechanics' Magazine*, vol. liii., p. 335.)

(469.) D. L. Williams (1850), patent for hollow furnace bars. (*Mechanics' Magazine*, vol. lv., p. 1.)

(470.) T. S. Prideaux (1850), patent for air-valve closing gradually to regulate the quantity of air which passes into the furnace through perforations in the inner door; the air becoming heated by passing through them. (*Mechanics' Magazine*, vol. lv., p. 18.)

(471.) George Anstey (1851), patent for passing smoke through a series of apertures to keep it in contact with flame. (*Mechanics' Magazine*, vol. lv., p. 75.)

(472.) Johann Stierba (1852), patent for feeding hopper, and air tubes for delivering air at the bridge. (*Mechanics' Magazine*, vol. lvii., p. 422.)

(473.) Sorrell's (1852) patent for oscillating furnace bars worked by cams; a feeding hopper worked by a roller; and for admitting air at the bridge of furnace. (*Mechanics' Magazine*, vol. lix., p. 26.)

(474.) Green's (1853) patent for double fires, to be fed alternately, together with peculiar system of flues. (*Mechanics' Magazine*, vol. lxi., p. 2.)

(475.) Jearrard (1853) hollow perforated fire-brick sides, &c., for admission of air to furnace. (*Repertory of Arts*, July, 1854.)

(476.) Gilbertson (1854), a perforated iron tube over the fuel to deliver air to furnace. (*Repertory of Arts*, March, 1855.)

(477.) Simpson (1854), stops or inverted bridges in the furnace to deflect air on to the fuel. (*Repertory of Arts*, April, 1855.)

(478.) Yates (1854), mechanical feeding apparatus for furnaces. (*Repertory of Arts*, March, 1855.)

(479.) Taylor (1854), furnace supplied with air above the fuel and not through furnace bars. (*Repertory of Arts*, September, 1855.)

(480.) Manley's (1854) method of descending stream of water, in the shaft or upright flue. (*Mechanics' Magazine*, vol. lxi., p. 205.)

(481.) W. Woodcock (1854), patent for air-tubes at sides of furnace, above the furnace bars, delivering heated air beyond the bridge of furnace in jets. (*Mechanics' Magazine*, vol. lxi., p. 410.)

(482.) Parker's (1854) patent for a loose or movable box or bridge, pierced with holes, and placed in front of the ordinary bridge, on the bars ; by which heated air is delivered to the gases as they pass over the furnace bridge. (*Mechanics' Magazine*, vol. lxi., p. 445.)

(483.) Galloway's (1854) patent for pipes placed below the furnace bars, and delivering air into a hollow or double bridge. (*Mechanics' Magazine*, vol. lxi., p. 458.)

(484.) It appears unnecessary to describe · more of these inventions for burning smoke. The recent plans have been far less numerous than they were a few years ago, partly because almost every conceivable form has been patented repeatedly before, and also because since the passing of the Act for the consumption of smoke in large towns, people appear to have become aware that there are many ways of accomplishing this object, and that no patent process whatever is necessary for the purpose: The process of smoke burning is indeed extremely simple in itself, and, with the exception of some very few of the preceding inventions, which are clearly founded on a misconception of the chemistry of combustion, they all attempt to effect the same object ; namely, to bring a larger quantity of atmospheric air into contact with the fuel, or with the products of the first combustion. It will be remarked that all the later plans, without exception, are mere modifications of the very earliest inventions; and that the first six or eight plans described in this list contain the gist of the many subsequent inventions for this purpose. In fact, it will be perceived that the same identical plans have been again and again patented by different individuals: and it is extremely doubtful whether any patent of the present day is really valid in law, and whether it cannot be shown to have been long . previously known to the world, and free for general use.

In most of these plans the design of bringing a larger quantity of air into the furnace is obvious. But in some few this principle is not so plainly developed. Among these latter may be noticed the inventions in which a jet of steam is thrown into the furnace, and two other inventions in which a shower of water falls down the flue with considerable velocity. In all these cases the great velocity of the steam and the water give an additional impetus to the motion of the gaseous bodies within their immediate sphere ; and these again communicate their velocity to those which are more distant, and thus draw into the furnace an additional quantity of air. There are practical difficulties, however, attending the use of a shower of water used in this way, which must effectually prevent these inventions coming into very general use.

(485.) A second class of inventions are those in which the fuel is gradually coked, before it enters into combustion. When coal is suddenly exposed to a high temperature, a very large quantity of gaseous matter is given off, which quantity gradually diminishes as the coal becomes more nearly converted into coke. By slowly heating the coal, a more equable evolution of gas is produced.

2 F

This result is very effectually accomplished by
using a large dead or dumb plate in front of the
furnace. The coals being placed on this plate are
gradually warmed, and at last arrive at a state of
incandescence, when all the gas has been driven off;
by which gradual distillation of the volatile gases,
the necessary quantity of air to produce combustion
is more easily obtained, as the demand for it thus
continues nearly uniform throughout the whole of
the process of combustion, instead of the very large
additional quantity required when a fresh charge of
coal is thrown on the fire in the ordinary method
of supplying fuel. The use of a large dead plate
in this way is quite sufficient of itself, without any
additional contrivance, to consume all the smoke
of a furnace, provided the fire on the bars be kept
thin, so as to allow a more ready entrance for the
air through the fuel. Considerable attention is,
however, required on the part of the fireman, that
the fuel on the dead plate be gradually pushed
forward into the furnace as it becomes heated.
When the fuel is supplied through a hopper, placed
in front of the furnace, the necessity for opening
the furnace door becomes much less frequent, and
the unnecessary cooling of the furnace is thus prevented. Where no air is admitted except through
the furnace bars, there will, however, always be
a considerable quantity of carbonic oxide formed
during the combustion; and although, under these
circumstances, there is no smoke, the greatest
effect of the fuel is not, by this means, produced.
A moderate quantity of air ought to be introduced
into the furnace, to mix with the gases distilled
from the coal, and also to convert the carbonic
oxide formed by the upper strata of coal into carbonic
acid. This air ought to be heated before it enters
the furnace, as it thus more readily inflames, mixes
more easily with the heated gases of the furnace,

and prevents injury by the unequal action of cold currents impinging against the bottom of the boiler. It perhaps matters but little in what part of the furnace this air be introduced; but the more it is diffused the better, as the heat is then less likely to become too intense on one particular part. Various simple methods of introducing heated air may be used; and the plan of using two pipes in the manner originally employed at the Royal Mint at Paris (Art. 406), as long ago as the year 1809, answers the purpose extremely well. The apertures to the pipes should be furnished with covers which can be partially closed; and by the experiments of Mr. Houldsworth,* it appears that an aperture for the air, varying from $1\frac{1}{2}$ to three square inches, for each square foot of the area of the furnace bars, will be sufficient for this purpose, the size of the pipes varying with the nature and quality of the coals.

(486.) The gradual coking of the fuel is likewise effected by such plans as those of Drew, Godson, and Coupland. Godson's plan effectually cokes the coal, and is perfectly compatible with any method of introducing additional air above the fuel to consume the carbonic oxide. The plans of Losh, Rodda, Thomas Hall, Collier, and some others, in which the flame from one fire passes over a second fire of clear bright burning fuel, is another mode of accomplishing the same object, but apparently less simple in its operation; and the mechanical means of continually feeding the fire, used by Stanley and by Brunton, produce nearly the same effect as the method of coking the coals would do; as by these means the evolution of the gases from the coal is equalised throughout the combustion, and the extraordinary demand for air, when fresh charges of

* "Report of the Select Committee of the House of Commons on the Prevention of Smoke," p. 105.

2 F 2

coal are supplied in the common mode of firing, is thereby avoided.

(487.) The method of supplying heated air through a split bridge has been repeatedly patented. Mr. Sheffield, in 1812, was undoubtedly the first to propose this plan, and his method of making the aperture deliver the heated air horizontally into the furnace, is perfectly correct in principle. The defect of these plans has frequently been that too large a quantity of air has thus been brought into the furnace, and the effect has been to lower its temperature. Mr. C. W. Williams' plan of diffusion would be very good, if it were used with hot air, instead of cold air, which latter is specially directed by the patent to be used; but by the former method, it would approach very near to the prior patent of Mr. Samuel Hall, differing only in being more simple and inexpensive. It should not, however, be overlooked, that it is not so easy to heat large quantities of atmospheric air to a high temperature as some persons imagine. When hot air is applied to blast furnaces,* it is found that to heat the air to about 600° Fahrenheit, it is necessary for it to traverse a surface of cast-iron pipes at nearly a red heat, for a distance of about $5\frac{1}{4}$ feet. The mere instantaneous passing of air through a heated metallic perforated plate, would therefore add but little to its temperature, unless it were also made to travel through heated pipes for some considerable distance.

(488.) Probably one of the most effectual methods of burning smoke is the plan of Mr. March (Art. 450), by blowing air downwards upon the fuel by a fan, and dispensing with the use of furnace bars. There can be no question that a most perfect combustion of the fuel may be thus produced; but it is very doubtful whether the

* Dufrenoy's Report on Hot Air in Iron Works of England, London, 1836, p. 77.

additional trouble which this method would cause, and the necessity for a mechanical power to produce the requisite blast of air, will not prevent its adoption to any considerable extent. A similar plan to this is stated to have been tried experimentally in some of the furnaces used in the manufacture of iron; and as the whole of the carbonic oxide must, by this plan, be consumed, there will necessarily be a considerable saving of fuel.*

(489.) It has been objected to the various plans for the admission of air to the gases above the fuel of the furnace, that the air, when thus admitted, prevents to a certain extent the admission of the air through the furnace bars, and thus reduces the rate of the combustion of the fuel on the furnace bars. This to a certain extent is true, but it can be no argument against the plan; for it can only be when the air is improperly admitted, and escapes through the flues in an uncombined state, that the total combustion of the furnace can be reduced by the admission of air above the fuel. And in general it will follow that the heat of the furnace being increased by the perfect combustion of the gases on the top of the fuel, the draught of the furnace will be increased, and therefore there will be a greater tendency to the influx of air through the furnace bars as well as through the other apertures. The use of heated air, however, in preference to cold air, is far more likely to prevent any loss by the passing of air in an uncombined state through the flues. When cold air is used, this result is not unlikely to

* Some interesting researches by M. Ebelmen, on the application of the carbonic oxide from blast furnaces to useful purposes, have shown that the loss of effect by the escape of the carbonic oxide amounts to 62 per cent. of the total quantity of fuel consumed in blast furnaces.—*Repertory of Arts*, vol. xviii. (1842), pp. 116-313.

occur; for gases, at temperatures differing considerably from each other, mix together very slowly; and therefore it may often happen that by introducing cold air into a furnace, the mixture of the air with the gases will not take place until they have passed into the flues, and the temperature becomes too much reduced to cause their accension.

(490.) The practical result of these remarks is, that there are many effectual contrivances for the combustion of smoke, combining the advantage of great economy of fuel. For this purpose, the more simple the apparatus the better; and with a very slight degree of attention on the part of the firemen, several of the plans which have been described would be certain to succeed in abating the nuisance of smoke entirely, and with considerable economy in the consumption of fuel. The saving in fuel would necessarily be very considerable. In very few furnaces the saving would be not less than twenty-five per cent.; and in many which now produce large volumes of smoke the saving would be considerably greater. A large amount of the present evil of smoke arises from most furnaces being overworked, in consequence of their being too small for the duty required from them. But, with a moderate degree of attention on the part of the firemen, the present furnaces, with a very slight alteration, could be made effectually to burn their smoke; and the most effectual way would be to combine the plan of delivering air through numerous small openings made in the inner plate of the furnace door (supplied by one or two large openings in the outer door covered with a slide); and also by allowing a further quantity of air to pass to the bridge through pipes laid along the furnace bed. For small furnaces either of these plans singly will suffice; but for larger furnaces the best effect will result from the two modes conjointly. Means must be taken to stop off a portion of the air at certain stages of the combustion, or there will be more air admitted than will be desirable. Both these plans have long been in use, many years prior to the date of any existing patents; and the necessary alterations to adapt them to ordinary furnaces need not involve anything beyond a very moderate expense.

(491.) Before concluding these remarks, a few words on artificial fuels may not be amiss. A great number of patents have been obtained for forming artificial fuel, the principle of them all being to combine the small and refuse coal into a solid body. As early as 1799, a patent was obtained by M. Chabannes for this purpose, and it is difficult to discover in what this patent differs from the various subsequent and recent ones for the same object. The principal ingredients used in all these compositions are coal-dust, coke, peat, bark, saw-dust, tan, clay, sand, pitch, coal-tar, alum, nitre, vegetable matter, and animal excrement. Different persons combine these substances in different proportions, and some omit altogether certain of these ingredients. A very powerful and efficient fuel can be composed by mixtures of these substances; and it appears by some experiments reported by Dr. Buckland to the British Scientific Association,[*] that when tried against Welsh coal, Pontop coal, and Wylam Main coal, the artificial compound was found to be very considerably more powerful in heating effect than either of these coals. These compound fuels, however, are subject to one inconvenience when used by themselves in furnaces; that the coal tar is very liable to distil from the fuel without being consumed, in which case it clogs up the furnace bars, and partially stops the due admission of air. When it is used with a certain propor-

* *Report of British Scientific Association,* vol. vii. (1838), p. 85.

tion of ordinary coal, this inconvenience is less likely to occur, and probably with moderate care it may be avoided. And these methods of combining refuse coal, which must otherwise be nearly valueless, may in many places be most efficiently applied to obtaining a powerful and useful fuel at a moderate expense.

APPENDIX.

TABLE I.

Table of the Expansive Force of Steam in Atmospheres, and in lbs. per square inch; for Temperatures above 212° of Fahrenheit.

N.B.—The steam is supposed to be in contact with the water from which it is formed, and the water and steam to be alike in temperature.

Heat in Degrees of Fahrenheit.	Pressure.		Heat in Degrees of Fahrenheit.	Pressure.		Heat in Degrees of Fahrenheit.	Pressure.		Heat in Degrees of Fahrenheit.	Pressure.	
	Atmospheres.	lbs.		Atmospheres.	lbs.		Atmospheres.	lbs.		Atmospheres.	lbs.
212	1	15	351	9	135	487	40	600	663	170	2550
216	—	16·5	359	10	150	499	45	675	671	180	2700
220	—	17·7	367	11	165	511	50	750	679	190	2850
225	—	19·5	374	12	180	521	55	825	686	200	3000
230	—	21·5	381	13	195	531	60	900	694	210	3150
235	—	23·6	387	14	210	540	65	975	700	220	3300
240	—	25·8	393	15	225	549	70	1050	707	230	3450
245	—	28·1	399	16	240	557	75	1125	713	240	3600
250	2	30·9	404	17	255	565	80	1200	719	250	3750
255	—	33·6	409	18	270	572	85	1275	726	260	3900
260	—	36·1	414	19	285	579	90	1350	731	270	4050
265	—	39·0	418	20	300	586	95	1425	737	280	4200
270	—	43·1	423	21	315	592	100	1500	742	290	4350
275	3	45·0	427	22	330	605	110	1650	748	300	4500
294	4	60·	431	23	345	616	120	1800	753	310	4650
308	5	75	436	24	360	627	130	1950	758	320	4800
320	6	90	439	25	375	636	140	2100	763	330	4950
332	7	105	457	30	450	646	150	2250	768	340	5100
342	8	120	473	35	525	655	160	2400	772	350	5250

** The pressures above three atmospheres in the above Table are deduced from the experiments of MM. Dulong and Arago. Their calculations extend only as far as 50 atmospheres; from thence the pressures are now calculated to 350 atmospheres by their formula, viz. :—

$$t = \frac{\sqrt[5]{e} - 1}{\cdot 7153}$$

where e represents the pressure in atmospheres, and t the tem-

perature above 100° of Centigrade. In this equation each 100°
of Centigrade is represented by unity.

In reducing these temperatures from Centigrade to Fahrenheit's scale, where the fractions amount to ·5, they have been
taken as the next degree above, and all fractions below ·5 have
been rejected.

More than twenty different formulæ for this purpose are
given in the *Encyclopædia Britannica*, art. Steam.

TABLE II.

Table of the quantity of Vapour contained in Atmospheric Air,
at different Temperatures, when saturated.

Temperature of the Air.	Quantity of Vapour per Cubic Foot, in Grains Weight.	Temperature of the Air.	Quantity of Vapour per Cubic Foot, in Grains Weight.	Temperature of the Air.	Quantity of Vapour per Cubic Foot, in Grains Weight.
20°	1·52	48°	3·94	76°	9·38
22	1·64	50	4·19	78	9·99
24	1·76	52	4·46	80	10·59
26	1·89	54	4·77	82	11·29
28	2·03	56	5·06	84	11·98
30	2·16	58	5·40	86	12·68
32	2·31	60	5·76	88	13·36
34	2·43	62	6·12	90	14·15
36	2·62	64	6·50	92	14·93
38	2·80	66	6·91	94	15·81
40	2·99	68	7·31	96	16·76
42	3·21	70	7·77	98	17·83
44	3·45	72	8·27	100	19·00
46	3·69	74	8·80	—	—

₊ The above Table is computed from Dr. Dalton's experiments on the Elastic Force of Vapour.

The weight of a cubic foot of steam, at the pressure of
30 inches of mercury, is 257·119 grains ; therefore at any other
pressure p, the weight will be $\frac{p \times 257·119}{30} = x$. But as vapours expand $\frac{1}{480}$ for each degree of Fahrenheit, this equation must be
corrected for the difference in the expansion of the vapour at
the temperature of 212°, and the temperature p, in the pre-

ceding equation. Therefore, the volume of the vapour at the temperature p, will be $1 + \frac{p}{480}$; the volume at 212°, being $1 + \frac{212}{480} = 1\cdot441$, when the volume is assumed to be unity at zero. The weight of a cubic foot of vapour will therefore be

$$\frac{1\cdot441 \times x}{1 + \frac{p}{480}} = w.$$

TABLE III.

Table of the Expansion of Air and other Gases by Heat, when perfectly free from Vapour.

Temperature, Fahrenheit's Scale.	Expansion.	Temperature, Fahrenheit's Scale.	Expansion.
32°	1000	100°	1152
35	1007	110	1178
40	1021	120	1194
45	1032	130	1215
50	1043	140	1235
55	1055	150	1255
60	1066	160	1275
65	1077	170	1295
70	1089	180	1315
75	1099	190	1334
80	1110	200	1354
85	1121	210	1372
90	1132	212	1376
95	1142	—	—

₊ The above numbers are obtained from Dr. Dalton's experiments, which give an average of $\frac{1}{483}$ part, or ·00207 for the expansion by each degree of Fahrenheit. Gay Lussac found it to be equal to $\frac{1}{480}$ part, or ·002083 for each degree of Fahrenheit; and that the same law extends to condensible vapours when excluded from contact of the liquids which produce them. Professor Daniell (*Chemical Philosophy*, p. 90) makes the expansion of air equal to ·373 for 180° of Fahrenheit: and in the Parliamentary Report "On Warming and Ventilating Dwellings, 1857," the expansion of air by 180° Fahrenheit is stated to be ·366 of its volume. Regnault (*Ann. de Chimie*) gives the expansion at ·00203 for each degree of Fahrenheit, or ·3654 for 180°.

TABLE IV.

Table of the Specific Gravity and Expansion of Water at different Temperatures.

Temperature, Fahrenheit's Scale.	Expansion.	Specific Gravity.	Weight of 1 Cubic Inch, in Grains.	Temperature, Fahrenheit's Scale.	Expansion.	Specific Gravity.	Weight of 1 Cubic Inch, in Grains.
30°	·00017	·9998	252·714	121°	·01236	·9878	249·677
32	·00010	·9999	252·734	124	·01319	·9870	249·473
34	·00005	·9999	252·745	127	·01403	·9861	249·265
36	·00004	·9999	252·753	130	·01490	·9853	249·053
38	·000002	·9999	252·758	133	·01578	·9844	248·836
39	·00000	1·0000	252·759	136	·01668	·9836	248·615
43	·00003	·9999	252·750	139	·01760	·9827	248·391
46	·00010	·9999	252·734	142	·01853	·9818	248·163
49	·00021	·9997	252·704	145	·01947	·9809	247·931
52	·00036	·9996	252·667	148	·02043	·9799	247·697
55	·00054	·9994	252·621	151	·02141	·9790	247·459
58	·00076	·9992	252·566	154	·02240	·9780	247·219
61	·00101	·9989	252·502	157	·02340	·9771	246·976
64	·00130	·9986	252·429	160	·02441	·9760	246·707
67	·00163	·9983	252·349	163	·02543	·9751	246·483
70	·00198	·9981	252·285	166	·02647	·9741	246·233
73	·00237	·9976	252·162	169	·02751	·9731	245·982
76	·00278	·9972	252·058	172	·02856	·9721	245·729
79	·00323	·9967	251·945	175	·02962	·9711	245·474
82	·00371	·9963	251·825	178	·03068	·9701	245·218
85	·00422	·9958	251·698	181	.03176	·9691	244·962
88	·00476	·9952	251·564	184	·03284	·9681	244·704
91	·00533	·9947	251·422	187	·03392	·9671	244·446
94	·00592	·9941	251·275	190	·03501	·9660	244·187
97	·00654	·9935	251·121	193	·03610	·9650	243·928
100	·00718	·9928	250·960	196	·03720	·9640	243·669
103	·00785	·9922	250·794	199	·03829	·9630	243·410
106	·00855	·9915	250·621	202	·03939	·9619	243·151
109	·00927	·9908	250·443	205	·04049	·9609	242·893
112	·01001	·9901	250·259	208	·04159	·9599	242·635
115	·01077	·9893	250·070	212	·04306	·9585	242·293
118	·01156	·9885	249·876				

₊ In the above Table the expansions are calculated by Dr. Young's formula, $22 f^2 (1 - ·002 f)$ in 10 millionths. The diminution of specific gravity is calculated by this equation: $·0000022 f^2 - ·00000000472 f^3$. In both equations f represents the number of degrees above or below 39° of Fahrenheit. The absolute weight of a cubic inch of water, at any temperature, may be found by multiplying the weight of a cubic inch at 39°, by the specific gravity at the required temperature. Water is 829 times the weight of air; and 13·2 cubic feet of air weigh 1 lb. There are 437·5 grains in an ounce, and 7000 grains in a pound.

TABLE V.

Table of the Specific Heat, Specific Gravity, and Expansion by Heat of different Bodies.

Barometer 30 Inches.—Thermometer 60°.

	Specific Heat		Specific Gravity	Weight of 100 Cubic Inches. Barometer 30 Inches. Thermometer 60°.	Linear Expansion by 180° of Heat, from 32° to 212°.
	Of equal Weights, by Berard, Delaroche.	and Petit and Dulong.			
				Grains.	
Air (atmospheric) ..	·2669	..	1·000	30·519	
—— (dry) .. *Apjohn*	·2767	..	..	..	
Aqueous vapour	·8470	..	·633	19·058*	
Azote	·2754	..	·9722	29·65	
—— oxide of	·2369	..	1·5277	46·596	
Carbonic acid	·2210	..	1·5277	46·596	
—————— oxide	·2884	..	·9722	29·65	
Hydrogen	3·2936	..	·0694	2·118	
Olefiant Gas	·4207	..	·9722	29·65	
Oxygen	·2361	..	1·1111	33·888	
				Ounces.	
Water	..	1·000	1·000	57·87	
Bismuth	..	·0288	9·880	571·7	
Brass	..	..	7·824	452·77	·00186671 = $\frac{1}{535}$
—— wire	..	..	8·396	485·87	·00193000 = $\frac{1}{518}$
Cobalt	..	·1498	8·600	497·6	
Copper	..	·0949	8·900	515·0	·00172244 = $\frac{1}{581}$
Gold	..	·0298	19·250	1114·0	·00146606 = $\frac{1}{682}$
Glass (flint)	..	..	2·760	159·72	·00081166 = $\frac{1}{1232}$
—— (tube)	..	..	2·520	145·83	·00087572 = $\frac{1}{1142}$
Iron (cast)	..	..	7·248	418·9	·00111111 = $\frac{1}{900}$
—— (bar)	..	·1100	7·788	450·2	·00122045 = $\frac{1}{819}$
Lead	..	·0293	11·350	656·8	·00284836 = $\frac{1}{351}$
Nickel	..	·1035	8·279	478·5	
Pewter (fine)	..	..	..	..	·00228300 = $\frac{1}{438}$
Platinum	..	·0314	21·470	1242·4	·00099180 = $\frac{1}{1008}$
Silver	..	·0557	10·470	605·8	·00208260 = $\frac{1}{480}$
Solder (lead 2 + tin 1)	..	..	..	..	·00250800 = $\frac{1}{398}$
Spelter(brass2+zinc 1)	..	..	..	..	·00205800 = $\frac{1}{485}$
Steel (untempered) ..	..	..	7·840	453·7	·00107875 = $\frac{1}{927}$
——(yellow tempered)	..	..	7·816	452·31	·00136900 = $\frac{1}{730}$
Sulphur	..	·1880	1·990	115·1	
Tellurium	..	·0912	6·115	353·5	
Tin	..	·0514	7·291	421·9	·00217298 = $\frac{1}{460}$
Zinc	..	·0927	7·191	416·0	·00294200 = $\frac{1}{339}$

. Air is taken as the standard for the specific gravity of the gases, and water as the standard for the solids.

The specific heat of gases has been recently investigated by Dr. Apjohn and a somewhat different result obtained. (See *London and Edinburgh Philosophical Magazine*, vol. xii. 102; xiii. 261, 339.)

The expansion in volume may be obtained without sensible error, by trebling the number which expresses the increase in length, where the fraction of its length is small.

* Specific gravity of steam at 212° = ·481. Weight of 100 cubic inches, 14·879 grains.

TABLE VI.

TABLE OF THE EFFECTS OF HEAT.

	Fahrenheit's Scale.
Soft iron melts (Clement and Desormes) . . .	3945°
Maximum temperature by Daniell's Pyrometer . .	3280
Cast iron melts	2786
Gold melts	2016
Copper melts	1996
Silver melts	1873
Bronze melts (copper 15 parts, tin 1 part) . . .	1750
Brass melts (copper 3 parts, zinc 1 part) . . .	1690
„ (copper 2 parts, zinc 2 parts) . . .	1672
Diamond burns	1552
Bronze melts (copper 7 parts, tin 1 part) . . .	1534
„ (copper 3 parts, tin 1 part) . . .	1446
Enamel colours burnt	1392
Iron red-hot in daylight	1272
„ in the twilight	884
„ in the dark	800
Charcoal burns	802
Heat of a common fire	790
Zinc melts (Davy 680°) (Daniell)	773
Mercury boils (Black 600°) (Crichton 655°) (Dalton 660°) (Petit and Dulong 656°) (Irving 672°) (Secondat 644°)	660
Linseed oil boils	640
Lowest ignition of iron in the dark	635
Lead melts (Guyton and Irving 594°) (Crichton) . .	612
Steel becomes dark blue, verging on black . . .	600
„ a full blue	560
Sulphur burns	560
Steel becomes blue	550
„ purple	530
„ brown, with purple spots . . .	510
„ brown	490
Bismuth melts	476
Steel becomes a full yellow	470
„ a pale straw colour	450
Tin melts	442
Steel becomes a very faint yellow	430
Tin 3 + lead 2 + bismuth 1, melts	334
Tin and bismuth, equal parts, melts	283
Sulphur melts	218
Bismuth 5 + tin 3 + lead 2, melts	212

Table of the Effects of Heat (*continued*).

	Fahrenheit's Scale.
Water boils (barometer 30 inches)	212°
Wax melts	149
Spermaceti melts	112
Tallow melts (Nicholson 127°)	92
Acetic acid congeals	50
Olive oil congeals	36
Water freezes	32
Milk freezes	30
Vinegar freezes	28
Sea-water freezes	28
Strong wine freezes	20
Oil of turpentine freezes	14
Mercury congeals	−39
Sulphuric æther congeals	−47
Natural temperature of Hudson's Bay . . .	−51
Greatest artificial cold	−91

₄ See also many other effects of heat in Chapter XII. Part 1.

Table VII.

Table of the Quantity of Water contained in 100 Feet of Pipe, of different Diameters.

Diameter of Pipe.	Contents of 100 Feet in length.
Inches.	Gallons.
½	·84
1	3·39
1½	7·64
2	13·58
2½	21·22
3	30·56
4	54·33
5	84·90
6	122·26

Table VIII.

Table of the Strength, or Cohesive Force, of different Substances. By Mr. George Rennie.

Bars of six inches long and a quarter of an inch square will break with the following weight suspended lengthways :—

	lbs.		per square inch.
Cast Iron (horizontal) . . .	1166	equal to	18,656 lbs.
Ditto (vertical) . . .	1218	„	19,488 „
Cast Steel (tilted) . . .	8391	„	134,256 „
Blistered Steel (hammered) . .	8322	„	133,152 „
Shear Steel (ditto) .	7977	„	127,632 „
Swedish Iron	4504	„	72,064 „
English Iron	3492	„	55,872 „
Hard Gun-metal . . .	2273	„	36,368 „
Wrought Copper (hammered) .	2112	„	33,792 „
Cast Copper . . .	1192	„	19,072 „
Fine Yellow Brass . . .	1123	„	17,968 „
Cast Tin	296	„	4,736 „
Cast Lead	114	„	1,824 „

Per Quetelet.

Iron Wire, ·0769 inches diameter, bears . 432 to 615 lbs.
Copper wire ditto . . 302 to 386 „

Per Committee of the Franklin Institute.*

Iron Wire. 1-3rd inch diameter, bears 81,387 lbs. per square inch, and 14 per cent. less when annealed.

Best Cable Iron . . .	59,105 lbs. per square inch.
Ditto ditto (hammer hardened)	71,000 lbs. „
Russian Iron	76,069 lbs. „

Table of the Relative Cohesive Strength of Metals.
By Sickenger.

Gold	150,955
Silver	190,771
Platinum	262,361
Copper	304,696
Soft Iron	362,927
Hard Iron	559,880

By Muschenbroek.

Copper 6 + Tin 1	41,000
Swedish Copper 6 + Malacca Tin 1 .	64,000
Brass	51,000
Block Tin 3 + Lead 1 . .	10,200
Ditto 8 + Zinc 1 . . .	10,000
Tin 4 + Regulus of Antimony 1 .	12,000
Lead 8 + Zinc 1	4,500
Tin 4 + Lead 1 + Zinc 1 . .	13,000

* For strength of iron, &c., at various temperatures, see Chap. XII., Part 1.

Table IX.

The following Table of the relative values of various sorts of fuel is compiled from Marcus Bull's Experiments on Fuel. In these experiments all the smoke was consumed.

Name.	Specific Gravity.	Weight per Bushel, lbs.	Relative Heating Value.
Hickory Charcoal . . .	·625	32·89	166
Cannel Coal	1·240	65·25	230
Liverpool Coal	1·331	70·04	215
Newcastle Coal	1·204	63·35	198
Scotch Coal	1·140	59·99	191
Karthaus Coal	1·263	66·46	208
Richmond Coal	1·246	65·56	205
Stony Creek Coal . . .	1·396	73·46	243
Maple Charcoal	·431	22·68	114
Oak Charcoal	·401	21·10	116
Pine Charcoal	·285	15·0	75
Coke	·557	29·31	126

The bushel measure in this Table is much smaller than the English Imperial bushel.

TABLE X.

Velocity of chimney draught at different temperatures, the external air being at 32° Fahrenheit. From Peclet's "Traité de la Chaleur," p. 79.

Temperature of Warm Air, Fahrenheit.	Relative Velocity.	Temperature of Warm Air, Fahrenheit.	Relative Velocity.	Temperature of Warm Air, Fahrenheit.	Relative Velocity.
86	4·93	356	8·09	608	8·25
104	5·51	374	8·14	662	8·21
122	5·98	392	8·17	752	8·13
140	6·35	410	8·21	842	8·03
158	6·66	428	8·23	932	7·92
176	6·92	446	8·25	1022	7·80
194	7·13	464	8·26	1112	7·62
212	7·33	482	8·27	1202	7·56
230	7·48	500	8·273	1292	7·44
248	7·62	518	8·278	1382	7·33
266	7·73	527	8·279	1472	7·22
284	7·83	536	8·276	1562	7·11
302	7·92	554	8·275	1652	7·00
320	7·98	572	8·27	1742	6·90
338	8·05	590	8·26	1832	6·80

It will be seen from this Table that the velocity of chimney draught diminishes at the extremely high temperatures, in consequence of the very great expansion of the air.

INDEX.

2 G 2

LONDON: PRINTED BY WILLIAM CLOWES AND SONS, LIMITED, STAMFORD STREET AND CHARING CROSS.

www.ingramcontent.com/pod-product-compliance
Lightning Source LLC
Chambersburg PA
CBHW052338110726
47901CB00005B/1276